CAMBRIDGE TRACTS IN MATHEMATICS

General Editors

---

174   Forcing Idealized

FORCING IDEALIZED

JINDŘICH ZAPLETAL
University of Florida

CAMBRIDGE UNIVERSITY PRESS
Cambridge, New York, Melbourne, Madrid, Cape Town, Singapore, São Paulo

Cambridge University Press
The Edinburgh Building, Cambridge CB2 8RU, UK

Published in the United States of America by Cambridge University Press, New York

www.cambridge.org
Information on this title: www.cambridge.org/9780521874267

First published 2008

Printed in the United Kingdom at the University Press, Cambridge

A catalogue record for this publication is available from the British Library

ISBN 978-0-521-87426-7 hardback

Contents

1
Introduction

1.1 Welcome

This book reports on the state of a research program that I initiated in 1999. It connects the practice of proper forcing introduced by Shelah [64] with the study of various σ-ideals on Polish spaces from the point of view of abstract analysis, descriptive set theory, measure theory, etc. It turns out that the connection is far richer than I dared to imagine in the beginning. Its benefits include theorems about methodology of forcing as well as isolation of new concepts in measure theory or abstract analysis. It is my sincere hope that this presentation will help to draw attention from experts from these fields and to bring set theory and forcing closer to the more traditional parts of mathematics.

The book uses several theorems and proofs from my earlier papers; in several cases I coauthored these papers with others. The first treatment of the subject in [83] is superseded here on many accounts, but several basic theorems and proofs remain unchanged. The papers [18], [67], [82], [86], and [87] are incorporated into the text, in all cases reorganized and with significant improvements.

Many mathematicians helped to make this book what it is. Thanks should go in the first place to Bohuslav Balcar for his patient listening and enlightening perspective of the subject. Vladimir Kanovei introduced me to effective descriptive set theory. Ilijas Farah helped me with many discussions on measure theory. Joerg Brendle and Peter Koepke allowed me to present the subject matter in several courses, and that greatly helped organize my thoughts and results. Last but not least, the influence of the mathematicians I consider my teachers (Thomas Jech, Hugh Woodin, and Alexander Kechris) is certainly apparent in the text.

I enjoyed financial support through NSF grant DMS 0300201 and grant GA ČR 201-03-0933 of the Grant Agency of Czech Republic as I wrote this book.

1.2 Navigation

This is not a textbook. The complexity of the subject is such that it is impossible to avoid forward references and multiple statements of closely related results, and to keep the book organized in a logical structure at the same time. As a result, the linear reading of the book will be necessarily interspersed with some page flipping. This section should help the reader to find the subjects he is most interested in.

Chapter 2 provides the basic definitions, restatements of properness, and basic implications of properness, such as the reading of reals in the generic extension as images of the generic point under ground model coded Borel functions. Every reader should start with this chapter. A sample theorem:

Theorem 1.2.1. *Suppose that I is a σ-ideal on a Polish space X. The forcing P_I of I-positive Borel sets ordered by inclusion adds a single point $\dot{x}_{gen} \in X$ such that a set B belongs to the generic filter if and only if it contains the generic point $\dot{x}_{gen}$.*

Chapter 3 investigates the possible finer forcing properties of the forcings of the form P_I. These divide into three basic groups. The first group is that of Fubini forcing properties, introduced in Section 3.2. These correspond to the classical preservation properties such as the bounding property or preservation of outer Lebesgue measure. A sample theorem:

Theorem 1.2.2. *Suppose that I is a σ-ideal on a Polish space X such that the forcing P_I is proper. The following are equivalent:*

1. *P_I is bounding;*
2. *for every Polish topology τ on the space X that yields the same Borel structure as the original one, every Borel I-positive set contains a τ-compact I-positive subset.*

The second group of properties is entirely absent in the combinatorial treatment of forcings. These are the descriptive set theoretic properties of the ideals, represented by the various dichotomies of Section 3.9 and the $\mathbf{\Pi}^1_1$ on $\mathbf{\Sigma}^1_1$ property. The dichotomies are constantly invoked in the proofs of absoluteness theorems and preservation theorems. The $\mathbf{\Pi}^1_1$ on $\mathbf{\Sigma}^1_1$ property of ideals allows ZFC treatment of such operations as the countable support iteration, product, and illfounded iteration, with a more definite understanding of the underlying issues. A sample theorem:

Theorem 1.2.3. *(LC+CH) Suppose that I is a σ-ideal generated by a universally Baire collection of analytic sets such that every I-positive $\mathbf{\Sigma}^1_2$ set has an I-positive Borel subset. If the forcing P_I is ω-proper then every function $f \in 2^{\omega_1}$ in the extension either is in the ground model or has a countable initial segment which is not in the ground model.*

Here, LC denotes a suitable large cardinal assumptions, as explained in the next section.

The third group of properties is connected with determinacy of games on Boolean algebras. A number of forcing properties can be expressed in terms of infinitary games of the poset P_I which are determined in the definable context. The games are usually variations on standard fusion arguments, and the winning strategies are a necessary tool in the treatment of product forcing, illfounded iteration, and other subjects. A sample application:

Theorem 1.2.4. *(LC) Suppose that I is a universally Baire σ-ideal on a Polish space X such that the forcing P_I is proper. The following are equivalent:*

1. *P_I preserves Baire category;*
2. *there is a collection T of Polish topologies on the space X such that I is the collection of all sets which are τ-meager for every topology $\tau \in T$.*

Chapter 4 gives a number of classes of σ-ideals I for which I can prove that the forcing P_I is proper. While the presentation is based on a joint paper with Ilijas Farah [18], it is nevertheless greatly expanded. There are two very distinct groups of ideals in this respect: the ideals satisfying the first dichotomy, whose treatment occupies almost the whole chapter, and the ideals that do not satisfy the first dichotomy, treated in Section 4.7. It seems that the former group is much larger. Its treatment is divided into several very populous subgroups, each treated in its own section. These subgroups are typically connected with a basic underlying idea from abstract analysis, such as capacities or Hausdorff measures. The sections are all very much alike: first comes the definition of the class of ideals, then the properness theorem, then the dichotomy theorem (which, mysteriously, is always proved in the same way as properness), then several general theorems regarding the finer forcing properties of the ideals. The section closes with a list of examples. A sample result:

Theorem 1.2.5. *Suppose that ϕ is an outer regular subadditive capacity on a Polish space X. Let $I = \{A \subset X : \phi(A) = 0\}$. Then:*

1. *if the capacity is stable then the forcing P_I is proper;*
2. *if the forcing P_I is proper and the capacity is strongly subadditive then the forcing P_I preserves outer Lebesgue measure;*
3. *if the forcing P_I is proper and the capacity is Ramsey then the forcing does not add splitting reals;*
4. *every capacity used in potential theory is stable.*

My original hope that the idealization of forcings would closely relate to the creature forcing technology [58] proved to be naive; the symmetric difference of the two approaches turned out to be quite large. Nevertheless, in several cases I could identify a precise correspondence between a class of ideals and a class of creature forcings.

Chapter 5 relates operations on ideals with operations on forcings. The key case here is that of the countable support iteration which corresponds to a transfinite Fubini product of ideals, Section 5.1. The other operations I can handle are side-by-side product with a great help from determinacy of games on Boolean algebras, the illfounded iteration, which provides a treatment dual to and more general than that of [43], the towers of ideals which is a method of obtaining forcings adding objects more complex than just reals, and the union of ideals, which forcingwise is an entirely mysterious operation. A sample theorem:

Theorem 1.2.6. *(LC) Suppose that $I_\alpha : \alpha \in \kappa$ is a collection of universally Baire σ-ideals on some Polish spaces such that the forcings P_{I_α} are all proper and preserve Baire category bases. Then the countable support side-by-side product of these forcings is proper as well and preserves Baire category bases. In addition, the ideals satisfy a rectangular Ramsey property.*

Chapter 6 is probably the primary reason why a forcing practitioner may want to read this book; however its methods are entirely incomprehensible without the reading of the previous chapters. There are several separate sections.

Section 6.1 contains the absoluteness results which originally motivated the work on the subject of this book. There are many theorems varying in the exact large cardinal strength necessary and in the class of problems they can handle, but on the heuristic level they all say the same thing. If χ is a simply definable cardinal invariant and I is a σ-ideal such that the forcing P_I is proper, then if the inequality $\chi < \mathrm{cov}^*(I)$ holds in some extension then it holds in the iterated P_I extension. Moreover, there is a forcing axiom CPA(I) which holds in the iterated P_I extension and which then must directly imply the inequality $\chi < \mathrm{cov}^*(I)$. The CPA-type axioms have been defined independently in the work of Ciesielski and Pawlikowski [9] in an effort to axiomatize the iterated Sacks model. A sample theorem:

Theorem 1.2.7. *(LC) Suppose that χ is a tame cardinal invariant and $\chi < \mathfrak{c}$ holds in some forcing extension. Then $\aleph_1 = \chi < \mathfrak{c}$ holds in every forcing extension satisfying CPA; in particular it holds in the iterated Sacks model.*

Section 6.2 considers the duality theorems. These are theorems that partially confirm the old duality heuristic: if I, J are σ-ideals and the inequality $\mathrm{cov}(I) \leq \mathrm{add}(J)$ is provable in ZFC, then so should be its dual inequality $\mathrm{non}(I) \geq \mathrm{cof}(J)$. This is really completely false, but several theorems can be proved that rescue nontrivial pieces of this unrealistic expectation. This is the one part of this book where the combinatorics of uncountable cardinals actually enters the computation of inequalities between cardinal invariants, with considerations involving various pcf and club guessing structures. A sample theorem:

Theorem 1.2.8. *Suppose that J is a σ-ideal on a Polish space generated by a universally Baire collection of analytic sets. If ZFC+LC proves $\mathrm{cov}(I) = \mathfrak{c}$ then ZFC+LC proves $\mathrm{non}(I) \leq \aleph_2$.*

Section 6.3 gives a long list of preservation theorems for the countable support iteration of definable forcings. Compared to the combinatorial approach of Shelah [64], these theorems have several advantages: they connect well with the motivating problems in abstract analysis, and they have an optimal statement. Among their disadvantages I must mention the restriction to definable forcings and the necessity of large cardinal assumptions for a full strength version. Many of the preservation theorems of this section have no combinatorial counterpart. A sample result:

Theorem 1.2.9. *(LC) Suppose that I is a universally Baire σ-ideal on a Polish space X such that the forcing P_I is proper. Suppose that ϕ is a strongly subadditive capacity. If P_I forces every set to have the same ϕ-mass in the ground model as it has in the extension, then even the countable support iterations of the forcing P_I have the same property.*

1.3 Notation

My notation follows the set theoretic standard of [29]. If T is a tree of finite sequences ordered by extension then $[T]$ denotes the set of all infinite paths through that tree; if $T \subset 2^{<\omega}$ then $[T]$ is a closed subset of the space 2^ω. If X, Y are Polish spaces and $A \subset X \times Y$ is a set then the expression $\mathrm{proj}(A)$ denotes the set $\{x \in X : \exists y \in Y \ \langle x, y \rangle \in A\}$, for a point $x \in X$ the expression A_x stands for the vertical section $\{y \in Y : \langle x, y \rangle \in A\}$, and for a point $y \in Y$ the expression A^y stands for the horizontal section $\{x \in X : \langle x, y \rangle \in A\}$. For a Polish space X, $K(X)$ is the hyperspace of its compact subsets with the Vietoris topology and $P(X)$ is the space of probability Borel measures on X. The expression $\mathcal{B}(X)$ denotes the collection of all Borel subsets of the space X. The word "measure" refers to a σ-additive Borel measure. If a set function is σ-subadditive rather than σ-additive then I use the word "submeasure." The value of a measure (submeasure, capacity) ϕ at a set B is referred to as the ϕ-mass of the set B. A tower of models is a sequence $\langle M_\alpha : \alpha \in \beta \rangle$ where β is an ordinal and M_α's are elementary submodels of some large structure (typically $\langle H_\theta, \in \rangle$ for a suitable large cardinal θ) such that

$\alpha' \in \alpha \in \beta$ implies $M_{\alpha'} \in M_\alpha$. The tower is continuous if for limit ordinals $\alpha \in \beta$, $M_\alpha = \bigcup_{\gamma \in \alpha} M_\gamma$.

One important deviation from the standard set theoretical usage is the liberal use of large cardinal assumptions. In order to prove suitably general theorems of a statement that is easy to understand and refer to, I frequently have to resort to a large cardinal assumption of this or that kind. There are only three classes of applications of large cardinal assumptions in this book–absoluteness, determinacy of (long and complex) games, and definable uniformization. The minimum large cardinal necessary for each of these applications is different, sometimes difficult to state, sometimes unknown, and invariably completely irrelevant for the goals of this book; the existence of a supercompact cardinal is always sufficient. As a result, I decided to denote the use of large cardinal assumptions by a simple (LC) preceding the statement of the theorems. For most but not all *specific* applications of the general theorems in this book the large cardinal assumption can be eliminated by manual construction of all the winning strategies and uniformization functions necessary. At least in one case (the countable support iteration of Laver forcing) I made an effort to show that the key dichotomy requires a large cardinal assumption, and in the rather restrictive case of Π^1_1 on Σ^1_1 ideals almost all general theorems in this book are proved in ZFC.

The labeling of the various claims in this book is indicative of their position and function. Facts are statements that are proved elsewhere, and I will not restate their proofs. Theorems are quotable self-standing statements, ready for use in the reader's work. Propositions are self-standing statements referred to at some other, possibly quite distant, place in the book. Finally, claims and lemmas appear in the proofs of theorems and propositions, and they are not referred to in any other place.

1.4 Background

The subject of this book demands the reader to be proficient in several areas of set theory and willing to ask at least the basic questions about several other fields of mathematics. This section sums up the basic definitions and results which are taken for granted in the text.

1.4.1 Polish spaces

A *Polish space* is a separable completely metrizable topological space. Many Polish spaces occur in this book. If T is a countably branching tree without endnodes, then the set $[T]$ of all infinite branches through the tree T equipped with the topology generated by the sets $O_t = \{x \in [T] : t \subset x\}$ is a Polish space, with important special cases the Cantor space 2^ω and the Baire space ω^ω.

I will make use of basic theory of Polish spaces as exposed in [40]. Every uncountable Polish space X is a Borel bijective image of the Cantor space and it is a continuous bijective image of a closed subset of the Baire space. A G_δ subset of a Polish space is again Polish in the inherited topology. Every Polish space is homeomorphic to a G_δ subset of the Hilbert cube.

There are several useful operations on Polish spaces. If X, Y are Polish spaces then their product is again Polish; even a product of countably many Polish spaces is still Polish. If X is a Polish space then $K(X)$ denotes the space of all compact subsets of X equipped with *Vietoris topology* generated by sets of the form $\{K \in K(X) : K \subset O\}$ and $\{K \in K(X) : K \cap O \neq 0\}$ for open sets $O \subset X$. The space $K(X)$ is referred to as the *hyperspace* of X; it is Polish and if X is compact then $K(X)$ is compact as well.

It is possible to change the topology on a Polish space to a new, more convenient one. Whenever X is Polish with topology τ and $B_n : n \in \omega$ are τ-Borel subsets of X then there is a Polish topology η extending τ such that the sets $B_n : n \in \omega$ are η-clopen and the η-Borel sets are exactly the τ-Borel sets.

1.4.2 Definable subsets of Polish spaces

Definability of subsets of Polish spaces plays a critical role. Let X be a Polish space, with a countable topology basis $\mathcal{O}$. *Borel sets* are those sets which can be obtained from the basic open sets by a repeated application of countable union, countable intersection, and taking a complement. This is a class of sets closed under continuous preimages and continuous one-to-one images, but not under arbitrary continuous images. *Analytic sets* are those that can be obtained as continuous images of Borel sets. This is a class of sets containing the Borel sets, closed under continuous images, countable unions and intersections, but not under complements. Every analytic set $A \subset X$ is a projection of a closed subset $C \subset X \times \omega^\omega$, $A = \text{proj}(C)$. Every analytic subset of the Baire space is of the form $\text{proj}[T]$. Every analytic set whose complement is analytic is in fact Borel.

The paper [20] isolated an important and very practical broad definability class of subsets of Polish spaces. A set $A \subset 2^\omega$ is *universally Baire* if there are class trees $S, T \subset (2 \times \text{Ord})^{<\omega}$ which in all set generic extensions project into complementary subsets of 2^ω and $A = \text{proj}[T]$. A subset of another Polish space is universally Baire if it is in Borel bijective correspondence with a universally Baire subset of the Cantor space. Equivalently, a set is universally Baire if all of its continuous preimages have the property of Baire.

In ZFC, analytic sets and coanalytic sets are universally Baire, and consistently the class of universally Baire sets does not reach far beyond that. However, under large cardinal assumptions the class of universally Baire sets expands considerably. If there is a proper class of Woodin cardinals then the class of universally Baire sets is closed under complementation and continuous images and preimages, and every set of reals in the model $L(\mathbb{R})$ is universally Baire.

1.4.3 Measure theory

Let X be a Polish space. A *submeasure* on X is a map $\phi : \mathcal{P}(X) \to \mathbb{R}^+$ such that $\phi(0) = 0$, $A \subset B \to \phi(A) \leq \phi(B)$ and $\phi(\bigcup_n A_n) \leq \Sigma_n \phi(A_n)$ whenever $A_n : n \in \omega$ is a countable collection of subsets of the space X. The submeasures on uncountable Polish spaces in this book will always be countably subadditive in this sense. The submeasure ϕ is *outer regular* if $\phi(A) = \inf\{\phi(O) : A \subset O, O \text{ open}\}$ and it is *outer* if $\phi(A) = \inf\{\phi(B) : A \subset B : B \text{ Borel}\}$.

A Borel measure (or *measure*) is a map $\phi : \mathcal{B}(X) \to \mathbb{R}^+$ such that $\phi(0) = 0$, $A \subset B \to \phi(A) \leq \phi(B)$ and $\phi(\bigcup_n A_n) = \Sigma_n \phi(A_n)$ if $A_n : n \in \omega$ is a countable collection of pairwise disjoint Borel sets. Finite Borel measures on Polish spaces are outer regular and tight: $\phi(A) = \inf\{\phi(O) : A \subset O, O \text{ open}\} = \sup\{\phi(K) : K \subset A, K \text{ compact}\}$. I will need a criterion for the restriction of a submeasure ϕ on X to the Borel subsets of X to be a measure. If d is a complete separable metric on X and for every pair of closed sets $C_0, C_1 \subset X$ which are nonzero distance apart, $\phi(C_0 \cup C_1) = \phi(C_0) + \phi(C_1)$ then indeed $\phi \restriction \mathcal{B}(X)$ is a measure. In this situation I will say that ϕ is a *metric measure*.

A capacity on a Polish space X is a map $\phi : \mathcal{P}(X) \to \mathbb{R}^+$ such that $\phi(0) = 0$, $A \subset B \to \phi(A) \leq \phi(B)$, $\phi(\bigcup_n A_n) = \sup_n \phi(A_n)$ whenever $A_n : n \in \omega$ is a countable inclusion-increasing sequence of subsets of the space X, and $\phi(K) = \inf\{\phi(O) : K \subset O, O \text{ open}\}$ for compact sets $K \subset X$. Capacities are tight on analytic sets: if $A \subset X$ is analytic then $\phi(A) = \sup\{\phi(K) : K \subset A : K \text{ compact}\}$.

1.4.4 Determinacy

Infinitary games of all kinds, lengths, and complexities are a basic feature of this book. The key problem always is whether one of the players must have a winning strategy, an issue referred to as the *determinacy* of the game in question.

An *integer game* of length ω is specified by the *payoff set* $A \subset \omega^\omega$. In the game, Players I and II alternate infinitely many times, each playing an integer in his turn. Player I wins if the infinite sequence they obtained belongs to the set A, otherwise Player II wins. Insignificant variations of this concept, which are nevertheless much more intuitive and easier to use, obtain when Players I and II can use moves from some other countable set in place of ω.

Fact 1.4.1. *[49] Games with Borel payoff set are determined. [20] If large cardinals exist then games with universally Baire payoff set are determined.*

A significant variation occurs if the players are allowed to choose their moves from a set larger than countable. Let U be an arbitrary set, and let $A \subset U^\omega$ be a set. The associated game with payoff A of length ω is played just as in the previous paragraph. To state the determinacy theorems, consider U^ω as a topological space with basic open neighborhoods of the form $O_t = \{\vec{u} \in U^\omega : t \subset \vec{u}\}$ as t varies over all finite sequences of elements of the set U.

Fact 1.4.2. *[48] Games with Borel payoff set are determined. Suppose that large cardinals exist, $A \subset U^\omega$ is a Borel set, $f : A \to X$ is a continuous function into a Polish space, and $B \subset X$ is a universally Baire set. The game with payoff set $f^{-1}B$ is determined, and moreover there is a winning strategy which remains winning in all set generic extensions.*

Still another significant variation occurs if the moves of the two players come from some fixed Polish space X and the game has α many rounds for some countable ordinal α. Consider the space X^α equipped with the standard Polish product topology.

Fact 1.4.3. *[55] (LC) Games with real entries, countable length, and universally Baire payoff set are determined.*

The games of longer than countable length are important and interesting, and in this book they appear in Section 6.1. However, I will never be concerned with their determinacy.

In numerous places I will refer to the Axiom of Determinacy (AD) and its variations, such as AD+, and the natural models for these axioms.

Definition 1.4.4. *The Axiom of Determinacy (AD) is the statement that integer games with arbitrary payoff set are determined. AD+ is the statement: every set of reals is ∞-Borel and games with ordinal entries, length ω, and payoff sets which are preimages of subsets of ω^ω under continuous maps $\mathrm{Ord}^\omega \to \omega^\omega$ are determined.*

Happily, I will never have to delve into the subtleties of AD+. Let me just state that it is an open question whether AD is in fact equivalent to AD+. In this book, I will need the following two pieces of information about the axiom AD+:

Fact 1.4.5. *Suppose that suitable large cardinals exist. Then $L(\mathbb{R}) \models AD+$. If Γ is a class of universally Baire sets closed under continuous preimages then $L(\mathbb{R})(\Gamma) \models AD+$.*

Fact 1.4.6. *[28] (ZF+DC+AD+) If $\kappa \in \Theta$ is a regular uncountable cardinal then there is a set $A \subset \omega^\omega$ and a prewellordering $\leq$ on A of length κ such that every analytic subset of A meets fewer than κ many classes.*

Here as usual Θ is the supremum of lengths of prewellorderings of the real numbers.

1.4.5 Forcing

The standard reference book for forcing terminology and basic facts is [29]. Suppose that $P, \leq$ is a partially ordered set, a *poset* for short. P is *separative* if for every $p, q \in P$, if every $r \leq p$ is compatible with q then $p \leq q$. The separative quotient of P is the partially ordered set of E-equivalence classes on P where pEq if every extension of p is compatible with q and vice versa, every extension of q is compatible with p, with the ordering inherited from the poset P. The separative quotient of P is separative. The posets considered in this book are generally not separative, and no effort is wasted on considering their separative quotients instead. Every separative poset P is isomorphic to a dense subset of a unique complete Boolean algebra denoted by $RO(P)$.

There is a historically and mathematically important forcing model mentioned in many places in the book, the *choiceless Solovay model*. Let me briefly outline its construction and basic features. Let κ be an inaccessible cardinal and $G \subset \text{Coll}(\omega, < \kappa)$ be a generic filter. Consider the submodel $M \subset V[G]$ consisting of those sets hereditarily definable in $V[G]$ from real parameters and parameters in the ground model. This is the definition of the choiceless Solovay model.

Fact 1.4.7. *The basic features of the Solovay model include*

1. *for every real number $r \in M$ the model M is a choiceless Solovay model over the model $V[r]$;*
2. *every set of reals is a wellordered union of length $\kappa = \omega_1^M$ of Borel sets.*

The book contains several isolated references to the *nonstationary tower forcing* Q_δ discovered by Woodin [79], recently exposed in [45]. If δ is a Woodin cardinal and $G \subset Q_\delta$ is a generic filter, then in $V[G]$ there is an elementary embedding $j: V \to M$ such that the model M is transitive, contains the same countable sequences of ordinals as $V[G]$, and $\omega_1^M = \delta$.

On several occasions I will refer to the Gandy–Harrington forcing [47]. This is the countable forcing of all nonempty lightface Σ_1^1 subsets of some fixed Polish space. As a countable forcing, this is similar to Cohen forcing; its worth derives from its particular representation. The forcing adds a single point in the Polish space which belongs to all sets in the generic filter. Note that there are some atoms in the forcing–the Σ^1_1 singletons, but there are nonatomic parts too. I will also consider the obvious relativized variations of the Gandy–Harrington forcing.

Throughout the book, I will use a trick commonplace in the literature. Let P be a partial ordering, M a countable elementary submodel of some large structure (the structure is typically H_θ for some large ordinal θ, never to be exactly spelled out) containing all the necessary information (the objects previously named in the argument, including the poset P). An *M-generic filter* $g \subset P$ is a filter on $P \cap M$ which intersects every dense subset of P which happens to be an element of the model M. The expression $M[g]$ describes the generic extension of the transitive collapse of the model M by the collapsed image of the filter g. If $\dot{x}$ is a P-name for an element of ω^ω then $\dot{x}/g$ is the element of ω^ω defined by $\dot{x}/g(n) = m \leftrightarrow \exists p \in g\ p \Vdash \dot{x}(\check{n}) = \check{m}$. The complexity of this operation is recorded in the following fact.

Fact 1.4.8. *Suppose that P is a forcing, $\dot{x}$ a P-name for an element of ω^ω, and M is a countable elementary submodel of a large enough structure. The set $A = \{y \in \omega^\omega : \exists g \subset M \cap P\ g$ is M-generic and $y = \dot{x}/g\}$ is Borel.*

Proof. Let $Q \subset r.o.(P)$ be the complete Boolean algebra generated by the name $\dot{x}$. Then $A = \{y \in \omega^\omega : \exists g \subset M \cap Q\ g$ is M-generic and $y = \dot{x}/g\}$. Let N by the transitive collapse of the model M, and consider the Polish space X of all N-generic filters on $\pi(Q)$ with the usual topology. Then A is the image of the space X under the continuous injection $g \mapsto \dot{x}/g$, and so A is Borel by a classical theorem of Lusin [40], 15.1. □

1.4.6 Absoluteness

The universally Baire sets (in particular, the analytic and coanalytic sets) have a natural interpretation in forcing extensions. Suppose $A \subset 2^\omega$ is universally Baire, as witnessed by trees $T, S \subset (2 \times \text{Ord})^{<\omega}$ which project to complements in all set forcing extensions and $A = \text{proj}[T]$. If $V[G], G \subset P$ is an arbitrary set forcing extension then $A^{V[G]}$, the interpretation of the set A in the model $V[G]$, is defined as $(\text{proj}[T])^{V[G]}$. A wellfoundedness argument shows that the interpretation does not depend on the choice of the witness trees T, S. I will use this feature to denote by $\dot{A}$ the P-name for the interpretation of the set A in the extension, and when speaking about this extended interpretation, I will frequently omit the superscript in the expression $A^{V[G]}$. This usage is commonplace throughout the book.

The following facts connecting the validity of certain sentences in generic extensions and the ground model are indispensable throughout the book.

Fact 1.4.9. *(Analytic absoluteness) Suppose that M is a transitive model of set theory, $\vec{x} \in M \cap \omega^\omega$ is a sequence of parameters, and ϕ is a Σ_1^1 formula with free variables. Then $\phi(\vec{x})$ holds if and only if $M \models \phi(\vec{x})$ holds.*

This is typically used in a situation where M is a generic extension of the transitive collapse of some countable elementary submodel of a large enough structure.

Fact 1.4.10. *(Shoenfield absoluteness) Suppose that M is a transitive model of set theory containing all countable ordinals, $\vec{x} \in M \cap \omega^\omega$ is a sequence of parameters, and ϕ is a Σ_2^1 formula with free variables. Then $\phi(\vec{x})$ holds if and only if $M \models \phi(\vec{x})$ holds.*

This is typically used in a generic extension with M equal to the ground model.

Fact 1.4.11. *[79] (Universally Baire absoluteness) (LC) Suppose that $\vec{A}$ is a finite sequence of universally Baire sets and M is a countable elementary submodel of some large structure containing $\vec{A}$. Suppose that $M[g]$ is a generic extension of the transitive collapse of the model M and $\vec{x} \in \omega$ is a finite sequence of parameters in the model $M[g]$. Suppose that ϕ is a formula quantifying over reals only. Then $\phi(\vec{x}, \vec{A})$ holds if and only if $M[g] \models \phi(\vec{x}, \vec{A})$ holds.*

Fact 1.4.12. *[45] (Σ_1^2 absoluteness) (LC+CH) Suppose that $\vec{A}$ is a finite sequence of universally Baire sets and ϕ is a formula of the form $\exists B \subset \omega^\omega \, \psi$ where ψ quantifies only over real numbers. If $\phi(\vec{A})$ holds in some generic extension, then $\phi(\vec{A})$ holds.*

1.4.7 Cardinal invariants of the continuum

The original motivation for the work contained in this book were the problems associated with comparison of cardinals defined in various ways from Polish spaces. I use [2] as a canonical reference.

Among the cardinal invariants that frequently occur in this book, let me quote $\mathfrak{a} =$ the least size of a maximal almost disjoint family of subsets of ω, $\mathfrak{b} =$ the least size of modulo finite unbounded subset of ω^ω, $\mathfrak{c} =$ the size of the continuum, $\mathfrak{d} =$ the least size of modulo finite dominating subset of ω^ω.

Given a σ-ideal I on a Polish space X, I will consider the cardinals $\mathrm{cov}(I) =$ the least number of sets in the ideal I necessary to cover the whole space, $\mathrm{non}(I) =$ the smallest possible size of an I-positive set, $\mathrm{add}(I) =$ the smallest size of a family of I-small sets whose union is not I-small, and $\mathrm{cof}(I) =$ the smallest possible size of a basis for the ideal I. It will be of advantage to consider starred variations of these cardinals: $\mathrm{cov}^*(I) = $ the least number of sets in the ideal I necessary to cover some Borel I-positive set, $\mathrm{non}^*(I) = $ the least cardinal such that every Borel I-positive set contains an I-positive subset of this size, and similarly for add^* and cof^*.

2

Basics

2.1 Forcing with ideals

2.1.1 The key definition

Definition 2.1.1. *Suppose that X is a Polish space and I is a σ-ideal on the space X. The symbol P_I denotes the partial order of I-positive Borel sets ordered by inclusion.*

I will always tacitly assume that the Polish space X is uncountable and the ideal I contains all singletons. There are several cases in which this will not hold, and they will be pointed out explicitly. Note that the poset P_I depends only on the membership of Borel sets in the ideal I, but it will frequently be of interest to look at the membership of non-Borel sets in I.

It is clear that the partial order P_I is not separative, and its separative quotient is the σ-algebra $\mathcal{B}(X)$ mod I. There is exactly one property all partial orders of this kind share.

Proposition 2.1.2. *The poset P_I adds an element $\dot{x}_{gen}$ of the Polish space X such that for every Borel set $B \subset X$ coded in the ground model, $B \in G$ iff $\dot{x}_{gen} \in B$.*

Proof. It is easy to see that the closed sets contained in the generic filter form a collection closed under intersection which contains sets of arbitrarily small diameter. A completeness argument shows that such a collection has a nonempty intersection containing a single point, and $\dot{x}_{gen}$ is a name for the single point in the intersection. Another way to describe the generic point is to say that it is the unique element in all basic open sets in the generic filter.

By induction on the complexity of the Borel set B prove that $B \Vdash \dot{x}_{gen} \in \dot{B}$. For closed sets this follows from the definition of the name $\dot{x}_{gen}$. Suppose that $B = \bigcup_n C_n$ and we already know that each set C_n forces $\dot{x}_{gen} \in \dot{C}_n$. Whenever $D \subset B$ is an I-positive Borel set then for some number n, $D \cap C_n$ is I-positive, $D \cap C_n \subset C_n$

and $D \cap C_n \Vdash \dot{x}_{gen} \in \dot{C}_n \subset \dot{B}$. By the genericity, $B \Vdash \dot{x}_{gen} \in \dot{B}$. Now suppose that $B = \bigcap_n C_n$ and we already know that each set C_n forces $\dot{x}_{gen} \in \dot{C}_n$. Then for every number n, $B \Vdash \dot{x}_{gen} \in \dot{C}_n$ since $B \subset C_n$. In other words, $B \Vdash \dot{x}_{gen} \in \bigcap_n C_n = \dot{B}$ as desired. Since the Borel sets in Polish spaces are obtained from closed sets by a repeated application of countable union and intersection, the induction is complete.

Now it is not difficult to prove that $C \Vdash \dot{x}_{gen} \in \dot{B}$ iff $C \setminus B \in I$. On one hand, if $C \setminus B \in I$ then every strengthening of the condition C is compatible with B and the previous paragraph applies to show that $C \Vdash \dot{x}_{gen} \in \dot{B}$. On the other hand, if $C \setminus B \notin I$, then $C \setminus B \subset C$ is a condition strengthening C which forces $\dot{x}_{gen} \in \dot{C} \setminus \dot{B}$ by the previous paragraph again, in particular $\dot{x}_{gen} \notin \dot{B}$.

The proposition follows. □

Note the key role played by the closure of the ideal I under countable unions in the argument. An important observation is that the forcings of the form P_I can be presented in various forms.

Definition 2.1.3. *Suppose I is a σ-ideal on a Polish space X. A different presentation of the poset P_I is a Borel bijection $f : X \to Y$ between X and another Polish space Y, the σ-ideal J on the space Y given by $A \in J \leftrightarrow f^{-1}A \in I$, and the resulting poset P_J.*

If f, J constitute a different presentation of the forcing P_I then the function f extends to a bijection $\hat{f} : P_I \to P_J$ given by $\hat{f}(A) = f''A$. Note that one-to-one Borel images of Borel sets are Borel by a theorem of Lusin [40], 15.1, and therefore the image of the function $\hat{f}$ indeed consists of Borel sets.

While a given forcing P_I can have many presentations, it is true that some presentations are more natural than others. In fact, I will frequently derive some forcing properties of the poset P_I from the topological features of a certain natural presentation. The forcing properties of P_I then persist through different presentations while the topological features may not. Note that there is a Borel bijection between any two uncountable Polish spaces, and so the nature of the Polish space does not restrict the kind of partial orders that can live on it. It may be occasionally difficult to decide whether a given presentation is the simplest possible one or the one most suitable to study.

2.1.2 Representation theorems

The study of the partial orders of the form P_I does entail a certain restriction in generality, but not too great a restriction. The following results show that many forcings encountered in practice can be presented as P_I for a suitable σ-ideal I on a Polish space.

Fact 2.1.4. *[68] Suppose that B is a σ-complete countably σ-generated Boolean algebra. Then there is a σ-ideal I on the Cantor space such that B is isomorphic to $\mathcal{B}(2^\omega)$ mod I.*

Corollary 2.1.5. *Suppose that P is a partially ordered set consisting of binary trees ordered by inclusion, such that for every tree $T \in P$ and every node $t \in T$ the tree $T \upharpoonright t$ is in P as well. Then P is in the forcing sense equivalent to a forcing of the form P_I.*

Proof. If $\dot{G}$ is a name for the generic filter write $\dot{x}_{gen}$ for the generic real: $\dot{x}_{gen} = \bigcup \bigcap \dot{G} \in 2^\omega$. Let $P \subset B$ be the complete Boolean algebra generated by the poset P. I will show that the σ-algebra $C \subset B$ σ-generated by the elements $b_t = |\check{t} \subset \dot{x}_{gen}|$: $t \in 2^{<\omega}$ is dense. By the previous fact the algebra C is isomorphic to some P_I and at the same time poset P is equivalent to it.

It is enough to show that for every tree $T \in P$ it is the case that $T = c_T$ where $c_T = \bigwedge_n \bigvee_{t \in 2^n \cap T} b_t$. It is clear that $T \leq c_T$. And if $c_T \not\leq T$ then there would be a tree $S \not\subset T$ such that $S \leq c_T$ and a node $s \in S \setminus T$ of length $n \in \omega$. Then $S \upharpoonright s \in P$ and clearly $S \upharpoonright s \Vdash \dot{x}_{gen} \upharpoonright n \notin \check{T}$, contradicting the assumption that $S \leq c_T$. $\quad\square$

There is frequently a more direct way of deriving the σ-ideal from the tree forcing in question.

Proposition 2.1.6. *Suppose that P is a partially ordered set consisting of binary trees ordered by inclusion such that for every tree $T \in P$ and every node $t \in T$ the tree $T \upharpoonright t$ is in P as well. Suppose moreover that P has the continuous reading of names. Then the collection $I = \{A \subset 2^\omega : A$ analytic and for no condition $T \in P$ it is the case that $[T] \subset A\}$ is a σ-ideal and the forcing P is in the forcing sense equivalent to P_I.*

Here the *continuous reading of names* is the statement that for every condition $T \in P$ and every name $\dot{f} \in \omega^\omega$ there is a condition $S \subset P$, natural numbers $n_0 \in n_1 \in \ldots$ and a function $g : \bigcup_m (S \cap 2^{n_m}) \to \omega$ such that for every number m and every sequence $t \in S \cap 2^{n_m}$ it is the case that $S \upharpoonright t \Vdash \dot{f}(\check{m}) = \check{g}(\check{t})$. This is a property frequently found in practice; consult Section 3.1 for a topological restatement of it.

Proof. Suppose that $A = \bigcup A_n : n \in \omega$ are analytic sets such that A contains all branches of some tree $T \in P$. I will produce a tree $S \subset T$ and a number $n \in \omega$ such that all branches of the tree S belong to the set A_n. This will prove the proposition.

Note that the forcing P adds a canonical generic point $\dot{x}_{gen} \in 2^{<\omega}$ which is a branch of all trees in the generic filter. Use a Shoenfield absoluteness argument to show that $T \Vdash \dot{x}_{gen} \in \dot{A}$ and therefore there is a condition $T' \subset T$ and a number n such that $T' \Vdash \dot{x}_{gen} \in \dot{A}_n$.

Let $U \subset (2 \times \omega)^{<\omega}$ be a tree such that $A_n = \mathrm{proj}[U]$. There is a name $\dot{f}$ for an element of the Baire space ω^ω such that $T' \Vdash \langle \dot{x}_{gen}, \dot{f} \rangle$ forms a branch through the tree $\check{U}$. Use the continuous reading of names to find a tree $S \subset T'$, natural numbers $n_0 \in n_1 \in \ldots$ and a function $g : \bigcup_m (S \cap 2^{n_m}) \to \omega$ such that for every number m and every node $t \in S \cap 2^{n_m}$ the condition $S \upharpoonright t$ forces $\dot{f}\check{m}) = \check{g}(\check{t})$. Then for every branch b through the tree S it must be the case that b together with the function $m \mapsto g(b \upharpoonright n_m)$ forms a branch through the tree U and therefore $b \in A_n$. I have just proved that $[S] \subset A_n$ as desired. $\qquad \square$

Partial orders for adding a real which do not consist of trees and the previous proposition cannot be applied to them are fairly rare in the practice of definable forcing. Nevertheless, many of them can be obtained through the methods of this book. The following is a characterization theorem which does not depend on the specific combinatorial form of the forcing.

Definition 2.1.7. *A forcing P is a* universally Baire real forcing *if*

1. *its conditions are elements of some Polish space Y;*
2. *there is a name $\dot{x}_{gen}$ for an element of some Polish space X;*
3. *there is a universally Baire set $A \subset X \times Y$ such that for every condition $p \in P$*
   $$P \Vdash \check{p} \in \dot{G} \leftrightarrow \langle \dot{x}_{gen}, \check{p} \rangle \in \dot{A};$$
4. *for every basic open set $O \subset X$ there is a condition $p \in P$ such that $P \Vdash \dot{x}_{gen} \in \dot{O} \leftrightarrow \check{p} \in \dot{G}$.*

Proposition 2.1.8. *[83] (LC) Every proper universally Baire real forcing is in the forcing sense equivalent to one of the form P_I.*

Proof. I claim that $I = \{B \subset X : B \text{ universally Baire and } P \Vdash \dot{x}_{gen} \notin \check{B}\}$ is the σ-ideal with the required properties. It is clear that I is closed under countable unions. Write $\dot{y}_{gen}$ for the P_I-name for its generic point in the space X, and let $\dot{G}$ be the P_I-name for the set $\{p \in \check{P} : \langle \dot{y}_{gen}, \check{p} \rangle \in \dot{A}\}$. It will be enough to show that $P_I \Vdash \dot{G} \subset \check{P}$ is a V-generic filter; the proposition then follows by standard abstract forcing considerations. Suppose that $B \in P_I$ is a condition, $p, q \in P$ are conditions such that $B \Vdash \check{p}, \check{q} \in \dot{G}$ and $D \subset P$ is open dense. I must find a condition $B' \in P_I$ and a condition $r \in P$ such that $B' \subset B$, $r \leq p, q$, $r \in D$, and $B' \Vdash \check{r} \in \dot{G}$.

Let M be a countable elementary submodel of a large enough structure, let Z be the Polish space of all M-generic filters on P with the usual topology, let $f : Z \to X$ be a map defined by $f(g) = \dot{x}_{gen}/g$. This map is continuous by (4) and injective by (3) of the definition of universally Baire real forcing. Thus the range $f''Z$ is Borel by a classical theorem of Luzin [40], 15.1. Write $C = B \cap \mathrm{rng}(f)$ and for every condition $r \in P \cap M$ write $C_r = C \cap f''O_r$ and $\bar{C}_r = C \setminus f''O_r$, where O_r is the open set of all filters in the space Z containing the condition r. Now,

- $C \notin I$. To see this, note that as $B \notin I$, there must be a condition $r \in P$ such that $r \Vdash \dot{x}_{gen} \in \dot{B}$. By elementarity, there must be such a condition in the model M. Any M-master condition below r forces $\dot{x}_{gen} \in \dot{C}$, and so $C \notin I$ as required.

- For every condition $r \in P \cap M$, $C_r \Vdash \check{r} \in \dot{G}$ and $\bar{C}_r \Vdash \check{r} \notin \dot{G}$ if these sets are I-positive. To see this, note $\forall x \in C_r$ $M[x] \models \langle x, r \rangle \in A$, by an absoluteness argument $\forall x \in C_r$ $\langle x, r \rangle \in A$, and by the universally Baire absoluteness this statement will still be true in the P_I extension, in particular $C_r \Vdash \dot{y}_{gen} \in \dot{C}_r$ and $\langle \dot{y}_{gen}, \check{r} \rangle \in \dot{A}$.

- $\bar{C}_p, \bar{C}_q \in I$ and so $C_p \cap C_q \notin I$. This follows from the previous item.

- The sets $C_r : r \in D \cap M$ is a lower bound of p, q cover the I-positive set $C_p \cap C_q$, therefore one of them is I-positive, and $C_r = B' \subset B$ is the required condition.

This completes the proof. □

Example 2.1.9. Consider the Sacks forcing P of all perfect binary trees ordered by inclusion. Corollary 2.1.5, Proposition 2.1.6 and Proposition 2.1.8 all can be used to show that $P = P_I$ for some σ-ideal I. None of this abstract reasoning can replace the information obtained from the perfect set theorem: the σ-ideal I is the ideal of countable subsets of 2^ω.

2.1.3 Generalizations

There are several ways in which the previous ideas can be generalized, each of them important and deserving a thorough discussion.

First, one can consider forcing with analytic (projective, universally Baire, etc.) sets positive with respect to a given σ-ideal I. For most of the forcings considered in this book it will be the case that every I-positive universally Baire set has an I-positive Borel subset, and so the poset P_I is dense in all of these variations, and under large cardinal assumptions it is dense in the poset $(\mathcal{P}(X) \mod I)^{L(\mathbb{R})}$ – refer to Section 3.9 for a thorough discussion. Nevertheless, I will have to enter situations in which this property has not been verified yet, and then the following definition and proposition will be important.

Definition 2.1.10. *Suppose that I is a σ-ideal on a Polish space X. The symbol Q_I stands for the poset of I-positive analytic sets ordered by inclusion.*

Proposition 2.1.11. *Suppose that the σ-ideal I is generated by coanalytic sets. There is a Q_I-name $\dot{x}_{gen}$ for an element of the Polish space X such that an analytic set belongs to the generic filter if and only if it contains the point $\dot{x}_{gen}$ in the extension.*

Proof. I will handle the case of $X = 2^\omega$, the other spaces being Borel bijective images of 2^ω. As in the P_I case, let $\dot{x}_{gen}$ be the unique point in the intersection of all basic open sets in the generic filter.

First note that any set $A \in Q_I$ forces $\dot{x}_{gen} \in \dot{A}$. To see this, let $T \subset (2 \times \omega)^{<\omega}$ be a tree such that $A = \text{proj}[T]$, let $G \subset Q_I$ be a generic filter containing the condition A and in the generic extension let $S \subset \omega^{<\omega}$ be the tree consisting of all nodes $t \in \omega^{<\omega}$ such that $\text{proj}[T \restriction \langle \dot{x}_{gen} \restriction |t|, t \rangle] \in \dot{G}$. Clearly $0 \in S$ and it will be enough to show that S contains no terminal nodes. Well, if $t \in S$ is a node and $B \subset A$ is a condition forcing $\check{t} \in \dot{S}$ then strengthening the condition B if necessary I may assume that there is a binary sequence s such that $B \subset \text{proj}[T \restriction \langle s, t \rangle]$. By the σ-additivity of the ideal I there must be a number $n \in \omega$ and a bit $b \in 2$ such that $C = B \cap \text{proj}[T \restriction \langle s^\frown b, t^\frown n \rangle] \notin I$. Clearly, $C \Vdash \check{t}^\frown \check{n} \in \dot{S}$ as required.

Second, if $A, B \in Q_I$ are sets and $A \Vdash \dot{x}_{gen} \in \dot{B}$ then $A \cap B \notin I$: if $A \cap B \in I$ then let $C \in I$ be a coanalytic set including it as a subset, and $A \setminus C \in Q_I$ is a condition which forces the point $\dot{x}_{gen}$ into itself by the previous paragraph, and by the analytic absoluteness it forces $\dot{x}_{gen} \notin B$, contradicting the assumption. But now $A \cap B \in Q_I$ is a common lower bound of A, B, forcing $B \in \dot{G}$. $\square$

It is remarkable that in all cases when I need to use the forcing Q_I it is only to show that in fact $P_I \subset Q_I$ is dense. However, the statement that every I-positive analytic set has an I-positive Borel subset seems to be interesting in its own right. See Section 3.9 on this and similar dichotomies.

The second way to generalize the forcings of the form P_I is to consider spaces of the form Y^ω with an uncountable set Y and the standard tree topology instead of Polish spaces X. For a sequence $t \in Y^{<\omega}$ let O_t be the basic open set determined by t, $O_t = \{x \in Y^\omega : t \subset x\}$.

Proposition 2.1.12. *Suppose that Y is a set and I is a σ-ideal on the space Y^ω with the following closure property:*

(*) *if $A_t : t \in Y^{<\omega}$ are sets in the ideal with $A_t \subset O_t$, then $\bigcup_t A_t \in I$.*

There is a name $\dot{x}_{gen}$ for an element of the space Y^ω such that in the generic extension by the poset P_I, a Borel set $B \subset Y^\omega$ belongs to the generic ultrafilter if an only if $\dot{x}_{gen} \in \dot{B}$.

Proof. Let $\dot{x}_{gen} = \bigcup \{t \in Y^{<\omega} : O_t \in G\}$ where G is the P_I-generic filter. It is clear that the sequences in the union are linearly ordered. Moreover if $n \in \omega$ is a number and $B \in P_I$ is a condition the one of the sets $B \cap O_t : t \in Y^n$ is I-positive by the property (*) and forces the corresponding sequence into the union defining the sequence $\dot{x}_{gen}$. Thus $P_I \Vdash \dot{x}_{gen} \in Y^\omega$.

Before the remainder of the proof note that similarly to Borel subsets of 2^ω the Borel subsets of Y^ω have natural interpretations in every generic extension which does not depend on the particular Borel definition of the set.

By induction on the complexity of the Borel set $B \in P_I$ I will show that $B \Vdash \dot{x}_{gen} \in \dot{B}$, where $\dot{B}$ denotes the interpretation of the Borel set B in the extension. Suppose first that B is open. If $C \subset B$ is any condition then by the property (*) there must be a sequence $t \in Y^{<\omega}$ such that $O_t \subset B$ and $C \cap O_t \notin I$. Clearly $C \cap O_t \subset C$ is a condition forcing $\dot{x}_{gen} \in \dot{B}$ and so $B \Vdash \dot{x}_{gen} \in \dot{B}$. The remaining steps in the induction are the same as in Proposition 2.1.2.

Now suppose that $B, C \in P_I$ are sets such that $B \Vdash \dot{x}_{gen} \in \dot{C}$. I must show that $B \cap C \notin I$; then $B \cap C$ is the required lower bound of the conditions B, C. Suppose $B \cap C \in I$. Then $B \setminus C$ is a condition in P_I which by the previous paragraph forces $\dot{x}_{gen}$ into $B \setminus C$ and outside of the set C, contradicting the choice of the set B. □

Example 2.1.13. Namba forcing [54]. Let $Y = \omega_2$ and let I be the ideal of sets $B \subset Y^\omega$ such that there is a map $f : Y^{<\omega} \to \omega_2$ such that $B \subset B_f = \{y \in Y^\omega : \exists^\infty n \; y(n) \in f(y \restriction n)\}$. It is not difficult to see that the ideal I has the closure property (*) from the previous proposition. I will show that a Borel set $B \subset Y^\omega$ is I-positive if and only if it contains all branches of some Namba tree, that is an infinite tree $T \subset Y^{<\omega}$ such that all but finitely many of its nodes have $\aleph_2$ many immediate successors. This means that the Namba forcing is in a natural sense isomorphic to a dense subset of the poset R_I.

Let $B \subset Y^\omega$ be a Borel set, and consider the game G between Player I and II. Player I produces a sequence of ordinals $\alpha_n \in \omega_2 : n \in \omega$ and Player II in response produces a sequence of ordinals $\beta_n \in \omega_2 : n \in \omega$. Moreover Player II must raise a flag at some round m and for all $n > m$ it must be the case that $\alpha_n \in \beta_n$. Player II wins if his sequence of answers belongs to the Borel set B. The payoff set of the game G is Borel and therefore determined by Fact 1.4.2. I will be finished if I show that Player I has a winning strategy iff $B \in I$ and Player II has a winning strategy iff B contains all branches of some Namba tree.

Suppose first that Player I has a winning strategy σ. For every sequence $t \in Y^{<\omega}$ there are at most $|t|$ many ways how the play could reach a position in which Player I followed his strategy σ and Player II produced the sequence t, depending on where and if Player II decided to raise the flag. Let $f(t) = $ maximum of all the possible answers by the strategy σ in that position. It is easy to see that $B \subset B_f$.

On the other hand, suppose that Player II has a winning strategy σ, and let t be some sequence for which there is a position in which Player II followed his strategy

σ, produced the sequence t and raised the flag at that point. It is easy to see that then $[T] \subset B$ for some Namba tree T with trunk t.

Another generalization is to consider spaces X^Y for a Polish space X and an uncountable set Y with the standard product topology, a σ-ideal I on it and a partial order R_I of I-positive Baire sets ordered by inclusion. Here the Baire sets are those subsets of the space X^Y obtained from basic open sets by countable repetition of countable unions, countable intersections, and complementation. The basic open sets are those of the form $\{\vec{x} \in X^Y : \vec{x}(y) \in O\}$ for some basic open set $O \subset X$ and an index $y \in Y$. Such partial orders are the results of the countable support iterations or products or the tower technology of Section 5.5. Let me include the basic property here, and defer the detailed treatment to that section.

Proposition 2.1.14. *There is a R_I-name $\vec{x}_{gen}$ for a function from Y to X such that a Baire set $A \subset X^Y$ belongs to the generic filter if and only if it contains the function $\vec{x}_{gen}$.*

Still another generalization is to consider partial orders $\mathcal{P}(Y)$ mod I for a suitable set Y and an ideal I on it. These partial orders lack the basic feature of the previously considered cases: the canonical generic object as an element of some ground model coded simple space. The case $Y = \omega$ has been extensively studied [15], [74], [87]. The case $Y = \omega_1$ and $I =$ the nonstationary ideal has been the subject of the precipitousness and saturation considerations. The general case of a σ-ideal I has been studied by Gitik and Shelah [24], [25] who showed that the resulting partial orders cannot be in the forcing sense equivalent to most of the forcings of the form P_J, where J is a σ-ideal on a Polish space.

2.1.4 Basic definability issues

This book deals with *suitably definable* σ-ideals on Polish spaces, with very few exceptions. The demands on definability vary depending on the large cardinal axioms one is willing to use. This section spells out several definitions used throughout the book.

In the presence of large cardinal axioms such as the existence of a supercompact cardinal, the following definability restriction is used.

Definition 2.1.15. *A σ-ideal I on a Polish space X is* universally Baire *if for every universally Baire set $A \subset 2^\omega \times X$ the set $\{y \in 2^\omega : A_y \in I\}$ is universally Baire.*

Without large cardinals more sophisticated notions of definability and absoluteness are needed.

Definition 2.1.16. *A σ-ideal I on a Polish space X is* ZFC-correct *if it is defined by a formula ϕ with a possible real parameter r (so that $I = \{A \subset X : \phi(A, r)\}$) and every transitive model M of a large fragment of ZFC containing r is correct about I on its analytic sets (so that if $s \in M$ is a code for an analytic set A_s then $\phi(A_s, r) \leftrightarrow M \models \phi(A_s, r)$).*

Note that this definition speaks really about the formula defining the ideal rather than the ideal itself. It turns out that nearly all definitions of σ-ideals considered in this book are ZFC-correct in this sense. This assertion is never completely trivial though and its proof is surprisingly close to the determinacy dichotomy and properness arguments used for other purposes.

Example 2.1.17. The ideals associated with Hausdorff submeasures as in Definition 4.4.1 are ZFC-correct. To see this, fix a Hausdorff submeasure ψ on a Polish space X with the associated σ-ideal I generated by sets of finite ψ-mass. Given an analytic set $A \subset X$ let $C \subset X \times \omega^\omega$ be a closed set which projects to A, and consider the integer game $G(C)$ as in the proof of Theorem 4.4.5. Player I has a winning strategy in the game $G(C)$ if and only if $A \in I$. Now given a transitive model M containing the set C, $M \models G(C)$ is determined. The winning strategy the model M finds is still a winning strategy in V since the nonexistence of a successful counterplay is a wellfoundedness statement. Thus the statement $A \in I$ is absolute between M and V.

A measure-theoretic counterpart of the above definition is the following.

Definition 2.1.18. *A submeasure ψ on a Polish space X is* ZFC-correct *if it is defined by a formula ϕ with a possible real parameter r (so that $\psi(A) < q \leftrightarrow \phi(A, q, r)$ for every set $A \subset X$) and every transitive model M of a large fragment of ZFC evaluates ψ-mass correctly (so that if $s \in M$ is a code for an analytic set A_s then $\psi(A_s, r, q) \leftrightarrow M \models \psi(A_s, r, q)$ for every rational number q).*

Example 2.1.19. Every pavement submeasure defined from a countable set of Borel pavers is ZFC-correct. Let ψ be the pavement submeasure on a Polish space X, let $A \subset X$ be an analytic set, let $C \subset X \times \omega^\omega$ be a closed set projecting to A, and let q be a rational number. $\psi(A) < q$ if and only if there is a rational number $q' < q$ such that Player I has a winning strategy in the game $G(C, q')$ as in the proof of Theorem 4.5.6. As in the previous arguments, whenever M is a transitive model containing the set C then it finds a winning strategy for one of the players in the games $G(C, q')$ for all rationals q', and these winning strategies of the model M stay winning in V. Thus M evaluates the ψ-mass of the set A correctly.

Example 2.1.20. Every outer regular strongly subadditive capacity is ZFC-correct. It is possible to supply the same argument as above using the integer game from

Theorem 4.3.6, however here I can use an argument which at least on the surface has no game theoretic content. Let ψ be a strongly subadditive capacity on a Polish space X, determined by its values on the sets from some fixed countable basis $\mathcal{O}$ closed under finite unions. The key fact: Fact 4.3.5, showing that the capacity ψ is simply derivable from its values on basic open sets. Now let $A \subset X$ be an analytic set, a projection of a closed subset $C \subset X \times \omega^\omega$. Let M be a transitive model containing the code for the set C, and let $q > 0$ be a rational number. By the definitions, if $M \models \psi(A) < q$ then $M \models \exists O \subset X$ O is open, $\psi(O) < q$ and $A \subset O$, this set O maintains these properties in V by a wellfoundedness argument, and therefore even in V, $\phi(A) < q$. What happens though if $M \models \psi(A) > q$? The key fact mentioned above implies that $M \models \psi$ is a capacity, and by the Choquet's capacitability theorem $M \models \exists K \subset X \times \omega^\omega$ K compact, $K \subset C$ and $\psi(\text{proj}(K)) > q$. Now the set K maintains these properties in V by a wellfoundedness argument. Note that $p(K) \subset X$ is a compact set, and therefore its ϕ-mass is the infimum of $\phi(O) : O \in \mathcal{O}, K \subset O$, a computation which works the same in the model M as in V by a wellfoundedness argument again.

The ZFC-correctness is a useful tool in a number of situations such as in the statement of ZFC-provable preservation theorems. Nevertheless, I will need a more sophisticated and more restrictive notion as well. Unlike the ZFC-correctness, it can be stated without a reference to models of ZFC and it has been studied in descriptive set theory for at least a century.

Definition 2.1.21. *A σ-ideal I on a Polish space X is $\mathbf{\Pi}^1_1$ on $\mathbf{\Sigma}^1_1$ if for every analytic set $A \subset 2^\omega \times X$ the set $\{y \in 2^\omega : A_y \in I\}$ is coanalytic.*

Unlike the ZFC-correctness which places no significant restrictions on the forcing properties of the poset P_I, the $\mathbf{\Pi}^1_1$ on $\mathbf{\Sigma}^1_1$ condition does have important forcing consequences – its associated forcing can never add dominating reals. This notion is studied in detail in Section 3.8. Here, let me just include two connections with ZFC-correctness.

Proposition 2.1.22. *If a σ-ideal I on a Polish space X is provably $\mathbf{\Pi}^1_1$ on $\mathbf{\Sigma}^1_1$ then it has a ZFC-correct definition.*

Proof. Let $A \subset 2^\omega \times X$ be a universal analytic set and $C \subset 2^\omega$ be a coanalytic set such that ZFC proves $\forall y \in 2^\omega$ $A_y \in I \leftrightarrow y \in C$. Every transitive model M evaluates the membership of a point $y \in 2^\omega$ in the set C correctly by a wellfoundedness argument. Thus M evaluates the membership in the ideal I correctly as well. $\square$

Proposition 2.1.23. *Every ZFC-correct ideal is $\mathbf{\Delta}^1_2$ on $\mathbf{\Sigma}^1_1$. Every ZFC-correct submeasure is $\mathbf{\Delta}^1_2$ on $\mathbf{\Sigma}^1_1$.*

Proof. Let I be a ZFC-correct σ-ideal on a Polish space X and let $A \subset 2^\omega \times X$ be an analytic set. I must show that the set $\{y \in 2^\omega : A_y \in I\}$ is $\mathbf{\Delta}_2^1$ on $\mathbf{\Sigma}_1^1$. To see this note that $A_y \in I \leftrightarrow$ for every countable model M containing the real y, either M is illfounded or $M \models A_y \in I$, and $A_y \notin I \leftrightarrow$ for every countable model M containing the real y, either M is illfounded or $M \models A_y \notin I$.

Let ϕ be a ZFC-correct submeasure on a Polish space X, let $\varepsilon \in \mathbb{R}^+$ be a real number, and let $A \subset 2^\omega \times X$ be an analytic set. I must show that the set $\{y \in 2^\omega : \phi(A_y) < \varepsilon\}$ is $\mathbf{\Delta}_2^1$ on $\mathbf{\Sigma}_1^1$. This is proved in the same way as in the previous paragraph. $\qquad\square$

2.2 Properness

The following definition has been central to the development of the forcing theory in the last several decades.

Definition 2.2.1. *[64] A forcing notion P is* proper *if for every set X and every stationary set $S \subset [X]^{\aleph_0}$ it is the case that $P \Vdash \check{S}$ is stationary. Another equivalent restatement is the following. The forcing P is* proper *if for every large enough cardinal θ, every countable elementary submodel $M \prec H_\theta$ containing P and every condition $p \in P \cap M$ there is an M-master condition $q \leq p$; that is, a condition q forcing $\dot{G} \cap \check{M}$ meets every dense subset of P which is an element of M, where $\dot{G}$ is the name for the P-generic filter.*

It turns out that in the context of definable forcing this is exactly the right notion. In its presence there is a rich structure and extensive theory, in its absence there is collapse. I will first restate it in the terms of σ-ideals:

Proposition 2.2.2. *Suppose that I is a σ-ideal on a Polish space X. The following are equivalent:*

1. *the forcing P_I is proper;*
2. *for every countable elementary submodel M of a large enough structure and every condition $B \in M \cap P_I$ the set $C = \{x \in B : x$ is M-generic$\}$ is not in the ideal I.*

Here, a point $x \in X$ is M-generic if the collection $\{A \in P_I \cap M : x \in A\}$ is a filter on $P_I \cap M$ which meets all open dense subsets of the poset P_I that are elements of the model M.

Proof. This is just a restatement of the definitions. First note that the set C is Borel: $C = B \cap \bigcap \{\bigcup(D \cap M) : D \in M$ is an open dense subset of the poset $P_I\}$. If $C \in I$

then $B \Vdash \dot{x}_{gen} \notin \dot{C}$ and so $\dot{G} \cap M$ is not M-generic by the definition of the set C; therefore there can be no M-master condition below the set B. On the other hand, if $C \notin I$ then $C \in P_I$ is a condition forcing $\dot{x}_{gen} \in \dot{C}$ and so $\dot{G} \cap M$ is M-generic by the definition of the set C; thus C is the required master condition. □

It is frequently a difficult job to decide the status of properness of a forcing of the form P_I, and a large part of this book is devoted to just that. Let me here include just three comparatively simple examples.

Example 2.2.3. Let $A \subset 2^\omega$ be an analytic non-Borel set and let I be the σ-ideal generated by Borel sets $B \subset 2^\omega$ such that the set $A \cap B$ is Borel. This is a nontrivial σ-ideal, the forcing P_I is not proper though. Let M be a countable elementary submodel of a large structure and consider the Borel set $B = \{x \in 2^\omega : x$ is M-generic$\}$. I will show that the set $A \cap B$ is Borel, so $B \in I$ and P_I is not proper. Suppose $x \in B$ is a point. We have $x \in A$ iff $M[x] \models x \in A$ (by analytic absoluteness) iff $\exists C \in P_I \cap M \ x \in C \wedge C \Vdash \dot{x}_{gen} \in \dot{A}$ (by the forcing theorem) iff $x \in \bigcup \{C \in P_I \cap M : C \Vdash \dot{x}_{gen} \in \dot{A}\}$. Thus $A \cap B$ is Borel as desired.

Example 2.2.4. Let X denote the Hilbert cube $[0, 1]^\omega$. It is a fundamental result of infinite-dimensional topology that X cannot be covered by countably many zero-dimensional sets [76], 4.8.5. Let I be the σ-ideal on X generated by the zero-dimensional sets. The forcing P_I adds a countable sequence $\vec{x}_{gen}$ of real numbers in $[0, 1]$. It turns out that the forcing collapses $\mathfrak{c}$ to $\aleph_0$. To see this, note that for every real $r \in \mathbb{R}$ the set $A_r \subset X$ which is the complement of the product of infinitely many copies of the set $\{r + q : q \in \mathbb{Q}\}$ is zero-dimensional; thus $P_I \Vdash \vec{x}_{gen} \notin \dot{A}_r$. Restated, in the generic extension, for every ground model real r there is a natural number n and a rational q such that $\vec{x}_{gen}(n) - q = r$. It follows that the set of ground model reals must be countable.

Example 2.2.5. Let I be the σ-ideal of countable subsets of 2^ω. Let M be a countable elementary submodel of a large enough structure and let $B \in M \cap P_I$ be a condition. To prove the properness of the forcing P_I it is enough to show that the set of M-generic points in the set B is uncountable. Let $y_n : n \in \omega$ be a countable list of infinite binary sequences; I must produce an M-generic sequence $x \in B$ which is not on the list. Let $D_n : n \in \omega$ be an enumeration of all open dense subsets of P_I that are elements of the model M. By induction on $n \in \omega$ build a descending chain $B = B_0 \supset B_1 \supset \ldots$ of conditions in $P_I \cap M$ such that for every number $n \in \omega$, $B_{n+1} \in D$ and $y_n \notin B_{n+1}$. To construct B_{n+1}, first note that since B_n is uncountable there must be a basic open set O such that $y_n \notin O$ and $B_n \cap O$ is still uncountable. Then use the elementarity of the model M to find a condition $B_{n+1} \subset B_n \cap O$ in the set $D_n \cap M$. By Proposition 2.1.2, there is a point $x \in \bigcap_n B_n$. A review of the construction shows that this is the desired M-generic point in the set B which is not among the points $y_n : n \in \omega$.

It is possible to formulate a host of various conjectures stating that more or less every forcing of the form P_I is proper. Most of them will be easily refuted by the following proposition. This is the richest source of ideals for which the quotient forcing is not proper.

Proposition 2.2.6. *Suppose that I is a σ-ideal on a Polish space X. The following are equivalent:*

1. *P_I adds a countable set of ground model reals which is not covered by any ground model countable set;*
2. *$I = \bigcap_n I_n$ where $I_n : n \in \omega$ is an inclusion decreasing nonstabilizing sequence of σ-ideals, that is for every Borel set $B \notin I$ and every number $n \in \omega$, $\{C \subset B : C \in I, C \text{ Borel}\} \neq \{C \subset B : C \in I_n, C \text{ Borel}\}$.*

In particular, in the presence of CH the second item is equivalent to the forcing P_I collapsing $\aleph_1$, and in the presence of CH and large cardinals (see below) the second item is equivalent to the improperness of the forcing P_I below every condition. Note that this gives a characterization of properness which does not mention forcing at all. One peculiar feature of this proposition is that using it I can produce a variety of σ-ideals I such that the forcing P_I is not proper, but I do not know how to prove in ZFC that they collapse any cardinals, see Question 7.1.2.

Proof. In (1)→(2) direction, choose a name $\dot{f}$ for a function from ω to $\check{\mathbb{R}}$ whose range is forced not to be included in any ground model countable set. For every number $n \in \omega$ let I_n be the σ-ideal generated by I and all Borel sets B for which there is a countable set $a \subset \mathbb{R}$ such that $B \Vdash \forall i \in n \ \dot{f}(i) \in \check{a}$. It is immediate that the σ-ideals $I_n : n \in \omega$ are as required in the second item.

For the (2)→(1) direction, suppose $I = \bigcap_n I_n$ as in (2). Note that the sets $P_I \cap I_n : n \in \omega$ are all dense in the poset P_I by the nonstabilizing property. Let $A_n \subset P_I \cap I_n$ be a maximal antichain for every number $n \in \omega$ and in the generic extension consider the set $a = \{B \in G : \exists n \in \omega \ B \in A_n\}$. It will be enough to show that no condition forces the set a to be covered by a countable ground model set, and to show this, it will be enough in turn to prove that for every condition $C \in P_I$ there is a number $n \in \omega$ such that C is compatible with uncountably many elements of the maximal antichain A_n. And indeed, there must be a number n such that $C \notin I_n$, and then for every countable collection $b \subset A_n$ the set $\bigcup b$ is I_n-small and the set $C \setminus \bigcup b$ is I_n-positive, and so a condition in the forcing P_I avoiding all the conditions in the set b. $\qquad\square$

As long as it seems to be difficult to verify the status of properness for a given forcing of the form P_I, a logician will attempt to tackle the two related questions: the complexity of the notion of properness and its absoluteness for definable partial orders. The main obstacle to their solution is the quantification over all countable elementary submodels of a large structure in the definition of properness. A forcing practitioner knows that in all known definable cases, the proof of properness goes through various fusion arguments and does not really consider the enormous structure H_θ. I cannot prove a completely general theorem to this effect, even though in certain classes of forcings there is some information, see Sections 3.10.9 and 3.10.10. The following proposition isolates the strongest statement that I know to be equivalent to properness for all definable forcings.

Proposition 2.2.7. *(LC) Suppose that P is a universally Baire forcing. Exactly one of the following is true:*

1. *P is proper;*
2. *there is a condition $p \in P$ which forces the set $([P]^{\aleph_0})^V$ to be nonstationary.*

This is an immediate corollary of the determinacy results in Section 3.10.2.

Corollary 2.2.8. *(LC + CH) If P is a universally Baire forcing on the reals then either P is proper or else P collapses $\aleph_1$ below some condition.*

It is impossible to remove the assumption of CH in this corollary. Consider the following example: If c is a Cohen real then by a result of Gitik the set $A = \{a \in [\omega_2]^{\aleph_0} : a \notin V\}$ is stationary in the extension $V[c]$. Let Q be the forcing shooting a closed unbounded set through the set A with countable conditions. It is not too difficult to show that if $\delta_2^1 = \omega_2$ then the iteration $P = \text{Cohen}^* \dot{Q}$ can be coded as a universally Baire forcing. Certainly P preserves $\aleph_1$ and P is not proper.

Corollary 2.2.9. *(LC + CH) Suppose that P is a proper universally Baire forcing on the reals. Then (the forcing with the definition of) P is proper in all set generic extensions.*

Proof. Suppose for contradiction that in some set generic extension $V[G]$ the forcing P is not proper. In a further σ-closed forcing extension $V[G][H]$, CH holds and $P^{V[G]} = P^{V[G][H]}$ is still not proper, and so by the previous corollary it collapses $\aleph_1$ below some condition. This is a Σ_1^2 statement with the universally Baire parameter P, and by the Σ_1^2 absoluteness 1.4.12, it must hold in V, contradicting the assumption of the properness of P in V. $\qquad\square$

I do not know if I can remove the CH assumption here, or perhaps the large cardinal assumption at the price of reducing the complexity of the partial orders to which the theorem applies.

2.3 Topological representation of names

The key feature of proper forcings of the form P_I is the way of topologizing the names for reals and Borel sets described in the following propositions.

Proposition 2.3.1. *Suppose that I is a σ-ideal on a Polish space X such that the forcing P_I is proper. Suppose Y is a Polish space, $B \in P_I$ is a condition and $B \Vdash \dot{y} \in Y$ is a point. Then there is a condition $C \subset B$ in the forcing P_I and a Borel function $f : C \to Y$ such that $C \Vdash \dot{y} = f(\dot{x}_{gen})$.*

Proof. Fix a countable base $\mathcal{O}$ for the topology of the space Y. Let M be a countable elementary submodel of a large enough structure and let $C = \{x \in B : x$ is M-generic$\}$. This is a Borel I-positive set. Consider the function $f : C \to Y$ assigning each point x the value $\dot{y}$ evaluated according to the filter generated by the point x.

First of all, f is a Borel function, since its graph can be written as $C \cap \bigcap_{O \in \mathcal{O}} A_O$ where $A_O = \bigcup\{z \times O : z \in P_I \cap M, z \Vdash \dot{y} \in \dot{O}\} \cup \bigcup\{z \times (X \setminus O) : z \in P_I \cap M, z \Vdash \dot{y} \notin \dot{O}\}$ by the forcing theorem. Second, $C \Vdash \dot{y} = f(\dot{x}_{gen})$ since $C \Vdash \dot{x}_{gen} \in \dot{C}$ is an M-generic point and so the point $\langle \dot{x}_{gen}, \dot{y} \rangle$ must satisfy the definition of the graph of the function f. The proposition follows. $\qquad\square$

Proposition 2.3.2. *Suppose that I is a σ-ideal on a Polish space X such that the forcing P_I is proper. Suppose that Y is a Polish space, $B \in P_I$ is a condition and $B \Vdash \dot{A} \subset \dot{Y}$ is a Borel set. Then there is a condition $C \subset B$ and a Borel set $D \subset C \times Y$ such that $C \Vdash \dot{A} = \dot{D}_{\dot{x}_{gen}}$.*

Proof. Since the forcing P_I preserves $\aleph_1$, strengthening the condition B if necessary I may assume that the Borel rank of $\dot{A}$ is forced to be $\leq \alpha$ for some fixed ordinal α. The proof of the proposition proceeds by induction on the ordinal α.

If $\dot{A}$ is forced to be closed, then again strengthening the condition B if necessary and using Proposition 2.3.1, I may assume that there is a Borel function $f : B \to \mathcal{P}(\mathcal{O})$ such that $B \Vdash f(\dot{x}_{gen}) = \{O \in \mathcal{O} : O \cap \dot{A} = 0\}$. Then the set $D \subset B \times Y, D = \{\langle x, y \rangle : y \notin \bigcup f(x)\}$ is the required Borel set. The proof for open sets is similar.

Suppose now that $\dot{A}$ is forced to be the union $\bigcup_n \dot{A}(n)$ of sets of lower complexity. Use the inductive assumption to find maximal antichains $Z(n) \subset P_I : n \in \omega$ such that for every condition $z \in Z(n)$ there is a Borel set $D(z, n) \subset z \times Y$ such that $z \Vdash \dot{A}(n) = \dot{D}(z, n)_{\dot{x}_{gen}}$. Let M be a countable elementary submodel of a large enough structure and consider the set $C = \{x \in B : x$ is M-generic$\}$. For every number $n \in \omega$ let $D(n) \subset C \times Y$ be the set $\bigcup\{D(z, n) : z \in M \cap Z(n)\} \cap C \times Y$, and let $D = \bigcup_n D(n)$. Clearly the set $D \subset C \times Y$ is Borel. It is not difficult to verify that $C \Vdash \dot{A} = \dot{D}_{\dot{x}_{gen}}$. Namely, the condition C forces that the generic point belongs to exactly one condition in the antichain $Z(n) \cap M$ for every number n, so then $\dot{A}(n) = \bigcup\{D(z, n) : z \in M \cap Z(n)\}_{\dot{x}_{gen}} = \dot{D}(n)_{\dot{x}_{gen}}$ and $\dot{A} = \bigcup_n \dot{D}(n)_{\dot{x}_{gen}}$ as desired. The countable intersection case lends itself to an identical argument. $\qquad\square$

Note that the argument does not give an ordinal correspondence, a statement of the kind "if $\dot{A}$ is a Borel set forced to be $\mathbf{\Sigma}^0_\alpha$ then it is equal to the generic section of a $\mathbf{\Sigma}^0_\alpha$ ground model set." Such a statement is true for bounding forcings though.

The identification of names for elements of Polish spaces with Borel functions has a number of consequences; I will state three that are used throughout the book.

Proposition 2.3.3. *Suppose that I is a σ-ideal on a Polish space X such that the forcing P_I is proper. Suppose that $B \in P_I$ is a Borel set, Y is a Polish space, $f : B \to Y$ is a Borel function, and $\{A_n : n \in \omega\}$ are analytic or coanalytic subsets of the space X. Then there is a Borel set $C \subset B$ such that for every number $n \in \omega$, the image $f''(C \cap A_n) \subset Y$ is a Borel set.*

Proof. The expression $\dot{y} = \dot{f}(\dot{x}_{gen})$ is a P_I-name for a point in the space Y. Write $P_I = P * \dot{Q}$ where P adds the point $\dot{y}$ and $\dot{Q}$ is the remainder forcing, adding the point $\dot{x}_{gen}$. For every number n let $p_n \in P$ be the Boolean value of the statement $\exists q \in \dot{Q} \; q \Vdash \dot{x}_{gen} \in \dot{A}_n$.

Let M be a countable elementary submodel of a large enough structure containing all the necessary information, and let $C = \{x \in B : x$ is M-generic for the poset $P_I\}$. I claim that this set has the required properties. It is certainly I-positive by the properness of the forcing P_I. For every number $n \in \omega$ let $D_n = \{y \in Y : y$ is M-generic for the poset P and the generic filter given by it meets the condition $p_n\}$. The set D_n is Borel by Fact 1.4.8. Moreover, $f''(C \cap A_n) = D_n$: whenever $y \in D_n$ is a point, it is possible to find a $M[y]$-generic x for the poset Q such that $M[x] \models x \in A_n$ by the choice of the condition p_n; but then $x \in A_n$ by analytic absoluteness, $y = f(x)$ and so $y \in f''(C \cap A_n)$. On the other hand, if $x \in C \cap A_n$ is a point, then $M[x] \models x \in A_n$ by analytic absoluteness, and $f(x) \in D_n$ by the forcing theorem applied in the model $M[f(x)]$. $\qquad\square$

Proposition 2.3.4. *(P_I-uniformization) Suppose that I is a σ-ideal on a Polish space X such that the forcing P_I is proper. Suppose that Y is a Polish space, $B \in P_I$ is a condition and $A \subset B \times Y$ is an analytic set with nonempty vertical sections. Then there is a condition $C \subset B$ in the forcing P_I and a Borel function*

$f : C \to Y$ *whose graph is a subset of the set A. With an appropriate large cardinal assumption it is possible to relax the demand on the set A to be universally Baire.*

Proof. By the Shoenfield absoluteness, $B \Vdash \exists y \in Y \ \langle \dot{x}_{gen}, y \rangle \in \dot{A}$. Let $\dot{y}$ be a name for the witness real. Let M be a countable elementary submodel of a large enough structure, let $C = \{x \in B : x$ is M-generic$\}$ and let $f : C \to Y$ be the Borel function defined by $f(x) = \dot{y}$ evaluated according to the filter generated by the point x. As in the previous proof, the set $C \subset B$ is Borel and I-positive and the function f is Borel. Moreover, for every point $x \in B$ it is the case that $M[x] \models \langle x, f(x) \rangle \in A$ by the forcing theorem, and $\langle x, f(x) \rangle \in A$ follows by the analytic absoluteness.

The universally Baire case argument follows the same lines with more absoluteness required. □

The ZFC variations of many theorems in this book constantly struggle with the necessity to uniformize sets that are more complex than analytic. There does not appear to be a general ZFC theorem which economically covers all the necessary cases.

Example 2.3.5. P_I uniformization of coanalytic sets by Borel functions fails in the constructible universe L. Consider the set $A \subset 2^\omega \times 2^\omega$ given by $\langle x, y \rangle \in A \leftrightarrow x \in L_{\omega_1^y}$. This is a coanalytic set with nonempty vertical sections. Whenever $B \subset 2^\omega$ is a Borel set and $f : B \to 2^\omega$ is a Borel function whose graph is a subset of the set A, it must be the case that B is countable. If B was uncountable, just choose a countable elementary submodel M of a large enough structure containing B, f, and choose an M-generic filter g for the Sacks forcing which contains some perfect subset of the set B. Writing $x \in 2^\omega$ for the resulting generic point we have $x \in B$, $f(x) \in M[x]$, and since $\omega_1^M = \omega_1^{M[x]}$ it is the case that $x \notin L_{\omega_1^M} \supset L_{\omega_1^{f(x)}}$ and so $\langle x, f(x) \rangle \notin A$, a contradiction.

Example 2.3.6. Suppose that ϕ is a ZFC-correct outer regular submeasure on a Polish space Y, and suppose that I is a σ-ideal on a Polish space X such that the forcing P_I is proper, let $B \in P_I$ be a condition, let $\varepsilon > 0$ be a real number and $A \subset B \times Y$ be an analytic set with vertical sections of ϕ-mass $< \varepsilon$. Then there is a condition $C \subset B$ in P_I and a Borel set $A' \subset C \times Y$ with open vertical sections of ϕ-mass $< \varepsilon$ which cover the corresponding vertical sections of the set A. A brief complexity computation reveals that this would require P_I uniformization of a coanalytic set, which may fail by the previous example. However, the special features of the current situation allow a ZFC argument.

Whenever M is a countable elementary submodel of a large enough structure containing the set B and x is an M-generic point then $M \models \phi(A_x) < \varepsilon$ by the ZFC correctness of the submeasure ϕ. By the forcing theorem then, there must be a name $\dot{O}$ for an open subset of the space Y which is forced to cover the set $A_{\dot{x}_{gen}}$ and has ϕ-mass $< \varepsilon$. Let $C = \{x \in B : x$ is M-generic$\}$ and let $A' \subset C \times Y$ be defined by $\langle x, y \rangle \in A' \leftrightarrow y \in \dot{O}/x$. The ZFC-correctness applied again shows that the vertical sections of the set A' have mass $< \varepsilon$, and since $M[x] \models A_x \subset \dot{O}/x$, a wellfoundedness argument shows that $V \models A_x \subset \dot{O}/x$ and the vertical sections of the set A' cover the vertical sections of the set A.

The last remark of this section concerns the homogeneity of the forcings P_I. Suppose that I is a σ-ideal on a Polish space X such that the forcing P_I is proper, and suppose that $B_0, B_1 \in P_I$ are conditions such that the forcings $P_I \upharpoonright B_0$ and $P_I \upharpoonright B_1$ are in the forcing sense equivalent. Then there is a $P_I \upharpoonright B_0$-name $\dot{y}$ for a generic element of the set B_1 such that $B_0 \Vdash \dot{x}_{gen}$ can be recovered from $\dot{y}$. Let M be a countable elementary submodel of a large enough structure, and let $C_0 \subset B_0$ be the set of all M-generic points. Let $f : C_0 \to B_1$ be the Borel function defined by $f(x) = \dot{y}/x$ and let $C_1 = \text{rng}(f)$. It is not difficult to see that the bijective function f transfers the ideal I below C_0 to the ideal I below C_1 at least as far as Borel sets are concerned. Thus the homogeneity of the forcing P_I is always witnessed by Borel functions. This helps to justify the following definition:

Definition 2.3.7. *An ideal I on a Polish space X is* homogeneous *if for every Borel set B there is a Borel function $f : X \to B$ such that f-preimages of I-small sets are I-small.*

It is clear that homogeneity of ideals does not imply the homogeneity of the resulting forcing. In all cases encountered in this book the homogeneity of the forcing and the underlying ideal always come together. Homogeneity is a rather special property. Many forcings in this book, such as the capacity forcings of Section 4.3, do not seem to be homogeneous, nevertheless I have not been able to find a definite proof of inhomogeneity in a single interesting instance.

3

Properties

3.1 Continuous reading of names

The continuous reading of names (CRN for short) is one of the more common and more slippery properties of forcing notions of the form P_I.

Definition 3.1.1. *Suppose that I is a σ-ideal on a Polish space X such that the forcing P_I is proper. The forcing P_I is said to have* continuous reading of names *(CRN) if for every Borel function $f : B \to 2^\omega$ with an I-positive Borel domain $B \subset X$ there is an I-positive Borel set $C \subset B$ such that the function $f \restriction C$ is continuous.*

How common this property is will be obvious from the list of examples below, and anyone acquainted with the combinatorial approach to forcing can appreciate it immediately. The slippery part is that the continuous reading of names may (or then again, may not) depend on the presentation of the forcing P_I. Note that extending the topology on the underlying Polish space without changing the Borel structure will make more functions continuous and so it can bring about the continuous reading of names in a forcing in which it was originally not present. This actually happens in the case of Steprāns forcing of Section 4.2.3. One simple observation is that every forcing with the CRN has a presentation on the Baire space with the CRN. If I is a σ-ideal on a Polish space X and $\pi : C \to X$ is a continuous bijection between a closed subset C of the Baire space and the space X as in [40], 7.9, define the σ-ideal J on the Baire space by $A \in J \leftrightarrow \pi''A \in I$ and observe that if P_I has the CRN then so does P_J. Deeper study will reveal that CRN has a natural game-theoretic restatement – Theorem 3.10.19 – and it is preserved under the countable support iteration – Theorem 6.3.16.

I will begin with several equivalent restatements of CRN.

Proposition 3.1.2. *[87] Let I be a σ-ideal on a Polish space X. The following are equivalent:*

33

1. *the forcing P_I has the continuous reading of names;*
2. *for every I-positive Borel set $B \subset X$ and a countable collection $\{D_n : n \in \omega\}$ of Borel subsets of X there is an I-positive Borel set $C \subset B$ such that all sets $D_n \cap C$ are relatively open in C;*
3. *for every I-positive Borel set B and every Borel function $f : B \to Y$ to a Polish space Y there is an I-positive Borel set $C \subset B$ such that $f \restriction C$ is continuous.*

Proof. (1)→(2). Fix sets $B, D_n : n \in \omega$ and define a Borel function $f : B \to 2^\omega$ by $f(r)(n) = 1$ if $r \in D_n$. By the continuous reading of names there is an I-positive Borel set $C \subset B$ such that $f \restriction C$ is continuous. It is immediate that the sets $D_n \cap C$ must be relatively open in C.

(2)→(3). Suppose that B is a Borel I-positive set and $f : B \to Y$ is a Borel function. For every basic open set O from some fixed countable basis for the space Y, let $D_O = f^{-1}O$. It is clear that D_O is a Borel set and if $C \subset B$ is any set such that all sets $D_O \cap C$ are relatively open in C, the function $f \restriction C$ must be continuous.

(3)→(1). Trivial. $\qquad\qquad\qquad\qquad\qquad\qquad\qquad\qquad\qquad\qquad\qquad\quad$ □

Example 3.1.3. Every bounding forcing of the form P_I has the continuous reading of names. This is the content of Theorem 3.3.2.

Example 3.1.4. If the σ-ideal I is σ-generated by closed sets then the forcing has the continuous reading of names. This is proved in Theorem 4.1.2.

Example 3.1.5. The Hechler forcing has the continuous reading of names. This is the forcing of all pairs $p = \langle t_p, f_p \rangle$ where $t_p \in \omega^{<\omega}$ and $f_p \in \omega^\omega$, ordered by $q \leq p \leftrightarrow t_p \subset t_q, t_q \setminus t_p$ dominates f_p on its domain, and for all numbers $n \notin \text{dom}(t_q)$, $f_p(n) \in f_q(n)$. This poset adds a single generic point in the Baire space which is the union of the first coordinates of the conditions in the generic filter.

Let I be the associated σ-ideal on the Baire space, let $B \in P_I$ be a positive set, and let $g : B \to 2^\omega$ be a Borel function. To find an I-positive set $C \subset B$ on which the function g is continuous, let M be a countable elementary submodel of a large enough structure, let $C' \subset B$ be the set of all M-generic points and let $h \in \omega^\omega$ be a function modulo finite dominating all functions in the model M. For some number $n \in \omega$, the set $C = \{f \in C' : \forall m > n \ f(m) > h(m)\}$ must be I-positive. To show that $g \restriction C$ is a continuous function, let $s \in 2^{<\omega}$ be a finite binary sequence and argue that $g^{-1}O_s \subset C$ is a relatively open set. Let $x \in C$ be a point such that $s \subset g(x)$. I must find an open neighborhood of the point x in the set C which is mapped into O_t. Use the forcing theorem to choose a condition $p = \langle t_p, f_p \rangle \in M$ which is in the filter determined by the point x and forces $\check{s} \subset \dot{g}(\dot{x}_{gen})$. Let $t \subset x$ be a finite initial segment such that $t_p \subset t, |t| > n$, and $\forall m \notin \text{dom}(t) \ h(m) > f_p(m)\}$. It follows from the definitions that for every point $y \in C \cap O_t$, its associated M-generic filter contains the condition p and by the forcing theorem applied in the model M, it must be the case that $s \subset g(y)$. In other words, $g''(O_l \cap C) \subset O_s$ as required.

Example 3.1.6. Let J be an ideal on ω and let $P(J)$ be the associated Prikry-type forcing. $P(J)$ has the continuous reading of names if and only if J is a P-ideal. The argument here follows the lines of the previous proof and Example 3.1.8.

Example 3.1.7. The Steprāns forcing of Section 4.2.3 has the continuous reading of names in one presentation but fails to have it in another one.

Example 3.1.8. The eventually different real forcing does not have the continuous reading of names in any presentation. The eventually different real forcing P is the set of all pairs $p = \langle t_p, f_p \rangle$ where t_p is a finite sequence of natural numbers and f_p is a finite set of functions in ω^ω. The ordering is defined by $q \leq p$ if $t_p \subset t_q$, $f_p \subset f_q$ and $(t_q \setminus t_p) \cap \bigcup f_p = 0$. The forcing P adds an element $\dot{x}_{gen}$ of the Baire space as the union of the first coordinates of the conditions in the generic filter. The function $\dot{x}_{gen}$ has finite intersection with every function in the ground model. The forcing P is clearly σ-centered since any two conditions with the same first coordinate are compatible. Let I be the σ-ideal of all Borel sets $B \subset \omega^\omega$ such that $P \Vdash \dot{x}_{gen} \notin \dot{B}$ so that P is in the forcing sense equivalent to the poset P_I.

It is enough to show that for no Polish topology τ on the Baire space extending the standard topology the forcing P_I has the τ-continuous reading of names. Let $B_n : n \in \omega$ enumerate a basis for the topology τ. These are all Borel subsets of the Baire space ω^ω and so there are countable antichains $A_n : n \in \omega$ in the forcing P such that every condition in A_n forces $\dot{x}_{gen} \in \dot{B}_n$ and the antichains are maximal with respect to this property.

A piece of notation and an easy construction: for a finite set $f \subset \omega^\omega$ of functions and a number $l \in \omega$ write $f(l) = \{x(l) : x \in f\}$. For every number $m \in \omega$ choose a set f_m of $m+1$ many functions in the Baire space which return mutually distinct values at every input and moreover such that for every number $k \in \omega$, every $n \in \omega$ and every condition $q \in A_n$ there is a number $l > k$ such that $f_m(l) \cap f_q(l) = 0$. Let $h : \omega^\omega \to \omega^\omega$ be the partial Borel function defined by $h(x)(n) =$ the least number k such that $x(l) \notin f_m(l)$ for all numbers $l > k$. Note that the function h is defined on all but I-many points in the Baire space. I claim that there is no Borel I-positive set $C \subset \omega^\omega$ such that $h \upharpoonright C$ is a t-continuous function.

Suppose there in fact is such a set $C \subset \omega^\omega$. Find a condition $p \in P$ such that $p \Vdash \dot{x}_{gen} \in \dot{C}$ and let $m = |f_p|$. The sets $C_k = \{x \in C : h(x)(m) = k\} : k \in \omega$ exhaust all of C and so one of them must be I-positive. This set C_k is relatively τ-open in the set C and there must be a collection $a \subset \omega$ such that $C_k = C \cap \bigcup_{n \in a} B_n$. Since the set C_k is I-positive, there must be a number $n \in a$ and a condition $q \in A_n$ such that p, q are compatible conditions. Now use the property of the finite set $f_m \subset \omega^\omega$

to find a number $l > k, |s_p|, |s_q|$ such that $f_m(l) \cap f_q(l) = 0$. Since there are $m + 1$ many functions in the finite set $f_m \subset \omega^\omega$ and only m many functions in the set f_p, there must be a function $y \in f_m$ such that $y(l) \notin f_p(l)$. It is now easy to find a finite sequence s extending both s_p and s_q such that the condition $r = \langle s, f_p \cup f_q \rangle$ is a lower bound of p, q and $s(l) = y(l)$. Since the condition r forces both $\dot{x}_{gen} \in \dot{C}$ and $\dot{x}_{gen} \in \dot{B}_n$, any sufficiently generic point $x \in \omega^\omega$ below the condition r will belong to the intersection $B_n \cap C$. However, for every such a point it is the case that $h(x)(m) > l > k$, contradicting the assumption that $B_n \cap C \subset C_k$!

The continuous reading of names has several consequences.

Proposition 3.1.9. *Suppose that I is a σ-ideal on a Polish space X with the continuous reading of names. Every I-positive Borel set has an I-positive G_δ subset.*

Proof. Suppose that $B \subset X$ is an I-positive Borel set and $D \subset X \times \omega^\omega$ is a closed set projecting to it. Since the poset P_I is proper, an application of P_I-uniformization shows that there must be an I-positive Borel set $C \subset B$ and a Borel function $f : C \to \omega^\omega$ whose graph is a subset of the set D. By the continuous reading of names we may assume that the function f is continuous on C. Every partial continuous function can be extended to a continuous function with a G_δ domain [40], 3.6. Let $C \subset C', f \subset g$ be such a G_δ set and a continuous extension, with C still dense in C'. Since the set $D \subset X \times \omega^\omega$ is closed, the graph of the function g is still a subset of it Then $C' \subset B$ is an I-positive G_δ subset of the set B.   □

The conclusion cannot be improved to the density in P_I of closed sets as many examples such as the Cohen forcing show.

Proposition 3.1.10. *Suppose that P_I is a proper forcing with the continuous reading of names and J is a c.c.c. ideal on a Polish space Y which does not have the continuous reading of names in any presentation below any condition. Then the forcing with P_I does not add a P_J-generic.*

Proof. Suppose the conclusion fails. By Proposition 2.3.1, there must be a Borel set $B \in P_I$ and a Borel function $f : B \to Y$ such that $B \Vdash \dot{f}(\dot{x}_{gen})$ is P_J-generic. Use Proposition 2.3.3 to find an I-positive Borel set $C \subset B$ such that the images $f''(C \cap O)$ are Borel for all basic open sets $O \subset X$ coming from some fixed countable basis for the space X. Find a Polish topology τ on the space Y extending the original one making all these image sets open. Since the forcing P_J does not have the continuous reading of names in the topology τ below the set $C' = f''C$ and it is c.c.c., there must be a Borel function $g : C' \to 2^\omega$ which is not τ-continuous on any Borel J-positive subset of E. Consider the Borel function $g \circ f : C \to 2^\omega$ and use the continuous reading of names of the poset P_I to find a Borel I-positive set $D \subset C$ on which this function is continuous. Now the set $f''D \subset E$ must have a Borel J-positive subset $D' \subset C'$. The function $g \restriction D'$ is τ-continuous by the choice of the topology τ and the set D. This contradicts the choice of the function g. $\square$

As a consequence, the generics for the eventually different real forcing cannot be obtained in the countable support iterations of forcings such as Laver, Cohen, Solovay, or Mathias, since these forcings all have the continuous reading of names and this property is preserved in the iteration by Theorem 6.3.16.

Fact 3.1.11. *[87] Suppose that I is a σ-ideal such that the forcing P_I is proper and has the continuous reading of names. There is an ideal J on ω such that the forcing $Q = \mathcal{P}(\omega) \mod J$ is proper and equal to the iteration of P_I followed with an $\aleph_0$-distributive forcing.*

To prove this it is enough to consider ideals I on the Baire space ω^ω. The ideal J is then defined on the underlying set $\omega^{<\omega}$ by $A \in J \leftrightarrow \{x \in \omega^\omega : \exists^\infty n \, x \restriction n \in A\} \in I\}$ for a set $A \subset \omega^{<\omega}$. I omit the proof.

3.2 Fubini properties of ideals

In a very restricted sense the whole field of preservation properties of definable forcings can be understood as the study of certain variations on the Fubini theorem.

3.2.1 Ideal vs. ideal

Definition 3.2.1. *Suppose that I, J are σ-ideals on the respective underlying Polish spaces X, Y. I will say that I, J are* perpendicular *$(I \perp J)$ if there are a Borel I-positive set $B \subset X$, a Borel J-positive set $C \subset Y$ and a Borel set $D \subset B \times C$ such that the vertical sections of the set D are J-small and the horizontal sections of its complement are I-small.*

It is part of the content of the classical Fubini theorems concerning the Lebesgue measure and Baire category that `null` $\not\perp$ `null` and `meager` $\not\perp$ `meager`. Incidentally, among the definable c.c.c. ideals these are essentially the only two instances of nonperpendicularity, see [17]. In the more general context, the perpendicularity turns out to have forcing content:

Proposition 3.2.2. *(LC) Suppose that P_I is a proper forcing and J is generated by a universally Baire collection of Borel sets. Then $I \perp J$ if and only if some condition in the poset P_I forces $\dot{C} \cap V \in J$ for some J-positive Borel set C. If the ideal J is ZFC-correct then the large cardinal assumption is not necessary.*

I will be frequently in the situation when I, J are homogeneous ideals, and then the conclusion clearly strengthens to the natural $P_I \Vdash \dot{X} \cap V \in J$.

Proof. On one hand, a review of the definitions shows that if B, C, D witness the statement $I \perp J$ then $B \Vdash \dot{C} \cap V \subset \dot{D}_{\dot{x}_{gen}}$ and the latter set is in the ideal J by an absoluteness argument. On the other hand, if $B \Vdash \dot{C} \cap V \in J$ then there is a name for a Borel set $\dot{A} \in J$ such that $B \Vdash \dot{C} \cap V \in I$. Using Proposition 2.3.2, thinning out the set B if necessary, I can find a Borel set $D \subset B \times C$ such that $B \Vdash \dot{A} = \dot{D}_{\dot{x}_{gen}}$. Using an absoluteness argument, thinning out the set B if necessary again, I can find the set D in such a way that its vertical sections are in the ideal J. Since $B \Vdash \dot{C} \cap V \subset \dot{D}_{\dot{x}_{gen}}$, it is clear that the horizontal sections of the complement of the set D must be in the ideal I.

For the ZFC version of the proposition, suppose on one hand that $P_I \Vdash \dot{C} \cap V \in J$ for some Borel set $C \subset Y$, and let $B \Vdash \dot{E} \in J$ is a Borel set such that $\dot{C} \cap V \subset \dot{E}$. Strengthening the condition B if necessary and using Proposition 2.3.2. I may assume that there is a Borel set $D \subset B \times Y$ such that $B \Vdash \dot{E} = \dot{D}_{\dot{x}_{gen}}$. Let M be a countable elementary submodel of a large enough structure and let $B' \subset B$ be the I-positive set of M-generic points in the set B. For every point $x \in B'$ it is the case that $M[x] \models D_x \in J$ by the forcing theorem and so $V \models D_x \in J$ by ZFC-correctness. It is clear that the sets $B', C, D \cap B' \times C$ witness the statement $I \perp J$. On the other hand, suppose that $I \perp J$ holds and is witnessed by some sets $B \subset X, C \subset Y$, and $D \subset B \times C$. I claim that $B \Vdash \dot{D}_{\dot{x}_{gen}} \in J$; since $B \Vdash \dot{C} \cap V \in \dot{D}_{\dot{x}_{gen}}$, this will show that the $\dot{C} \cap V$ is forced to belong to the ideal J. Suppose for contradiction that $B' \subset B$ is some condition forcing $\dot{D}_{\dot{x}_{gen}} \notin J$, let M be a countable submodel of a large enough structure and let $x \in B'$ be an M-generic point. By the forcing theorem it should be the case that $M \models D_x \notin J$. However, since $D_x \in J$, the ZFC-correctness implies that $M \models D_x \in J$, contradiction! $\square$

The reader may wonder whether the condition on the set $D \subset B \times C$ in the definition of perpendicularity can be relaxed to analytic or weaker while the proposition above remains true. The answer is yes in most cases, but is somewhat awkward to state precisely. This is used in several critical situations later in the book, notably in the proofs of preservation theorems for the countable support iteration.

Definition 3.2.3. *Suppose Γ is a class of sets and I, J are σ-ideals on Polish spaces X, Y respectively. The symbols $I \perp_\Gamma J$ stand for the statement that there are sets $B \subset X, C \subset Y$ and $D \subset B \times C$ in the class Γ such that $B \notin I$, $C \notin J$, all vertical sections of D are in J while all horizontal sections of its complement are in I. $I \perp_{uB} J$ is the perpendicularity with the class of universally Baire sets, $I \perp_a J$ is the perpendicularity with the class of analytic sets.*

Proposition 3.2.4. *(LC) Suppose that I, J are universally Baire σ-ideals on their respective domain Polish spaces X, Y, P_I is proper, I satisfies the second universally Baire dichotomy, and J satisfies the first universally Baire dichotomy. Then $I \perp J \leftrightarrow I \perp_{uB} J$.*

Proof. The left-to-right implication does not need an argument. For the right-to-left implication fix universally Baire sets $B_0 \subset X$, $C_0 \subset Y$ and $D_0 \subset X \times Y$ witnessing $I \perp_{uB} J$. First use the dichotomies to find I-positive Borel set $B_1 \subset B_0$ and a J-positive Borel set $C_1 \subset C_0$. By the first dichotomy applied to the σ-ideal J, every vertical section of the set D_0 is a subset of a J-small Borel set. By a universally Baire absoluteness argument, this will happen in the P_I generic extension at the generic vertical section as well. Use Proposition 2.3.2 to find a Borel set $B_2 \subset B_1$ in the poset P_I and a Borel set $D_1 \subset B_2 \times C_1$ such that $B_2 \Vdash C_1 \cap (\dot{D}_0)_{\dot{x}_{gen}} \subset (\dot{D}_1)_{\dot{x}_{gen}}$ and the latter set is in the σ-ideal J. Let M be a countable elementary submodel of a large enough structure and let $B_3 = \{x \in B_2 : x \text{ is } M\text{-generic}\} \subset B_2$. This is a Borel I-positive set, by the forcing theorem for every $x \in B_3$ it is the case that $M[x] \models C_1 \cap (D_0)_x \subset (D_1)_x$ and the latter set is in the σ-ideal I, and by the universally Baire absoluteness it is even true that $M[x] \models C_1 \cap (D_0)_x \subset (D_1)_x$. It is clear that the sets $B_3 \subset X$, $C_1 \subset Y$, and $D_2 = D_3 \cap (B_3 \times C_1)$ witness the statement $I \perp J$. □

Proposition 3.2.5. *Suppose that I, J are σ-ideals on their respective domain Polish spaces X, Y, both satisfying the third dichotomy, the forcing P_I is proper and the ideal J is ZFC-correct, generated by Borel sets. Then $I \perp_{uB} J \leftrightarrow I \perp_a J$.*

Proof. The left-to-right direction does not need an argument. For the opposite direction, suppose the analytic sets B, C, D are given. The third dichotomy yields a Borel I-positive subset $B_1 \subset B$. Let M be a countable elementary submodel of a large enough structure and $x \in B_1$ be an M-generic point. By the ZFC-correctness, $M[x] \models D_x \in J$ and therefore there is a name $\dot{E}$ for a Borel set such that $M[x] \models E/x \in J$ and $D_x \subset \dot{E}$. There must be a condition $B_2 \subset B_1$ and a Borel set $D_1 \subset B_2 \times Y$ such that $B_2 \Vdash \dot{E} = (\dot{D}_1)_{\dot{x}_{gen}}$ and $\dot{D}_{\dot{x}_{gen}} \subset \dot{E} \in J$. Let $B_3 \subset B_2$ be the set of all M-generic points in the set B_2. Let $C_3 \subset C$ be some J-positive Borel set obtained by an application of the third dichotomy. Let $D_3 = D_1 \cap B_3 \times C_3$. Another application of ZFC-correctness shows that the sets B_3, C_3, D_3 witness the relation $I \perp J$. □

3.2.2 Ideal vs. submeasure

Definition 3.2.6. *Suppose that I is a σ-ideal on a Polish space X and ϕ is a submeasure on a Polish space Y. Say that the ideal I is* not *perpendicular to the* submeasure ϕ *($I \not\perp \phi$) if for every real number $\varepsilon > 0$, every I-positive Borel set $B \subset X$ and every Borel set $D \subset B \times Y$ with vertical sections of submeasure $\leq \varepsilon$ the set*

$$\int_B D \, dI = \{y \in Y : \{x \in B : \langle x, y \rangle \notin D\} \in I\}$$

has mass $\leq \varepsilon$. In parallel to Definition 3.2.3, I will also introduce the notation $I \perp_r \phi$, $I \perp_{uB} \phi$, $I \perp_a \phi$.

It is a consequence of the classical Fubini theorem that the ideal of Lebesgue null sets is not perpendicular to the outer Lebesgue measure. In the more general context, the perpendicularity turns out to have forcing meaning again.

Definition 3.2.7. *Suppose that ϕ is an outer submeasure on a Polish space Y (that is, $w(A) = \inf\{w(B) : A \subset B \text{ and } B \subset Y \text{ Borel}\}$) which is universally Baire (that is, the function $x \mapsto \phi(A_x) : x \in 2^\omega$ where $A \subset 2^\omega \times Y$ is a universal analytic set, is universally Baire). Suppose that P is a forcing. I will say that P preserves the submeasure ϕ if $P \Vdash \check{\phi}(\dot{A}) = \dot{\phi}(\check{A})$ for every set $A \subset Y$.*

The following propositions have proofs analogous to those in the previous subsection, and I omit them.

Proposition 3.2.8. *(LC) Suppose that I is a σ-ideal on a Polish space X such that the forcing P_I is proper. Suppose that ϕ is an outer universally Baire submeasure on a Polish space Y. Then $I \perp \phi$ is equivalent to the failure of preservation of the submeasure ϕ by the forcing P_I. If the submeasure ϕ is ZFC-correct then the large cardinal assumption is not necessary.*

Proposition 3.2.9. *(LC) Suppose that I is a σ-ideal on a Polish space X satisfying the second dichotomy such that the forcing P_I is proper. Suppose that ϕ is a universally Baire submeasure such that every universally Baire set has a Borel subset of the same submeasure. Then $I \perp \phi \leftrightarrow I \perp_{uB} \phi$.*

Proposition 3.2.10. *The previous proposition holds in ZFC if ϕ is a ZFC-correct submeasure and $\perp_{uB}$ is replaced with $\perp_a$.*

In certain prominent cases, the perpendicularity with ϕ and its null ideal coincide. The following two propositions strengthen of [2].

Proposition 3.2.11. *(LC) Suppose that I is a universally Baire σ-ideal on a Polish space X such that the forcing P_I is proper. Let λ be the outer Lebesgue measure. Then $I \perp \lambda$ iff $I \perp \text{null}$; in other words P_I preserves the Lebesgue measure if and only if it does not make the ground model reals Lebesgue null. If the ideal I is $\mathbf{\Pi}_1^1$ on $\mathbf{\Sigma}_1^1$ then the large cardinal assumption can be omitted.*

Proof. I will first treat the large cardinal version. The right-to-left direction does not need an argument. For the other direction, suppose that P_I does not preserve outer Lebesgue measure. I will first show that it does not preserve the outer Lebesgue measure of some Borel set. Suppose that $A \subset 2^\omega$ is a set and $B \in P_I$ is a condition such that $B \Vdash \lambda(\check{A}) < \check{\lambda}(\check{A})$. There must be a Borel set $B_1 \subset B$ and a Borel set $D \subset B_1 \times 2^\omega$ such that the vertical sections of the set D have Lebesgue measure $< \varepsilon$ while $\lambda(A) > \varepsilon$, and at the same time $B_1 \Vdash \check{A} \subset \dot{D}_{\dot{x}_{gen}}$. Now consider the set $C = \{y \in 2^\omega : B_1 \Vdash \check{y} \in \dot{D}_{\dot{x}_{gen}}\}$. This is a universally Baire set, since $C = \{y \in 2^\omega : B_1 \setminus D^y \in I\}$, it contains the set A and therefore $\lambda(C) > \varepsilon$. Use the Lebesgue measurability of universally Baire sets to find a Borel set $C_1 \subset C$ of the same Lebesgue measure. Then $B_1 \Vdash \lambda(\check{C}_1) < \check{\lambda}(\check{C}_1)$ as desired.

By the measure isomorphism theorem [40], 14.71, I can then find in the P_I-generic extension an open set $O \subset 2^\omega$ such that O contains all the ground model reals and $\lambda(O) < 1$. Let $E = 2^\omega \setminus O$; this is a λ-positive closed set and since the Lebesgue measure is ergodic, the closure F of the set E under rational translations has full measure. However, the set F still contains no ground model reals and so $\lambda^*(2^\omega \cap V) = 0$.

The Π_1^1 on Σ_1^1 case is identical with the additional observation that the set C above is coanalytic and therefore Lebesgue measurable. $\qquad\square$

Proposition 3.2.12. *(LC) Suppose that I is a universally Baire σ-ideal on a Polish space X such that the quotient forcing P_I is homogeneous and proper. Suppose that ϕ is either a pavement submeasure or a strongly subadditive submeasure on a Polish space Y. Then $I \perp \phi \leftrightarrow I \perp I_\phi$.*

Proof. The right-to-left implication does not need a proof. For the opposite implication, I will treat the case of a pavement submeasure ϕ, obtained from some countable set U of pavers and a weight function w. Suppose that $I \perp \phi$. As in the previous argument this means that there are Borel sets $B \subset X$, $C \subset Y$ and $D \subset B \times C$ such that $B \notin I$, $C \notin I_\phi$, the vertical sections of D have ϕ-mass $< \varepsilon < \phi(C)$ and the horizontal sections of the complement of the set D are I-small. Using the ZFC-correctness of the submeasure ϕ, shrinking the set B if possible I can find a Borel function $f : B \to \mathcal{P}(U)$ such that $\forall x \in B \ \Sigma\{w(u) : u \in f(x)\} < \varepsilon$ and $D_x \subset \bigcup f(x)$. Using the σ-completeness of the ideal I repeatedly, I can find a descending collection of Borel I-positive sets $B_n : n \in \omega$ and inclusion-increasing sequence $a_n : n \in \omega$ of subsets of U such that $B_0 = B$, $a_0 = 0$, and for every number $n \in \omega$ and every point $x \in B_{n+1}$ it is the case that $a_{n+1} \subset f(x)$ and $\Sigma\{w(u) : u \in f(x) \setminus a_{n+1}\} \leq 2^{-n}$.

In the end, let $a = \bigcup_n a_n$ and note that $\Sigma\{w(u) : u \in a\} \leq \varepsilon$. Thus the set $C' = C \setminus \bigcup a$ is ϕ-positive. I claim that $P_I \Vdash \phi(C' \cap V) = 0$, which will complete the proof. To see this, note that for every number $n \in \omega$, $B_{n+1} \Vdash \phi(C' \cap V) < 2^{-n}$

since $C' \cap V$ must be covered by the set $\dot{f}(\dot{x}_{gen}) \setminus a_{n+1}$ sum of whose weights if smaller than 2^{-n}. By the homogeneity of the forcing P_I, $P_I \Vdash \phi(C' \cap V) < 2^{-n}$, and since this happens for every number $n \in \omega$, $P_I \Vdash \phi(C' \cap V) = 0$ as required. □

However, in other cases this feature is apparently absent and submeasure preservation takes a life on its own. The reader should note that while homogeneity is obvious in most popular forcings of the form P_I – such as the Sacks or Miller forcing – it is really quite difficult to argue for it in a more general setting and most likely in a "typical" σ-ideal I this property fails. It is apparently difficult to find forcings which preserve this submeasure and fail to preserve that submeasure, as outstanding mathematicians have been willing to write long papers on the issue [73].

Before I leave this subject let me point out that the concept of perpendicularity to submeasures makes sense and is interesting already for ideals on ω. Kanovei and Reeken defined the class of Fubini ideals as those ideals on ω which are not perpendicular to outer Lebesgue measure, produced a long list of Fubini ideals, and showed that they all have the Radon–Nikodym property.

3.3 Bounding forcings

The bounding property of forcings is one of the most commonly studied and used properties. This section offers several topological restatements of it.

Definition 3.3.1. *A forcing P is* bounding *(or weakly ω, ω-distributive) if for every function $f \in \omega^\omega$ in the P-extension there is a function $g \in \omega^\omega$ in the ground model which dominates f pointwise.*

Theorem 3.3.2. *Suppose that I is a σ-ideal on a Polish space X such that the forcing P_I is proper. The following are equivalent:*

1. *P_I is bounding;*
2. *every Borel I-positive set contains a compact I-positive subset* (compact sets are dense) *and every Borel function on an I-positive domain has a continuous restriction with a Borel I-positive domain* (continuous reading of names);
3. *for every Polish topology t on X producing the same Borel structure, every Borel I-positive set has a t-compact positive subset.*

Proof. For the (1)→(2) direction, I will first argue for the case $X = 2^\omega$ and then reduce the general case to this. Suppose $B \in P_I$ is a Borel set and $g : B \to 2^\omega$ be a Borel function. I must produce a compact I-positive set $C \subset B$ on which the function g is continuous. Let $T \subset (2 \times 2 \times \omega)^{<\omega}$ be a tree projecting into the graph of the function g. A Shoenfield absoluteness argument shows that $B \Vdash \exists y \in \omega^\omega \langle \dot{x}_{gen}, \dot{g}(\dot{x}_{gen}), y \rangle$ forms a branch through the tree $\check{T}$. Let $\dot{y}$ be a corresponding P_I name for an element of ω^ω. Use the bounding property to find a condition $B' \subset B$ and a function $z \in \omega^\omega$ such that $B' \Vdash \dot{y} < \check{z}$. Now consider the tree $S \subset T$ of all triples of sequences whose third coordinate is dominated by z. The tree S is finitely branching and so $C = \text{proj}(\text{proj}[S]) \subset B$ is compact. Since $B' \Vdash \dot{x}_{gen} \in \dot{C}$ it must be the case that $C \notin I$. Also the graph of the function $g \upharpoonright C$ is equal to the compact set $\text{proj}[S]$ and therefore $g \upharpoonright C$ is continuous as desired.

The general case easily follows. If X is a Polish space and I is a σ-ideal on it such that the forcing P_I is proper and bounding, find a Borel bijection $\pi : 2^\omega \to X$ and let J be the σ-ideal of those sets $A \subset 2^\omega$ such that $\pi''A \in I$. Clearly P_J is just a different presentation of P_I, therefore it is bounding and proper and I can apply the previous paragraph to it. Thus if $B \in P_I$ is a Borel I-positive subset of X and $g : B \to 2^\omega$ is a function, there is a compact J-positive set $C \subset 2^\omega$ such that $\pi''C \subset B$ and the functions $\pi, g \circ \pi$ are both continuous on C. Since C is a compact zero dimensional space and $\pi : C \to \pi''C$ is a continuous bijection, in fact $\pi \upharpoonright C$ is a homeomorphism of C and the compact set $\pi''C \subset B$. Thus $\pi''C$ is a compact I-positive subset of the set B on which the function g is continuous.

The $(2) \to (1)$ direction is easier. If $B \Vdash \dot{y} \in \omega^\omega$ is a function, then there is a condition $B' \subset B$ and a Borel function $g : B' \to B$ such that $B' \Vdash \dot{y} = \dot{g}(\dot{x}_{gen})$ by the results of Section 2.3. Let $C \subset B'$ be an I-positive compact set on which the function g is continuous. The set $g''C \subset \omega^\omega$ is compact and therefore bounded by some function $z \in \omega^\omega$. An analytic absoluteness argument shows that $C \Vdash \dot{y} < \check{z}$.

$(1) \to (3)$ is really subsumed in $(1) \to (2)$ since the proof of that implication did not depend on the topology of the space X. Finally, to see that $(3) \to (1)$, suppose for contradiction $(3) \wedge \neg(1)$ and find a Borel set $B \subset X$ and a Borel function $f : B \to \omega^\omega$ such that f-preimages of compact sets are I-small. As in [40], 13.11, find a Polish topology t on X that makes f a continuous function. Let $C \subset B$ be a t-compact I-positive set. The image $f''C \subset \omega^\omega$ is compact, contradicting the choice of the function f! $\qquad\qquad\qquad\qquad\qquad\qquad\qquad\qquad\qquad\qquad\qquad\qquad\quad\Box$

The theorem justifies the widespread use of tree forcings in the forcing practice. Note that in fact I proved that zero-dimensional compact sets are dense in P_I. The condition of compact sets being dense in P_I cannot be relaxed to the equivalence classes of compact sets being dense in the algebra $\mathcal{B}(X) \mod I$ as the example of the Cohen forcing shows. The density of compact sets does not imply the continuous reading of names and vice versa. An instructive example is that of Steprāns forcing in Section 4.2.3 which has two distinct presentations, in one compact sets are dense and in the other it has the continuous reading of names, but these two properties cannot hold in conjunction since the forcing is not bounding.

The following proposition shows that the bounding property is in fact a Fubini property.

Proposition 3.3.3. *Suppose that I is a σ-ideal on a Polish space X such that the forcing P_I is proper. The following are equivalent:*

1. *P_I is bounding;*
2. *$I \not\perp J$ where J is the Laver ideal on ω^ω.*

Here the Laver ideal J on ω^ω is generated by sets $A_g = \{f \in \omega^\omega :$ for infinitely many n, $f(n) \in g(f \upharpoonright n)\}$ as g varies through all functions from $\omega^{<\omega}$ to ω. It is well-known and proved in Section 4.5.2 that every analytic subset of ω^ω either is in the ideal J or contains all branches of some Laver tree. The ideal J is ZFC-correct as shown in Theorem 4.5.6.

Proof. Suppose that the poset P_I is bounding, $B \in P_I$ and $C \in P_J$ are Borel sets and $D \subset B \times C$ is a Borel set with J-small vertical sections. I must show that its complement contains an I-positive horizontal section. By the ZFC-correctness of the Laver ideal, thinning out the set B if necessary I may assume that there is a Borel function $G : B \to \omega^{\omega^{<\omega}}$ such that for every pair $\langle x, f \rangle \in C$ it is the case that for infinitely many numbers n, $f(n) \in G(x)(f \upharpoonright n)$. Since the poset P_I is bounding, there is an I-positive Borel set $B_1 \subset B$ and a function $h : \omega^{<\omega} \to \omega$ such that for every point $x \in B_1$ and every finite integer sequence s, $G(x)(s) \in h(s)$. Since the set C is J-positive, there must be a function $f \in C$ such that for all but finitely many numbers n, $h(f \upharpoonright n) \in f(n)$. It is clear that $B_1 \times \{f\}$ is the required I-positive horizontal section of the complement of the set C.

On the other hand assume that $\neg I \perp J$, and let $B \in P_I$ be a condition and $\dot{g}$ a P_I-name for an element of ω^ω. By Proposition 2.3.4, thinning out the set B if necessary I may assume that there is a Borel function $G : B \to \omega^\omega$ such that $B \Vdash \dot{g} = \dot{G}(\dot{x}_{gen})$. Let $D \subset B \times \omega^\omega$ be the Borel set consisting of all pairs $\langle x, f \rangle$ for which f is dominated by $G(x)$ at infinitely many values. It is clear that the vertical sections of the set D are J-small, and since $\neg I \perp J$ holds, there must be a Borel I-positive set $B' \subset B$ and a function $f \in \omega^\omega$ such that the horizontal section $B' \times \{f\}$ is a subset of the complement of the set D. An analytic absoluteness argument shows that B' forces the function $\dot{g}$ to be dominated by $\check{f}$. $\qquad\square$

The bounding property has a game-theoretic restatement. This is important since then a determinacy argument can be used to derive a stronger property than just bounding in the case of definable forcing.

Definition 3.3.4. *Let P be a partial ordering. In the* bounding game G *Player I first indicates an initial condition p_{ini} and then plays open dense sets $D_n : n \in \omega$. Player II in response to the set D_n chooses a finite set $d_n \subset D_n$. Player II wins if in the end the expression $p_{ini} \wedge \bigwedge_n \bigvee d_n$ describes a nonzero element of the completion algebra $RO(P)$.*

The following is proved in Section 3.10.3.

Proposition 3.3.5. *Suppose the forcing P_I is proper. The following are equivalent:*

1. *Player I has a winning strategy in the bounding game;*
2. *P_I is not bounding.*

If the σ-ideal I is $\mathbf{\Pi}_1^1$ on $\mathbf{\Sigma}_1^1$ then the bounding game is determined. If large cardinals exist, the bounding game is determined for every universally Baire σ-ideal I.

A spectacular application of the bounding game was found by Fremlin. It concerns the relationship between Maharam's and von Neumann's problem. Here Maharam's problem is whether every c.c.c. complete algebra carrying a continuous submeasure carries a measure. Von Neumann's problem asks whether every c.c.c. complete bounding algebra carries a measure. It turns out that it is consistent and in fact follows from the P-ideal dichotomy of Todorcevic that every c.c.c. complete bounding algebra carries a continuous submeasure, and so in such a context the two problems coincide. Fremlin showed that the determinacy of the bounding game is of critical concern here.

Fact 3.3.6. *(Fremlin) Suppose that B is a c.c.c. complete algebra. The following are equivalent:*

1. *Player II has a winning strategy in the bounding game;*
2. *there is a continuous submeasure on the algebra B.*

Corollary 3.3.7. *Suppose that I is a $\mathbf{\Pi}_1^1$ on $\mathbf{\Sigma}_1^1$ c.c.c. σ-ideal on a Polish space X. The following are equivalent:*

1. *the forcing P_I is bounding;*
2. *there is a continuous submeasure ϕ on the space X such that I contains the same Borel sets as the ideal $\{A \subset X : \phi(A) = 0\}$.*

If large cardinal assumptions are used then the definability condition on the ideal can be relaxed to universal Baireness.

Recently Talagrand answered Maharam's problem in the negative [75].

3.4 Bounding and not adding splitting real

Shelah introduced the notion of preservation of P-points [2] 6.2.1 – a forcing preserves P-points if every P-point ultrafilter in the ground model generates a P-point ultrafilter in the extension–and proved a number of beautiful results with it; in particular, the preservation of P-points is preserved under the countable support iteration. Curiously I find it impossible to restate this property as a Fubini property even in the case of definable forcing. Instead, the natural notion of preservation of Ramsey ultrafilters has a Fubini type restatement, and there is a corresponding countable support preservation Theorem 6.3.7. Recall that a Ramsey ultrafilter on ω is an ultrafilter which contains homogeneous sets for every partition $p : [\omega]^2 \to 2$.

Theorem 3.4.1. *(LC) Suppose that I is a universally Baire σ-ideal on a Polish space X such that the forcing P_I is proper, and let U be a Ramsey ultrafilter on ω. Writing J for the Mathias ideal of Section 4.7.7, the following are equivalent:*

1. *U generates a Ramsey ultrafilter in the P_I-extension;*
2. *MRR(I, J);*
3. *$I \not\perp J$;*
4. *P_I is bounding and does not add a splitting real.*

In the case of a Π^1_1 on Σ^1_1 ideal I the large cardinal assumption is not necessary.

It is not difficult to see that (2), (3), and (4) are equivalent even without the presence of a Ramsey ultrafilter.

Proof. Let Q be the Prikry-type forcing with the ultrafilter U. This is the set of all pairs $p = \langle t_p, a_p \rangle$ where $t_p \in 2^{<\omega}$ and $a_p \in U$ and $q \le p$ if $t_p \subset t_q$, $a_q \subset a_p$, and $\{n \in \text{dom}(t_q \setminus t_p) : t_q(n) = 1\} \subset a_p$. The union of first coordinates of conditions in the generic filter gives a characteristic function of the generic set $\dot{y}_{gen} \subset \omega$. I will rely on the following well-known fact.

Fact 3.4.2. *[66] Suppose M is a transitive model of ZFC and $y \subset \omega$ is an infinite set. Then Y is M-generic for Q if and only if it is modulo finite included in every set in $U \cap M$.*

For the (1)→(2) implication, assume (1) holds and $B \subset X$ and $C \subset \mathcal{P}(\omega)$ are I, J-positive Borel sets respectively and $B \times C = \bigcup_n D_n$ is a countable union of Borel sets. By the homogeneity of the Mathias ideal I may assume that $C = \mathcal{P}(\omega)$. Find conditions $B_0 \in P_I$, $B_0 \subset B$ and $q \in Q$ such that $(B_0, q) \in P_I \times Q$ decides the value of the number n such that $(\dot{x}_{gen}, \dot{y}_{gen}) \in \dot{D}_n$. Let M be a countable elementary submodel of some large structure, let $B_1 = \{x \in B_0 : x \text{ is } M\text{-generic}\}$ and $C_1 = \{y \subset \omega : y \text{ is } M\text{-generic for the poset } Q, \text{ meeting the condition } q\}$. Now observe:

- $B_1 \notin I$ since P_I is proper;
- $C_1 \notin J$ since by the above fact C_1 is closed under infinite subsets and therefore dense in $\mathcal{P}(\omega)$ mod fin below each of its elements;
- every pair $(x, y) \in B_1 \times C_1$ is M-generic for $P_I \times Q$, since in the model $M[x]$, U still generates an ultrafilter by the assumption and therefore the set y still satisfies the genericity criterion above for the model $M[x]$.

The forcing theorem applied in the model M to the poset $P_I \times Q$ then shows that $B_1 \times C_1 \subset D_n$, concluding the proof of (1).

(2)→(3) does not need an argument. For (3)→(4), argue in the contrapositive and assume that P_I under some condition adds a splitting real. By Proposition 2.3.1 this means that there is an I-positive Borel set $B \subset X$ and a Borel function $f : B \to \mathcal{P}(\omega)$ such that for every infinite set $b \subset \omega$ both sets $\{x \in B : b \subset f(x)\}$ and $\{x \in B : b \cap f(x) = 0\}$ are I-small. A review of definitions shows that the set $D \subset B \times \mathcal{P}(\omega)$, defined by $\langle x, y \rangle \in D$ iff $y \times f(x)$ is infinite, has J-small vertical sections and its complement has I-small horizontal sections as desired in the definition of perpendicularity. Now suppose that the forcing P_I adds an unbounded real and use Proposition 2.3.1 to find an I-positive Borel set $B \subset X$ and a Borel function $g : B \to \omega^\omega$ such that for every function $h \in \omega^\omega$ the set $\{x \in B : g(x) < h \text{ pointwise}\}$ is I-small. A review of definitions shows that the set $E \subset B \times \mathcal{P}(\omega)$, defined by $\langle x, y \rangle \in E$ iff the enumerating function of y does not dominate $g(x)$ modulo finite, again witnesses $I \perp J$.

Finally, for (4)→(1) I need the following claim of independent interest.

Claim 3.4.3. *(LC) Suppose that U is a Ramsey ultrafilter and $D \subset \mathcal{P}(\omega)$ is a universally Baire dense set in $\mathcal{P}(\omega)$ mod fin. Then $U \cap D \neq 0$. If D is in addition analytic then the large cardinal assumption is not necessary.*

Proof. First note that $Q \Vdash \dot{y}_{gen} \in \dot{D}$. To see this, observe that the set D is still dense in the Q-extension by universally Baire absoluteness in the large cardinal case, and by Shoenfield absoluteness in the analytic case. So in the extension, given any V-generic set $y \subset \omega$ meeting a condition $p \in Q$, it is possible to find its infinite subset $z \subset y$ in the set D still meeting the condition p. The set z satisfies the diagonalization condition from Fact 3.4.2 and therefore it is V-generic for the poset Q. Another absoluteness argument shows that $V[z] \models z \in D$ and therefore p could not have forced $\dot{y}_{gen} \notin \dot{D}$.

Now let M be a countable elementary submodel of a large enough structure containing both U and D. The ultrafilter U contains a set y which modulo finite diagonalizes all the countably many sets in $U \cap M$. The set y is M-generic for Q by Fact 3.4.2, and by the previous paragraph and the forcing theorem $M[y] \models y \in D$. Another absoluteness argument shows that $y \in D$ as required. □

Now assume that P_I does not add a splitting real. I will show that P_I preserves U as an ultrafilter. Suppose that $B \in P_I$ is a Borel set and $f : B \to [\omega]^{\aleph_0}$ is a Borel function. By Proposition 2.3.1, it will be enough to show that there is a set $y \in U$ such that one of the sets $\{x \in B : y \subset x\}$, $\{x \in B : y \cap x = 0\}$ is I-positive. For that, notice that the set $D = \{y \in [\omega]^{\aleph_0} : \text{one of the sets } \{x \in B : y \subset x\}, \{x \in B : y \cap x = 0\}$ is I-positive$\}$ is dense in $\mathcal{P}(\omega)$ mod fin since the forcing P_I does not add a splitting real. The set D is universally Baire (or analytic in the $\mathbf{\Pi}^1_1$ on $\mathbf{\Sigma}^1_1$ case) and therefore $D \cap U$ is nonempty, giving the desired set $y \in U$. The bounding condition can now be used to show that in fact U generates a Ramsey ultrafilter in the extension. □

There is an interesting model theoretic restatement. Looking forward to Section 3.9, write I^{**} for the collection of all sets which do not contain a Borel I-positive subset.

Proposition 3.4.4. *(LC) Suppose that I is a universally Baire σ-ideal on a Polish space X such that the forcing P_I is proper and bounding. Suppose that U is a Ramsey ultrafilter. The following are equivalent:*

1. *P_I is does not add splitting reals;*
2. *in the model $L(\mathbb{R})[I \cap B][U]$, I^{**} is a σ-ideal.*

This and especially the following two propositions should be compared to the results of DiPrisco and Todorcevic on the dichotomies in the model $L(\mathbb{R})[U]$ [13].

Proof. An instrumental well-known fact in this proof is that the ultrafilter U is a $L(\mathbb{R})[I \cap B]$-generic subset of $\mathcal{P}(\omega)$ mod fin. This immediately follows from Claim 3.4.3.

For the ease of notation assume that $I \cap B \in L(\mathbb{R})$. Suppose on one hand that P_I is bounding and does not add splitting reals and suppose that $A_n : n \in \omega$ are sets in the model $L(\mathbb{R})[U]$ whose union covers a Borel I-positive set $B \subset X$. I will show that one of the sets A_n contains a Borel I-positive subset. A genericity argument shows that there must be in $L(\mathbb{R})$ names $\dot{A}_n : n \in \omega$ and a set $a \subset \omega$ such that in the algebra $\mathcal{P}(\omega)$ mod fin the set a forces $\check{B} \subset \bigcup_n \dot{A}_n$, and it will be enough to find a condition $b \subset a$, Borel I-positive set $B' \subset B$ and a number $n \in \omega$ such that $b \Vdash \check{B}' \subset \dot{A}_n$. Consider the Mathias ideal J, the Mathias forcing P_J and the condition $C \in P_J$, $C = \{b \subset \omega : b \subset a \text{ modulo finite}\}$. Let $f : B \times C \to \omega + 1$ be the function defined by $f(x, b) = n$ if n is the least number such that $b \Vdash \check{x} \in \dot{A}_n$ if there is such a number n; otherwise let $f(x, b) = \omega$. The function g is in the model $L(\mathbb{R})$. As in the proof of Theorem 3.4.1, there must be a rectangle $B' \times C'$ with I, J-positive sides respectively on which the function f is constant. The constant value cannot be ω because given any point $x \in B'$, in the somewhere dense set $C' \subset \mathcal{P}(\omega)$ mod fin there must be a condition $b \subset \omega$ forcing $\check{x} \in \dot{A}_n$ for some fixed number n. Now write $n \in \omega$ for the constant value of the function f on the rectangle and find a pair $\langle s, b \rangle$ such that $s \subset \omega$ is finite, $b \subset \omega$ is infinite, and $\{c \subset \omega : s \subset c \wedge c \subset b\} \subset C'$. A review of the definitions now shows that $b \Vdash \check{B}' \subset \dot{A}_n$ as required.

On the other hand, suppose that P_I adds a splitting real. This means that there is an I-positive Borel set $B \subset X$ and a Borel function $f : B \to \mathcal{P}(\omega)$ such that for every infinite set $b \subset \omega$ both sets $\{x \in B : b \subset f(x)\}$ and $\{x \in B : b \cap f(x) = 0\}$ are I-small. Let $A_0 = \{x \in B : f(x) \in U\}$ and $A_1 = \{x \in B : \omega \setminus f(x) \in U\}$. Since $A_0 \cup A_1 = B$ and $A_0, A_1 \in L(\mathbb{R})[U]$, it will be enough to show that neither of these sets has a Borel I-positive subset. Suppose for contradiction that some condition $a \subset \omega$ in $\mathcal{P}(\omega)$ mod fin forces that (say) $\check{C} \subset \dot{A}_0$ for some Borel I-positive set $C \subset B$. Unraveling the definitions, this means that $a \subset f(x)$ modulo finite for every point $x \in C$, which contradicts the asumption on the function f! $\qquad\square$

The proposition opens the possibility that there are two ways of not adding a splitting real. The stronger one, in which $L(\mathbb{R})[I \cap B][U] \models I^{**}$ is generated by the σ-ideal I; and the weaker one, in which I^{**} is a σ-ideal in this model for a different reason. Related questions were studied by DiPrisco and Todorcevic who proved in particular that every set of reals in the model $L(\mathbb{R})[U]$ is either countable or contains a perfect subset. Which Borel generated σ-ideals have the property that every set of reals in the model $L(\mathbb{R})[I][U]$ has either Borel I-small superset or Borel I-positive subset? This question is related to the dichotomies of Section 3.9 and is interesting regardless of the properness of the forcing P_I. I conclude this section with the investigation of two particular cases which appear later in the book.

Proposition 3.4.5. *(LC) Suppose that I is a σ-ideal on a compact metric space X which is σ-generated by a σ-compact family of compact subsets of X. Let U be a Ramsey ultrafilter. Then every set in $L(\mathbb{R})[U]$ has either a Borel I-positive subset or a Borel I-small superset.*

Proof. For notational simplicity I will deal with the case $X = 2^\omega$. Recal that the ultrafilter U is an $L(\mathbb{R})$-generic subset of $\mathcal{P}(\omega)$ mod fin. Suppose that $A \subset X$ is a set in $L(\mathbb{R})[U]$. There must be a $\mathcal{P}(\omega)$ mod fin-name $\dot{A} \in L(\mathbb{R})$ such that $A = \dot{A}/U$. There are two separate cases.

Either there is a set $a \in U$ and an I-positive set $B \subset X$ such that $\forall x \in B \, a \Vdash \check{x} \in \dot{A}$. Since $B \in L(\mathbb{R})$, the dichotomy 4.1.3 shows that B has an I-positive Borel subset $B' \subset B$. In this case clearly $B' \subset A$ and as desired.

Or this fails and by a genericity argument then there must be a set $a \in U$ such that for all sets $b \subset a$ the set $A_b = \{x \in X : b \Vdash \check{x} \in \dot{A}\}$ is I-small. In $L(\mathbb{R})[U]$ consider the Prikry forcing P_U, its name $\dot{y}_{gen}$ for a generic subset of ω and the name $\dot{A}_{\dot{y}_{gen}}$ for an I-small set in the extension. Note that since the set $\dot{b}_{gen}$ seals the ultrafilter U, it is the case that $P_U \Vdash \dot{A} = \dot{A}_{\dot{y}_{gen}} \cap V$. I will now produce a Borel set $B \in I$ such that $A \subset B$.

Let $F_n \subset K(X) : n \in \omega$ be the countable collection of compact sets generating the ideal I. Without loss of generality $F_0 \subset F_1 \subset \dots$ and so there are P_U-names $\dot{T}_n : n \in \omega$ for binary trees such that P_U forces $[\dot{T}_n] \in [F_n]$ and $\dot{A}_{\dot{y}_{gen}} \subset \bigcup_n [\dot{T}_n]$. The forcing P_U is σ-filtered and therefore its Boolean algebra is a union of countably many ultrafilters $G_m : m \in \omega$. For numbers $m, n \in \omega$ let S_{nm} be the tree of all finite binary sequences t such that some condition in the ultrafilter G_m forces $\check{t} \in \dot{T}_n$. Note that since G_m is closed under finite conjunctions and the set $F_n \subset K(X)$ is compact, it must be the case that $[S_{nm}] \in F_n$. I will show that $A \subset \bigcup_{n,m} [S_{nm}]$, and that will complete the proof of the proposition.

Indeed, suppose that $x \in A$ is a binary sequence. There must be a condition $p \in P_U$ and a number $n \in \omega$ such that $p \Vdash \check{x} \in [\dot{T}_n]$. There is a number $m \in \omega$ such that $p \in G_m$. It then must be the case that $x \in [S_{nm}]$, because if $t \subset x$ were an initial segment not in the tree S_{nm} then the common lower bound of p and the condition in G_m forcing $\check{t} \notin \dot{T}_n$ would also force $\check{x} \notin [\dot{T}_n]$, contradicting the choice of the condition p. $\qquad \square$

Proposition 3.4.6. *(LC) Suppose that ϕ is a subadditive outer regular capacity on a Polish space X such that $L(\mathbb{R}) \models \phi$ is continuous in increasing wellordered unions. Let U be a Ramsey ultrafilter. The following are equivalent:*

1. *for every real number $\varepsilon > 0$ and every collection $A_n : n \in \omega$ of sets of capacity $\leq \varepsilon$ there is an infinite set $b \subset \omega$ such that $\phi(\{x \in X : \exists^\infty n \in b \; x \in A_n\}) \leq \varepsilon$;*
2. *$L(\mathbb{R})[U] \models \phi$ is continuous in increasing wellordered unions and every set has a Borel subset of the same capacity.*

Note that the first item is a condition on the capacity weaker than Ramseyness of Section 4.3.5. That section provides several examples of capacities satisfying the assumptions of the proposition as well as (1).

Proof. The key element of the proof is again the observation of that the Ramsey ultrafilter U is an $L(\mathbb{R})$-generic subset of the poset $P = \mathcal{P}(\omega)$ mod fin.

$(2) \rightarrow (1)$ is easier. Suppose that $A_n : n \in \omega$ is a collection of capacity $\leq \varepsilon$ such that for every infinite set $b \subset \omega$ the set $B_b = \{x \in X : \exists^\infty n \in b \; x \in A_n\}$ has capacity $> \varepsilon$. Going into a subsequence if necessary I can find a real number $\delta > 0$ such that $\phi(B_b) > \varepsilon + \delta$. Without loss of generality each set A_n is Borel and so the whole sequence is in $L(\mathbb{R})$. Consider the set $C = \int_\omega A_n \, dU = \{x \in X : \{n \in \omega : x \in A_n\} \in U\} \subset X$ in the model $L(\mathbb{R})[U]$. I will show that $\phi(C) \geq \varepsilon + \delta$ while C has no Borel subset of capacity $> \varepsilon$, proving (1).

Suppose that $O \subset X$ is an open set of capacity $< \varepsilon + \delta$ and $b \in P$ is a condition. Since $\phi(B_b) > \varepsilon + \delta$ the set $B_b \setminus O$ must be nonempty, containing some point $x \in X$. The infinite set $c = \{n \in b : x \in A_n\}$ then forces in P that $\check{x} \in C \setminus O$. The genericity of the ultrafilter U then shows that $\phi(C) \geq \varepsilon + \delta$. Now suppose that $B \subset X$ is a Borel set and $b \in P$ is a condition such that $b \Vdash \check{B} \subset \dot{C}$. Then for every point $x \in B$ the set $\{n \in b : x \notin A_n\}$ must be finite, since if infinite this would be a condition forcing $\check{x} \notin \dot{C}$. This means that $B = \bigcup_{n \in \omega} \bigcap_{m \in b \setminus n} A_m$, an increasing union of sets of capacity $\leq \varepsilon$, therefore $\phi(B) \leq \varepsilon$ as desired.

(1)$\rightarrow$(2) is considerably harder. I will first show that the capacity ϕ is still continuous in increasing wellordered unions in the model $L(\mathbb{R})[U]$. Suppose that $P \Vdash \dot{A}_\beta : \beta \in \alpha$ is an inclusion-increasing sequence of sets of mass $< \varepsilon$ for some ordinal β. I will find a condition q and a set $B \subset X$ of mass $\leq \varepsilon$ such that $q \Vdash \bigcup_\alpha \dot{A}_\beta \subset \check{B}$. Let $D_\beta = \{p \in P : \exists O \text{ open of } \phi\text{-mass} < \varepsilon \text{ such that } p \Vdash \dot{A}_\beta \subset \check{O}\}$, this for every ordinal $\beta \in \alpha$. These sets are all open dense. The forcing $P \in L(\mathbb{R})$ has the property that wellordered intersections of open dense sets are still open dense, so there is some condition $q \in \bigcap_\beta D_\beta$. For every ordinal $\beta \in \alpha$ let $B_\beta = \bigcap\{O \subset X : O \text{ open and } p \Vdash \dot{A}_\beta \subset \check{O}\}$. These sets form an inclusion increasing sequence of sets of mass $< \varepsilon$ in the model $L(\mathbb{R})$ and therefore $B = \bigcup_\beta B_\beta$ has mass $\leq \varepsilon$. It is clear that B, p are as required.

The second step in the proof of (1)$\rightarrow$(2) is to show that $U \not\perp \phi$, that is, if $\varepsilon > 0$ is a real number and $D \subset \omega \times X$ is a set whose vertical sections have ϕ-mass $\leq \varepsilon$ then even the set $\int_\omega D \, dU = \{x \in X : \{n \in \omega : \langle n, x \rangle \in U\}$ has ϕ-mass $\leq \varepsilon$. Since the capacity ϕ is outer regular, I may assume that the vertical sections of the set D are G_δ. In $L(\mathbb{R})$, condition (2) implies that the collection $\{a \subset \omega : \phi(E_a) \leq \varepsilon\}$ is dense in the algebra $\mathcal{P}(\omega)$ mod fin, where $E_a = \{x \in X : \exists^\infty n \in a \ x \in D_a\}$. By the genericity of the Ramsey ultrafilter U over $L(\mathbb{R})$, there is a set $a \in U$ such that $\phi(E_a) \leq \varepsilon$. Now clearly $\int_\omega D \, dU \subset E_a$ and so $\phi(\int_\omega D \, dU) \leq \varepsilon$ as required.

It now follows that the Laver-type forcing $P(U)$ with the ultrafilter U preserves the capacity ϕ using a proof similar to the argument for Theorem 3.6.11. Suppose $T \in P(U)$ is a tree and $T \Vdash \dot{O} \subset X$ is an open set of capacity $< \varepsilon$. I must show that $\phi(\{x \in X : T \Vdash \check{x} \in \dot{O}\}) \leq \varepsilon$. Thinning out the tree T if necessary I may assume that there is a function $f : T \rightarrow \mathcal{O}$ such that $\forall t \in T \ \phi(f(t)) < \varepsilon$, $\forall s \subset t \in T \ f(s) \subset f(t)$, and $T \Vdash \dot{O} = \bigcup_n f(\dot{y}_{gen} \upharpoonright n)$ where $\dot{y}_{gen}$ is the name for the $P(U)$-generic path through the tree T. To simplify the notation suppose that the trunk of the tree T is empty. By induction on an ordinal α build sets $A(\alpha, t) \subset X$ as follows:

- $A(0, t) = f(t)$;
- $A(\alpha, t) = \bigcup_{\beta \in \alpha} A(\beta, t)$ if the ordinal α is limit;
- $A(\alpha + 1, t) = \int_\omega D \, dU$ where $D \subset \omega \times X$ is defined by $D_n = A(\alpha, t^\frown n)$.

It is not difficult to argue by simultaneous induction on α such that $A(\alpha, s) \subset A(\alpha, t)$ whenever $s \subset t \in T$, $A(\beta, t) \subset A(\alpha, t)$ whenever $\beta \in \alpha$, and finally $\phi(A(\alpha, t)) \leq \varepsilon$. The latter statement uses the work in the previous two paragraphs. By the Replacement Axiom, the inductive process must stabilize at some ordinal Ω. I claim that $\{x \in X : T \Vdash \check{x} \in \dot{O}\} \subset A(\Omega, 0)$, which will complete the proof that $P(U)$ preserves the capacity ϕ. In fact these two sets are equal, but this is immaterial for the purpose here. Suppose $x \in X$, $x \notin A(\Omega, 0)$ is a point. The definition of the inductive process shows that there is a tree $S \subset T$ in $P(U)$ such that $x \notin A(\Omega, s)$ for every node $s \in S$. In particular, $x \notin f(s)$ for every node $s \in S$ and so $S \Vdash \check{x} \notin \dot{O}$ as required.

Now I am ready to conclude the proof. In $L(\mathbb{R})$, let $\dot{A}$ be a P-name for a set of capacity $> \varepsilon$. I must find a set $a \subset \omega$ such that $\phi(E_a) > \varepsilon$ where $E_a = \{x \in X : a \Vdash \check{x} \in \dot{A}\}$. Note that the latter set is in $L(\mathbb{R})$ and so by the assumptions on the capacity ϕ it has a Borel subset of the same capacity which in the P-extension will be a subset of the set $\dot{A}$. The implication $(1) \to (2)$ then follows by the genericity of the Ramsey ultrafilter U over the model $L(\mathbb{R})$. Let $a \subset \omega$ be a $P(U)$-generic set and in the model $V[a]$ consider the set E_a. Since y diagonalizes the ultrafilter U, it follows that $\dot{A}/U \subset E_a$. Now $\dot{A}/U \in V$ was a set of ϕ-mass $> \varepsilon$, the forcing $P(U)$ preserves the capacity ϕ, and so $V[a] \models \phi(E_a) > \varepsilon$. A universally Baire absoluteness argument now shows that already in the ground model there must be a set $a \subset \omega$ such that $\phi(E_a) > \varepsilon$ as desired. $\qquad\square$

3.5 Preservation of Baire category

Let P be a forcing. I will say that P preserves Baire category if for every nonmeager set $A \subset 2^\omega$, $P \Vdash \check{A}$ is nonmeager. I will also say that P collapses Baire category (under some condition) if there is a condition $p \in P$ such that $p \Vdash 2^\omega \cap V$ is a meager set. These two properties have been frequently investigated and used. This section offers many topological restatements of category preservation in the definable case. It follows from Proposition 3.2.2 and the homogeneity of the meager ideal that if I is a suitably definable σ-ideal on a Polish space X and the forcing P_I is proper then category preservation is equivalent to $I \not\perp \mathtt{meager}$.

It turns out that category preservation is a property that in universally Baire forcing case has many consequences that sound almost too good to be true. The key tool is the following game characterization.

Definition 3.5.1. *Suppose P is a partial order. The* category game *G between Players I and II proceeds as follows. First Player II indicates an initial condition $p_{ini} \in P$. After that, the moves alternate, Player I producing a condition p_n and Player II playing its strengthening $q_n \leq p_n$. Player I wins if the result of the play, the expression*

$$p_{ini} \wedge \bigwedge_m \bigvee_{n > m} q_n$$

denotes a non-zero element in the complete Boolean algebra RO(P).

The following is proved in Section 3.10.9, Theorem 3.10.21.

Theorem 3.5.2. *Suppose that the forcing P is proper. The following are equivalent:*

1. P below some condition makes the set of the ground model reals meager;
2. Player II has a winning strategy in the game G.

Moreover, if suitable large cardinals exist and the forcing P is universally Baire then the game G is determined.

I do not know if in the case of a forcing P_I associated with a $\mathbf{\Pi}_1^1$ on $\mathbf{\Sigma}_1^1$ σ-ideal I the category game must be determined in ZFC. This would be quite helpful given the fact that all category preserving ideals in this book are $\mathbf{\Pi}_1^1$ on $\mathbf{\Sigma}_1^1$.

There is a large number of corollaries.

Corollary 3.5.3. *(LC) Suppose that I is a universally Baire ideal such that the forcing P_I is proper and preserves category. Then the forcing P_I is such in all forcing extensions.*

Proof. To see this note that the large cardinal assumptions imply that there is a universally Baire winning strategy for Player I in the category game. An absoluteness argument shows that this strategy remains winning in every set forcing extension. Existence of such strategy implies that the forcing is proper and preserves category. $\qquad\square$

The following corollaries depend on an idea which I will use throughout the book in the case of category-preserving forcings. If I is a σ-ideal on a Polish space X and M is a countable elementary submodel of a large structure, consider the countable forcing $P_I \cap M$. A repetition of the proof of Proposition 2.1.2 will show that the forcing $P_I \cap M$ adds a single point $\dot{x}_{gen} \in \dot{X}$ which belongs to all sets in its generic filter and falls out of all sets in the model M which do not belong to the generic filter. Moreover, the name for the $P_I \cap M$-generic point is just the restriction of the name for the P_I-generic point to the model M. In general, the forcing $P_I \cap M$

is so different from P_I itself as to be useless to consider; however, in the category preserving case the situation changes.

Corollary 3.5.4. *(LC) Suppose that I is a universally Baire σ-ideal on a Polish space X such that the forcing P_I is proper. The following are equivalent:*

1. *P_I preserves category;*
2. *for every countable elementary submodel M of a large structure the forcing $P_I \cap M$ forces its generic point to fall out of all ground model coded I-small sets.*

Proof. For (1)→(2) suppose for contradiction that P_I preserves category and some condition $B \in P_I \cap M$ forces in $P_I \cap M$ that the $P_I \cap M$-generic point belongs to some Borel I-small set $C \in I$. By the Baire category theorem applied to the poset $P_I \cap M$ there are open dense sets $D_n \subset P_I \cap M : n \in \omega$ such that $B \cap \bigcap_n \bigcup D_n \subset C$. Choose a winning strategy σ in the category game G for Player I in the model M, and simulate a play against this strategy in which Player II indicates $B = p_{ini}$ and his n-th move is a condition in the model M and in the set $\bigcap_{m \in n} D_m$. The result of this play is a subset of the set C, and Player II won, a contradiction.

For the other direction suppose that (2) holds and $B \subset X$, $C \subset 2^\omega$ and $D \subset B \times C$ are a Borel I-positive set, a Borel nonmeager set, and a Borel set with meager vertical sections. I must find a point $y \in C$ with $B \setminus D^y \notin I$. Fix a countable elementary submodel M of a large enough structure. If all the sets $B \setminus D^y$ belonged to the ideal I, then the condition B in the forcing $P_I \cap M$ forced $C \cap V \subset D_{\dot{x}_{gen}}$ by the property (2). This is impossible since $P_I \cap M$ is a countable forcing, therefore equivalent to Cohen forcing, and Cohen forcing preserves Baire category by Kuratowski–Ulam theorem [40], 8.41. $\square$

Corollary 3.5.5. *(LC) Suppose that I is a universally Baire ideal on a Polish space X such that P_I is proper. The following are equivalent:*

1. *P_I preserves category;*
2. *there is a collection T of Polish topologies on the space X giving the same Borel structure such that $I = \{A \subset X : \forall t \in T \ A \text{ is meager in } t\}$.*

Proof. The direction (2)→(1) is an immediate corollary of the Kuratowski–Ulam theorem for category. If $B \in P_I$ is a Borel set, $C \subset 2^\omega$ is a Borel nonmeager set and $D \subset B \times C$ is a Borel set, find a Polish topology t on the space X such that the ideal I is a subideal of the t-meager ideal and B is not t-meager, and use the Fubini theorem for category to either find a vertical section of the set D which is not meager or a horizontal section of the complement of the set D which is not t-meager and therefore not in the ideal I.

For the opposite direction, let M be a countable elementary submodel of a large enough structure, consider the countable forcing $Q = P_I \cap M$, and consider the Stone space Y of the forcing Q and its G_δ subset $Z \subset Y$ consisting of M-generic filters. Since the forcing Q is countable, the space Y is Polish and so is the set Z. Let $\pi : Z \to X$ be the one-to-one continuous function defined by $\pi(G) = \dot{x}/G$; manipulating π on a nowhere dense set I can assume that π is a Borel bijection. Let t be the π-image of the Y-topology on the space X. This is the desired topology.

To see this note that the collection of filters containing B is non-meager in the Stone space Y and so the set B is not meager in the topology τ. On the other hand, if $C \in P_I \cap M$ and $D_n : n \in \omega$ is a collection of open dense subsets of the poset $P_I \cap M$ (possibly not elements of the model M) then there is a play observing the strategy σ with all moves in the model M such that $p_{ini} = C$ and q_n–the n-th move of Player II–is in the open dense set $\bigcap_{m \in n} D_m$. The result of such a play is an I-positive set. This, together with the Baire category theorem, shows that no Borel t-nonmeager set can be I-small, or in other words $I \subset t$-meager ideal as claimed. □

Note that the corresponding equivalence for the Lebesgue measure – P_I is bounding and preserves Lebesgue measure if and only if I is polar – is an open Question 7.2.3.

Corollary 3.5.6. *(LC) Suppose that I is a universally Baire ideal on a Polish space X such that P_I is proper. The following are equivalent:*

1. *P_I preserves Baire category;*
2. *For every I-positive Borel set $B \subset X$ the Cohen forcing adds an element of the set B which falls out of all ground model coded Borel I-small sets.*

This is immediate from the previous corollary.

Corollary 3.5.7. *(LC) Suppose that I is a universally Baire c.c.c. ideal such that the forcing preserves Baire category. Then P_I is in the forcing sense equivalent to the Cohen forcing.*

This has been proved earlier by Shelah [61] using entirely different methods. The corresponding implication for the case of Lebesgue measure – if I is a universally Baire c.c.c. σ-ideal and $I \not\perp \mathtt{null}$ then P_I is equivalent to the Solovay forcing – is true and its proof follows somewhat different lines [16], [17].

Proof. This is an immediate consequence of the previous corollary. It will be enough to show that the set $D = \{C \in P_I : P_I$ below C has a countable dense set$\} \subset P_I$ is dense–by the c.c.c. I will be then able to find a countable maximal antichain in D and get a countable dense subset of the whole forcing P_I. So fix a condition $B \in P_I$. By the previous corollary the Cohen forcing adds an element $\dot{x} \in \dot{B}$ which falls out of all I-small Borel sets coded in the ground model. By the c.c.c. such a point $\dot{x}$ is P_I-generic. A standard forcing theory argument then shows that there must be a condition $C \subset B$ and a complete embedding of the poset P_I below C to the Cohen forcing algebra. Since Cohen forcing has a countable dense subset so must the poset P_I below C, thus $C \in D$ and the proof is complete. $\square$

Corollary 3.5.8. *(LC) Suppose that I is a universally Baire σ-ideal on a Polish space X such that the forcing P_I is proper. The following are equivalent:*

1. *there is a Borel I-positive Borel set $B \in P_I$ and a Polish topology τ on B generating the same Borel structure such that every countable subset of the set B is covered by a $\tau - G_\delta$-set in the ideal I;*
2. *P_I fails to preserve Baire category.*

This shows that the forcings associated with Hausdorff measures of Section 4.4 and outer regular capacities of Section 4.3 make the set of ground model reals meager. In all the specific cases there is a quite simple proof of this statement which does not use any large cardinal assumption.

Proof. I will start with (1)→(2) direction. Extend the topology on some positive Borel set $B_0 \subset X$ to satisfy the property (1). Suppose for contradiction that $B_1 \in P_I$, $B_1 \subset B_0$ is a condition forcing that category is preserved. Fix the corresponding winning strategy σ for Player I in the category game below B_1, and choose a countable elementary submodel M of a large enough structure containing I, B_1, σ. Let $A \subset \{x \in B_1 : x \text{ is } M\text{-generic}\}$ be a countable set meeting every condition in $M \cap P_I$ below B_1. Use the assumption to find a G_δ-set $\bigcap_{m \in \omega} O_m$ in the ideal I covering A, where $O_m : m \in \omega$ are open sets. I claim that there is a counterplay against the strategy σ in which Player II uses only conditions in the model M, and at n-th round his move q_n is a set which is a subset of all the open sets $O_m : m \in n$. The result $\bigcap_k \bigcup_{n>k} q_n$ of such a play must be a subset of the set $O = \bigcap_m O_m \in I$, contradicting the assumption that σ is a winning strategy.

Suppose that the play has been constructed up to the move $p_n \in M$ of Player I. The set $\bigcap_{m \in n} O_m$ is open and it contains some M-generic point $x \in A \cap p_n$. So there must be a basic open neighborhood $P \subset X$ such that $x \in P \subset \bigcap_{m \in n} O_m$. Now $P, P \cap p_n \in M$, and since the latter set contains the M-generic point x, it must be I-positive. Player II indicates $q_n = P \cap p_n$ and the construction of the play can proceed.

The (2)→(1) direction uses no large cardinal or definability assumptions. Fix a winning strategy σ for Player II in the category game starting with some condition $p_{ini} \in P_I$, let M be a countable elementary submodel of a large enough structure containing the strategy σ, and let $B = \{x \in p_{ini} : x \text{ is } M\text{-generic}\} \in P_I$. Find a Polish topology τ on the Borel set B extending the original one in which all Borel sets in the model M are open, using [40], II.13.5. I claim that the topology τ has the required property. Suppose that $A \subset B$ is a countable set. I will find a play in the category game in which all moves belong to the model M, Player II follows the strategy σ and A is a subset of the result of the play. This will complete the proof since the result of the game is $\tau - G_\delta$ I-small set.

To find the desired play of the category game, first fix an enumeration $A = \{x_n : n \in \omega\}$ and a partition $\omega = \{a_m : m \in \omega\}$ into infinite sets. I will build the play in such a way that for every number $n \in \omega$ the n-th move q_n of the strategy σ contains the point x_m whenever $n \in a_m$. This will guarantee that the set A is indeed a subset of the result of the play. Well, suppose that the move q_{n-1} of the play has been obtained, and find a number m such that $n \in a_m$. The set $D = \{q \in P_I : \exists p \le p_{ini}\, p$ induces the strategy σ to answer with $q\} \subset P_I$ is dense below the condition P_I, and since the point $x_m \in A$ is M-generic, it is an element of some condition $q \in D \cap M$. This shows precisely that there is an extension of the play by some moves p_n and $q_n = q$ in the model M such that $x_m \in q_n$ and Player II still follows his strategy σ. $\qquad\square$

Corollary 3.5.9. *(LC) Suppose that P is a universally Baire proper category preserving forcing. Suppose that T is an ω_1-tree. Every uncountable subtree of T in the P-extension contains an uncountable subtree in the ground model.*

Proof. Suppose $\dot{S}$ is a name for an uncountable subtree of the tree T, assume that no condition in the forcing P forces an uncountable set of elements of T into $\dot{S}$, and work towards a contradiction.

Use Theorem 3.10.21 to find a winning strategy σ for Player I in the category game. Find a countable elementary submodel M containing all the necessary information, let $\alpha = M \cap \omega_1$, and choose an enumeration $t_m : m \in \omega$ of α-th level of the tree T. Find a play of the category game such that

- Player I follows the strategy σ;
- moves of Player II come from the model M;
- whenever $n \in \omega$ is a number and $m \in n$ then there is a node $u_m < t_m$ in the tree T such that the condition q_n, the answer of Player II at round n, forces $\check{u}_m \notin \dot{S}$.

This is not difficult to do: once strategy σ makes the move $p_n \in M$ by induction on $m \in n$ find a descending chain $q_n^m : m \in n$ of conditions in the model M below p_n such that for every $m \in n$ there is a node $u_m < t_m$ such that q_n^m forces $\check{u}_m \notin \dot{S}$. Suppose q_n^{m-1} has been found. Consider the set $\{u \in T : \exists r \le q_n^{m-1}\, r \Vdash \check{u} \notin \dot{S}\} \in M$. By the assumption on the name $\dot{S}$, this set is co-countable, in particular it contains some level T_β for $\beta \in \alpha$. Let $u_m \in T_\beta$ be the unique node smaller than t_m and find the condition $q_n^m \leq q_n^{m-1}$ in the model M forcing $\check{u}_m \notin \dot{S}$. After this subinduction has been performed, put $q_n = q_n^n$.

Let $q \leq p$ be the result of the play described in the previous items. Clearly, $q \Vdash \forall m \in \omega \; \check{t}_m \notin \dot{S}$ and therefore the tree $\dot{S} \subset \check{T}$ is countable. $\qquad \square$

As an immediate consequence the forcing P does not add new branches to ω_1-trees and preserves Suslin trees.

3.6 Preservation of outer Lebesgue measure

Let λ denote the outer Lebesgue measure on 2^ω. Let P be a forcing. I will say that P preserves outer Lebesgue measure if for every set $A \subset 2^\omega$, $P \Vdash (\lambda(A))^V = \lambda(\check{A})$. I will also say that P collapses outer Lebesgue measure (under some condition) if there is a condition $p \in P$ such that $p \Vdash 2^\omega \cap V$ is a Lebesgue null set. These two properties have been frequently investigated and used. It follows from the results of Section 3.2.2 that if I is a suitably definable σ-ideal on a Polish space X and the forcing P_I is proper then P_I collapses outer Lebesgue measure iff it does not preserve it iff $I \perp \lambda$ holds.

In this section I will provide several tools for proving outer Lebesgue measure preservation.

3.6.1 Polar ideals

By far the most powerful tool for proving Lebesgue measure preservation theorems is associated with polar ideals.

Definition 3.6.1. *A σ-ideal I on a Polish space X is* polar *if there is a collection M of countably additive probability measures on the space X such that $I = \{A \subset X : \forall \mu \in M \; \mu(A) = 0\}$.*

The terminology comes from a paper of Gabriel Debs [12] where he considers polar ideals of *compact* sets. While his work is highly relevant to our investigation here, most examples of polar ideals below are not generated by compact sets in any presentation. The basic theorem:

Theorem 3.6.2. *Suppose that I is a polar σ-ideal such that the forcing P_I is proper. Then the forcing P_I is bounding and preserves outer Lebesgue measure.*

Proof. To prove the bounding condition, let $B \in P_I$ be a Borel set and $f : B \to 2^\omega$ be a Borel function. I must produce a compact set $C \subset B$ such that $C \in P_I$ and $f \upharpoonright C$ is continuous. To do that, find a probability measure μ on the underlying space X such that $\mu(B) \neq 0$ and μ vanishes on all sets in the ideal I. Let $J = \{A \subset X : \mu(A) = 0\}$. Then P_J is isomorphic to the Solovay forcing by the measure isomorphism theorem [40], 17.41, it is bounding, and there must be a compact set $C \subset B$ such that $f \upharpoonright C$ is continuous and $C \notin J$. However, $I \subset J$ and so $C \notin I$ as desired.

The preservation of Lebesgue measure is similar. Let $B \in P_I$ be a Borel set and $D \subset B \times 2^\omega$ be a Borel set with vertical sections of Lebesgue measure $\leq \varepsilon$. Proceed as above and find a measure μ which vanishes on all sets in the ideal I and $\mu(B) \neq 0$. Let $J = \{A \subset X : \mu(A) = 0\}$. By the Fubini theorem, $\lambda(\int_B D \, dJ) \leq \varepsilon$. Since $I \subset J$, it is necessarily the case that $\int_B D \, dI \subset \int_B D \, dJ$, $\lambda(\int_B D \, dI) \leq \lambda(\int_B D \, dJ) \leq \varepsilon$, and the proof is complete. $\qquad\square$

One of the most pressing questions left open in this book is whether this implication can be reversed, Question 7.2.3. Note that a similar equivalence indeed is true on the category side as proved in Corollary 3.5.5. Proposition 3.6.10 below shows that the implication can be reversed in a rather extensive class of ideals, nevertheless the general case remains open. As long as this important piece is missing, I will just go on and give a list of examples of ideals which are or are not polar. It turns out that measure theorists have been involved in this type of investigation for a long time.

Example 3.6.3. If η is a strongly subadditive capacity on a Polish space X then η is an envelope of measures by a theorem of Choquet [6], and the associated null ideal $I_\eta = \{A \subset X : \eta(A) = 0\}$ is therefore polar. Here η is an envelope of measures if for every Borel set $B \subset X$ and every real $\varepsilon > 0$ there is a measure μ on X such that $\mu \leq \eta$ on Borel sets and $\mu(B) \geq \eta(B) - \varepsilon$. Many forcings of the form P_{I_η} are proper; this extensive subject is handled in Section 4.3.

Example 3.6.4. Suppose that ϕ is a (non-σ-finite) measure on a Polish space X, and consider the ideal $I_{\sigma\phi}$ σ-generated by sets of finite ϕ-measure. If the measure satisfies the condition

(*) every Borel ϕ-positive set contains a Borel subset of nonzero finite ϕ-mass,

then the ideal $I_{\sigma\phi}$ is polar. The verification of the condition (*) for various measures made measure theorists busy for decades. For instance, Howroyd [26] proved that r-dimensional Hausdorff measures satisfy it where $r > 0$ is a real number, and Preiss and Joyce [32] proved that r-dimensional packing measures satisfy it where $r > 0$ is a real number. In both cases, the associated forcings are proper and interesting.

The polarity of the ideal $I_{\sigma\phi}$ requires an argument. Let $B \in P_{I_{\sigma\phi}}$ be a Borel set. I must find a probability measure μ on the space X which vanishes on all sets in the ideal $I_{\sigma\phi}$ and assigns the set B a positive value. Consider the ideal J of sets of ϕ-mass zero. The property (*) together with the measure isomorphism theorem [40], 17.41, imply that the forcing P_J is densely isomorphic to Solovay forcing and therefore bounding. It also implies that there is an uncountable antichain in the forcing P_J below the set B: any maximal antichain in the dense set $\{C \in P_J : \phi(C) < \infty\}$ below B must be uncountable since B cannot be covered by countably many sets of finite ϕ-measure. The argument from 3.7.7 then shows that there must be a perfect collection $P \subset K(X)$ of mutually disjoint ϕ-positive perfect subsets of the set B.

Use the condition (*), the measure isomorphism theorem and the Sacks uniformization 2.3.4 to find a perfect set $Q \subset P$ and a Borel function $f : Q \times 2^\omega \to X$ such that for every compact set $C \in Q$ the range of the function f_C is a subset of C and f_C is a measure isomorphism between the Lebesgue measure λ on the Cantor set and the measure ϕ on $\mathrm{rng}(f_C)$. Let ψ be some probability measure on the perfect set Q and let μ be the probability measure on the space X given by $\mu(A) = (\psi \times \lambda)(f^{-1}(A))$. I claim that the measure μ has the desired properties.

First, it is clear that $\mu(B) = \mu(X) = 1$. Moreover, whenever $A \subset X$ is a Borel set in the σ-finite ideal $I_{\sigma\phi}$, the set $\{C \in Q : \phi(A \cap C) > 0\}$ must be countable by the countable additivity of the measure ϕ. It follows that the preimage $f^{-1}(A) \subset Q \times 2^\omega$ has Lebesgue null vertical sections with possibly countably many exceptions. By the Fubini theorem then, $\mu(A) = (\psi \times \lambda)(f^{-1}(A)) = 0$ as desired.

Example 3.6.5. Rogers and Davies [11] constructed a classical example of a Hausdorff measure ϕ which fails the condition (*). Necessarily it has to be a measure associated with a fast growing gauge function, since gauge functions like the exponentiation to the power r, $r > 0$, give rise to measures satisfying the condition (*) by a result of Howroyd [26]. The example gives a Hausdorff measure ϕ such that its only values are zero or infinity. Thus $I_{\sigma\phi} = \{A \subset X : \phi(A) = 0\}$ and the whole argument in [11] shows that $I_{\sigma\phi}$ is not a polar ideal. I investigate the forcing P_{I_ϕ} in depth in Section 4.4. It turns out that it is proper, bounding, does not add splitting reals, and collapses the outer Lebesgue measure.

Example 3.6.6. Let I be the σ-ideal of sets of extended uniqueness on the unit circle T in the complex plane. Then $I = \{A \subset T : \mu(T) = 0$ for every Rajchman measure $\mu\}$ and therefore I is a polar ideal. It turns out that the ideal I is σ-generated by closed sets, and therefore the associated forcing P_I is proper, bounding, and preserves category and outer Lebesgue measure.

Example 3.6.7. A number of polar ideals appears in the theory of Borel equivalence relations. Let E be a Borel equivalence relation on a Polish space X which is not smooth, and consider the ideal I generated by Borel sets $B \subset X$ such that $E \upharpoonright B$ is smooth. Then by [47], a Borel set $B \subset X$ is I-positive if and only if it supports a probability E-ergodic measure. Thus $I = \{B \subset X : \mu(B) = 0$ for every E-ergodic probability measure} and the ideal I is polar. A special case of this ideal appears in Section 4.7.1.

If E is a countable Borel equivalence on a Polish space X, call a set $B \subset X$ *compressible* if there is a Borel injection $f : B \to B$ such that $\forall x \in B\ f(x) Ex$ and the set $B \setminus f''B$ hits every equivalence class of B. The collection I of all Borel sets B such that the E-saturation of B is compressible is a σ-ideal. A theorem of Nadkarni [53] shows that $B \notin I$ if and only if there is an E-invariant probability measure μ such that $\mu(B) > 0$. Here a measure μ is E-invariant if it is invariant with respect to some (any) Borel action of a countable group generating the equivalence E as the orbit equivalence. Thus $I = \{B : \mu(B) = 0$ for every E-invariant probability measure $\mu\}$ and again, I is a polar ideal. The ergodic decomposition theorem shows that P_I, if nontrivial, is densely isomorphic to the Solovay forcing.

Example 3.6.8. A number of other σ-ideals are defined in such a way that they are polar. A typical example is the σ-ideal of Gauss null sets on any separable Banach space – these are the sets with zero mass for every Gauss measure on the space [10]. The problem is that such definitions do not yield any hint as to the possible proof of properness or improperness of the resulting quotient forcing.

Example 3.6.9. Let I be the σ-ideal of σ-porous sets on the real line. The forcing P_I is proper by the results of Section 4.2, and it is bounding by the results of [80]. Preiss and Humke [27] produced a Borel I-positive set $B \subset \mathbb{R}$ such that every measure on it concentrates on a σ-porous set, therefore the ideal I is not polar. I do not know if the forcing P_I preserves outer Lebesgue measure. In Section 4.2 I will show an example of another compact metric space for which the metric porosity forcing demonstrably makes the set of the ground model reals Lebesgue null.

Finally, there is the promised proposition showing than in a large number of cases polarity of ideals coincides with the preservation of outer Lebesgue measure and the bounding property.

Proposition 3.6.10. *Suppose that ϕ is a countably subadditive submeasure on a Polish space X which is outer regular on compact sets: $\phi(K) = \inf\{\phi(O) : K \subset O$ and O open$\}$ for every compact set $K \subset X$. Suppose that the forcing P_{I_ϕ} is proper, where $I_\phi = \{A \subset X : \phi(A) = 0\}$. The following are equivalent:*

1. *the forcing P_{I_ϕ} is bounding and preserves outer Lebesgue measure;*
2. *the ideal I_ϕ is polar.*

The proposition applies to all capacities as well as to all pavement submeasures, since changing the topology of the underlying space X it is possible to present any pavement submeasure as an outer regular submeasure.

Proof. The (2)→(1) direction is included in Theorem 3.6.2. The opposite direction is the heart of the matter. The proof is an elaboration of an argument of Christensen [7].

Suppose that the forcing P_{I_ϕ} is bounding and proper, and $B \notin I_\phi$ is a Borel set. Since the forcing P_{I_ϕ} is bounding, the compact sets are dense in it by Theorem 3.3.2 and I can assume that the set B is in fact compact. There are two distinct cases.

Either the submeasure $\phi \restriction B$ is not *pathological*, meaning that there is a bounded finitely additive measure ψ on the set X which is dominated by ϕ such that $\psi(B) > 0$. In this case I will use the standard Caratheodory construction to extend the measure $\psi \restriction K(X)$ to a countably subadditive measure μ. Then $\mu(B) = \psi(B) > 0$ and a tightness argument will show that $\mu \leq \phi$. For every set $A \subset X$, define $\mu(A) = \sup_{\delta>0} \inf\{\Sigma_k \psi(O_k) : A \subset \bigcup_k O_k$ and O_k is open of diameter $< \delta\}$. It is immediate that μ is a metric measure. I claim that μ is a nonzero countably additive measure dominated by ϕ as desired. The finite additivity of the measure ψ and a compactness argument show that $\mu(K) = \inf\{\psi(O) : K \subset O$ and O open$\}$ for every compact set $K \subset X$. In particular $\mu(B) = \psi(B) > 0$ and $\mu(K) \leq \phi(K)$ for every compact set $K \subset X$ by the outer regularity of the submeasure ϕ. For an arbitrary Borel set $A \subset B$ the tightness of the measure μ implies that $\mu(A) = \sup\{\mu(K) : K \subset A$ compact$\} \leq \sup\{\phi(K) : K \subset A$ compact$\} \leq \phi(A)$ as desired.

Or the submeasure $\phi \restriction B$ is pathological. In this case I will show that $B \Vdash$ the ground model reals form a Lebesgue null set. By a theorem of Christensen [7], for every number $n \in \omega$ there is a finite collection $\{A_i^n : i \in i_n\}$ of Borel subsets of B such that $\phi(A_i^n) \leq 2^{-n}$ and there are nonnegative numbers $c_i^n : i \in i_n$ with unit sum and $\Sigma_i c_i^n \cdot \chi(A_i^n) \geq 1/2$. Consider the space $Y = \prod_n i_n$ equipped with the probability measure μ which is the product of the measures μ_n on i_n defined by $\mu_n(a) = \Sigma_{i \in a} c_i^n$. Consider the set $D \subset B \times Y$, $D = \{\langle x, y \rangle \in B \times Y : \forall^\infty n \; x \notin A_{y(n)}^n\}$. It will be enough to show that the vertical sections of the set D have μ-mass 0 while the horizontal sections of its complement have ϕ-mass 0. Fix a point $x \in B$. Then $D_x = \bigcup_n E_x^n$ where $E_x^n = \{y \in Y : \forall k > n \; x \notin A_{y(k)}^k\}$. By the choice of the sets A_i^k, the numbers $\mu_k(\{i \in i_k : x \notin A_i^k\})$ are smaller than $1/2$ and therefore the μ-masses of the sets E_x^n and D_x are zero. On the other hand, fix a point $y \in Y$. The set $B \setminus D^y$ is the intersection of sets $F_y^n = \{x \in X : \exists k > n \; x \in A_{y(k)}^k\}$ which have the respective ϕ-masses $\leq \Sigma_{k>n} 2^{-k}$, numbers which tend to zero as n tends to ∞. Thus $\phi(B \setminus D^y) = 0$ as desired.

The proposition immediately follows from these two cases. $\square$

3.6.2 Other proofs

Of course, some forcings preserve outer Lebesgue measure without being polar. The archetype of such behavior is the Laver forcing; it is not polar since it is not bounding. Woodin [2], 7.3.36, proved that Laver forcing preserves outer Lebesgue measure. A more careful argument actually provides a much stronger result, and the method will be applied again in the proof of a powerful preservation theorem.

Theorem 3.6.11. *Suppose that ϕ is an outer regular capacity on a Polish space X such that every coanalytic set is capacitable for ϕ. Then Laver forcing preserves ϕ.*

Here, a set is capacitable for ϕ if it can be sandwiched between two Borel sets of the same ϕ-mass. It turns out that in the constructible universe for most capacities there is a coanalytic set which is not capacitable–Theorem 4.3.21. On the other hand, under assumptions such as the determinacy of coanalytic games or add(null) $> \aleph_1$, coanalytic sets are capacitable for very many capacities – Theorem 4.3.6. This subject is handled in detail in Section 4.3. Note that the theorem does not require the capacity to be subadditive.

Proof. Suppose $T \subset \omega^{<\omega}$ is a Laver tree forcing $\dot{O} \subset \dot{X}$ to be an open set of capacity $\leq \varepsilon$. I have to show that the set $\{x \in X : T \Vdash \check{x} \in \dot{O}\}$ has capacity $\leq \varepsilon$.

Let $\mathcal{O}$ be some countable topology basis for the space X closed under finite unions. A standard fusion argument gives a Laver tree $S \subset T$ and a function $f : S \to \mathcal{O}$ such that $\forall t \in S \; \phi(f(t)) \leq \varepsilon$, $s \subset t \to f(s) \subset f(t)$, and $S \Vdash \dot{O} = \bigcup_n f(\dot{x}_{gen} \upharpoonright n)$. To simplify the notation assume that $S = \omega^{<\omega}$.

Consider the space $Y = \omega^{<\omega} \times X$ and the operator $\Gamma : \mathcal{P}(Y) \to \mathcal{P}(Y)$ on it defined by $\langle s, x \rangle \in \Gamma(B) \leftrightarrow \langle s, x \rangle \in B \vee \forall^\infty n \; \langle s^\frown n, x \rangle \in B$. This is a monotone inductive coanalytic operator, and therefore by a theorem of Cenzer and Mauldin [5], 1.6, given a coanalytic set $A \subset Y$, the transfinite sequence given by the description $A = A_0$, $A_{\alpha+1} = \Gamma(A_\alpha)$ and $A_\alpha = \bigcup_{\beta \in \alpha} A_\beta$ for limit ordinals α, stabilizes at ω_1 in a coanalytic set A_{ω_1} such that for every analytic set $C \subset A_{\omega_1}$ there is an ordinal $\alpha \in \omega_1$ such that $C \subset A_\alpha$.

Now consider the set $A \subset Y$ defined by $\langle s, x \rangle \in A$ if $x \in f(s)$. It is not difficult to see that writing A^s for the set $\{x \in X : \langle s, x \rangle \in A\}$ it is the case that $s \subset t \to A^s \subset A^t$, these sets have capacity $\leq \varepsilon$ and this feature persists through the countable stages of the iteration. To see that $\phi(A^s_{\alpha+1}) \leq \varepsilon$ note that the set $A^s_{\alpha+1}$ is an increasing union of the sets $\bigcap_{m>n} A^{s^\frown m}_\alpha : n \in \omega$, each of them of ϕ-mass $\leq \varepsilon$, and use the continuity of the capacity under increasing unions. At limit stages, use the continuity of the capacity again to argue that $c(A^s_\alpha) \leq \varepsilon$.

Consider the coanalytic set A_{ω_1}, the fixed point of the operator Γ, and its first coordinate $B = A^0_{\omega_1}$. First note that $x \notin B$ means that $S \nVdash \check{x} \in \dot{O}$, since if $x \notin B$ then the tree $U = \{s \in S : x \notin A^s_{\omega_1}\}$ is a Laver tree by the definition of the operator Γ and it forces $\check{x} \notin \dot{O}$. In fact a transfinite induction argument will show that $B = \{x \in X : S \Vdash \check{x} \in \dot{O}\}$. Now it is enough to show that $\phi(B) \leq \varepsilon$. However, if $\phi(B) > \varepsilon$, then by the capacitability of the set B there is a compact set $C \subset B$

such that $\phi(C) > \varepsilon$, and such a set must be included in the set A_α^0 for some countable ordinal α. However, $\phi(A_\alpha^0) \leq \varepsilon$ as proved in the previous paragraph, a contradiction! $\qquad\square$

It is instructive to compare this argument with the original Woodin's proof for preservation of outer Lebesgue measure in the Laver extension in [2], 7.3.36.

Corollary 3.6.12. *Suppose that ϕ is an outer regular capacity such that every coanalytic set is capacitable for it. Then Miller forcing and Steprāns forcing preserve ϕ.*

To show this it is possible to literally repeat the Laver forcing proof. A parallel argument more in line with the doctrine presented in this book reduces the Miller case to the Laver case by noticing that Laver forcing adds a point falling out from all Borel sets in the ideal associated with the Miller forcing.

3.7 The countable chain condition

The countable chain condition is a rare guest in the realm of definable forcing.

3.7.1 Ergodicity

Definition 3.7.1. *A σ-ideal I on a Polish space X is called* ergodic *if there is a Borel equivalence relation E on X with countable classes such that every Borel E-invariant set is either in I or its complement is in I. A c.c.c. forcing P adding a single point $\dot{x} \in X$ is called* ergodic *if its associated ideal $\{B \subset X : P \Vdash \dot{x} \notin \dot{B}\}$ is ergodic.*

While ergodicity does not imply c.c.c. and vice versa, most c.c.c. forcings for adding a single real are ergodic:

Example 3.7.2. The Cohen forcing is ergodic. Consider the presentation as P_I where I is the ideal of meager subsets of the Cantor space 2^ω. Let E be the equivalence relation defined by xEy if $x \Delta y$ is finite. Now if $B \subset 2^\omega$ is a Borel nonmeager set, then it is comeager in some basic open set O_t for $t \in 2^n$ for some number $n \in \omega$, its E-saturation must be comeager in every set O_s for $s \in 2^n$, and therefore the E-saturation of the set B is comeager as required.

Example 3.7.3. The Solovay forcing is ergodic. This is the fundamental fact of ergodicity theory [78]; the equivalence used is the Vitali equivalence relation.

Example 3.7.4. The Hechler forcing is ergodic. Hechler forcing P is the set of all pairs $p \in \omega^{<\omega} \times \omega^{\omega}$, $p = \langle t_p, f_p \rangle$ ordered by $q \leq p$ if $t_p \subset t_q$, $f_p \leq f_q$ pointwise and $t_q \setminus t_p \geq f_p$ pointwise on its domain. The forcing adds a single point $\dot{f}_{gen} \in \omega^{\omega}$ as the union of the first coordinates in the generic filter. Let I be its associated ideal on the Baire space. Let E be the equivalence on the Baire space defined by $fEg \leftrightarrow |f\Delta g| < \aleph_0$. Suppose $B \subset \omega^{\omega}$ is a Borel E-invariant set which is I-positive, that is there is a condition $p \Vdash \dot{f}_{gen} \in B$. I will show that then $1 \Vdash \dot{f}_{gen} \in B$, in other words $\omega^{\omega} \setminus B \in I$ as required. Suppose for contradiction that there is a condition $q \Vdash \dot{f}_{gen} \notin B$. It is not difficult to extend the conditions p, q to p', q' such that $|t_{p'}| = |t_{q'}|$ and $f_{p'} = f_{q'}$. Consider the function $\pi : P \upharpoonright p' \to Q \upharpoonright q'$ where $\pi(r)$ is the condition obtained from r by replacing the appropriate initial segment of t_r with $t_{q'}$. The function π is clearly an isomorphism of the two partial orders. Thus if $G \subset P \upharpoonright p'$ is a generic filter then so is $\pi''G \subset P \upharpoonright q'$. By the forcing theorem, the generic point $x \in \omega^{\omega}$ associated with G is in the set B while the generic point $y \in \omega^{\omega}$ associated with $\pi''G$ is in the complement of B. However, a review of the definitions reveals that y is obtained from x by replacing its appropriate initial segment with $t_{q'}$. Thus xEy, contradicting the E-invariance of the set B.

I do not know an example of an ergodic forcing whose ergodicity would be witnessed by an equivalence relation in the Borel reducibility sense more complicated than the Vitali equivalence. The previous examples seem to indicate that the ergodicity is connected with c.c.c. and the homogeneity of the forcing, and this is in fact true.

Proposition 3.7.5. *Suppose that I is an ergodic σ-ideal on a Polish space X. Then:*

1. *either the forcing P_I is c.c.c.;*
2. *or the ground model coded I-small sets cover the space X in every $\aleph_1$-preserving extension.*

In other words, if it is at all possible to increase the invariant $\text{cov}(I)$ while preserving $\aleph_1$, then the forcing P_I is the only tool.

Proof. Clearly (1) implies that (2) fails. To see that $\neg(1)$ implies (2), let E be the countable Borel equivalence relation witnessing the ergodicity. If (1) fails, there is an uncountable collection $\{B_\alpha : \alpha \in \omega_1\}$ of I-positive Borel sets with I-small pairwise intersections. Use the σ-completeness of the ideal I inductively to thin out the sets B_α in such a way that they are actually pairwise disjoint. The sets C_α, complements of the E-saturations of the sets B_α, are in the ideal I, and I will show that $\bigcup_{\alpha \in \omega_1} C_\alpha = X$ in every $\aleph_1$-preserving extension. Well, if $x \in X$ is a point in some extension which is not in this union, then the equivalence class of the point x must have a nonempty intersection with every set $B_\alpha : \alpha \in \omega_1$. However, the equivalence class of the point x is countable while there are $\aleph_1^V$ many disjoint sets B_α; ergo, $\aleph_1^V$ is collapsed in the extension. $\square$

Proposition 3.7.6. *Suppose that I is an ergodic c.c.c. σ-ideal on a Polish space X. Then:*

1. *the countable Borel equivalence relation witnessing ergodicity can be chosen so that saturations of I-small Borel sets are small;*
2. *P_I is a homogeneous notion of forcing and if the ideal I is generated by Borel sets, then it is homogeneous;*
3. *the dichotomy in the definition of ergodicity holds even for universally Baire sets in place of Borel sets.*

Proof. For the first item, choose an arbitrary equivalence relation F witnessing the ergodicity. Note that there is a Borel set $A \subset X$ whose complement is in the ideal I and such that the F-saturation of every Borel I-small subset of A is F-small. Suppose this fails. Build an inclusion-decreasing sequence $\langle A_\alpha : \alpha \in \omega_1 \rangle$ of I-large subsets of the space X and sets $\langle B_\alpha : \alpha \in \omega_1 \rangle$ such that $B_\alpha \subset A_\alpha$ is I-small with I-positive (by the ergodicity, I-large) F-saturation, and such that $A_{\alpha+1} \cap B_\alpha = 0$. This is easy enough to do, at limit stages taking intersections of the sets A_α built so far and using the σ-additivity of the ideal I. Now consider the P_I-generic point. It falls into the I-large F-saturations of all the sets $\langle B_\alpha : \alpha \in \omega_1 \rangle$, and therefore its equivalence class visits all the sets B_α. But the equivalence class is countable while the sets B_α are disjoint, meaning that $\aleph_1$ was collapsed. Once the existence of the set $A \subset X$ has been established, it is clear that the Borel equivalence relation $E = (F \cap A^2) \cup (= \cap (X \setminus A)^2)$ satisfies the demands of the first item.

By the Feldman–Moore theorem [19], there is a Borel action of a countable group Γ such that the E-equivalence classes are exactly the orbits under the action. Since E-saturations of I-small Borel subsets of A are I-small, the action preserves the ideal I: whenever $\pi \in \Gamma$ and $B \subset A$ is a Borel set, $B \in I$ iff $\pi''B \in I$. Clearly, each member $\pi \in \Gamma$ induces an automorphism $\bar\pi$ of the poset P_I by setting $\bar\pi(B) = \pi''(B)$. Since the E-saturations of I-positive Borel sets are large, for every two positive Borel sets B_0, B_1 there must be a member $\pi \in \Gamma$ such that $\bar\pi(B_0) \cap B_1 \notin I$, which proves the homogeneity of the forcing P_I.

For the homogeneity of the σ-ideal I let $B \subset X$ be a Borel positive set. Enumerate the acting group $\Gamma = \{\pi_n : n \in \omega\}$ and note that for all but I-many points $x \in X$ there is an element $\pi \in \Gamma$ such that $\pi(x) \in B$, since the E-saturation of the set

B is I-large. This makes it possible to define $f : X \to B$ by setting $f(x) = \pi_n(x)$ whenever n is the least number m such that $\pi_m(x) \in B$, and $f(x) =$ some fixed element in the set B if no such number m exists. Since Γ-preimages of Borel I-small sets are small, it is also true that the f-preimages of Borel I-small sets are I-small.

The ergodicity dichotomy clearly extends to all universally Baire sets, since for every c.c.c. ideal, every universally Baire set either has a Borel superset in the ideal I or it has a Borel I-positive subset. In the former case the E-saturation of the universally Baire set is I-small, in the latter it is I-large. $\qquad\square$

3.7.2 Perfect antichains

A long time ago Woodin asked whether every suitably definable non-c.c.c. forcing must contain a perfect antichain. This question was answered in the negative in [33], where the authors showed that the standard Baumgartner forcing for adding a closed unbounded subset of ω_1 can be coded in such a way that no perfect antichains can exist. It nevertheless turns out that the dichotomy holds for definable *bounding* forcing notions and for $< \omega_1$-proper forcing notions of the form P_I:

Proposition 3.7.7. *(LC) Suppose that I is a universally Baire σ-ideal on a Polish space X such that the forcing P_I is proper and bounding. Then exactly one of the following holds:*

1. P_I is c.c.c.;
2. there is a perfect collection of mutually disjoint I-positive compact sets.

If the ideal I is $\mathbf{\Pi}^1_1$ on $\mathbf{\Sigma}^1_1$ then the large cardinal assumption can be omitted.

Here the word "perfect" refers to the hyperspace $K(X)$ of compact subsets of the Polish space X.

Proof. Assume that P_I is not c.c.c. Then it is possible to find a collection of ω_1 many mutually disjoint I-positive Borel sets, and since the compact sets are dense in the forcing P_I, there is a collection $\vec{C} = \langle C_\alpha : \alpha \in \omega_1 \rangle$ of mutually disjoint I-positive compact sets.

Now assume that I is a universally Baire σ-ideal, and consider a Woodin cardinal δ with the associated stationary tower Q and the Q-name $j : V \to N$ for the generic elementary embedding. Clearly, $Q \Vdash j\vec{C}_{\omega_1} \in P_I$. Choose a countable elementary submodel M of a large enough structure. If $g \subset M \cap Q$ is an M-generic filter then the models $M[g]$ and N/g are both correct about the membership in the ideal I and therefore the compact set $C_g = j\vec{C}_{\omega_1}/g$ is I-positive. I will complete the proof by finding a perfect collection P of M-generic filters on $M \cap Q$ such that the collection

$\{C_g : g \in P\}$ consists of mutually disjoint sets. Fix a countable topology basis $\mathcal{O}$ of the underlying Polish space.

Claim 3.7.8. *If $S \in Q$ is a stationary set then there are disjoint basic open sets O_0, O_1 such that both sets $S_0 = \{x \in S : C_{x \cap \omega_1} \subset O_0\}$ and $S_1 = \{x \in S : C_{x \cap \omega_1} \subset O_1\}$ are stationary.*

Proof. If this failed then for every pair O_0, O_1 of disjoint basic open sets one of the sets S_0, S_1 is nonstationary. Collect all of these countably many nonstationary sets and subtract them from the stationary set S, obtaining a stationary set R. Let $x, y \in R$ be countable sets such that $x \cap \omega_1$ and $y \cap \omega_1$ are distinct countable ordinals. Since the compact sets $C_{x \cap \omega_1}$ and $C_{y \cap \omega_1}$ are disjoint, there are disjoint basic open sets O_0, O_1 separating them. This contradicts the construction of the set R though. $\qquad\square$

Note that the sets S_0, S_1 force the compact set $j\vec{C}_{\omega^y}$ to be included in O_0 or O_1 respectively. Now let $D_n : n \in \omega$ enumerate all open dense subsets of the forcing Q in the model M and by induction on sequences $u \in 2^{<\omega}$ build conditions $p_u \in Q \cap M$ and basic open sets O_u so that

- $p_u \in D_{|u|}$, $v \subset u \to p_v \geq p_u$;
- $O_{u^\frown 0} \cap O_{u^\frown 1} = 0$;
- $C_{x \cap \omega_1} \subset O_u$ for every set $x \in p_u$.

This is not difficult to arrange using the previous claim repeatedly. In the end, for every infinite binary sequence $v \in 2^\omega$ the conditions $p_u : u \subset v$ generate an M-generic filter $g_v \subset Q \cap M$ and the I-positive compact sets $C_{g_v} : v \in 2^\omega$ are mutually disjoint as desired.

The ZFC case of a $\mathbf{\Pi}_1^1$ on $\mathbf{\Sigma}_1^1$ ideal I is just a version of the previous argument. To simplify the notation assume that the underlying Polish space is just the Cantor space 2^ω. The set $\{C \in K(2^\omega) : C \notin I\}$ is analytic, so find a tree $T \subset (\omega \times 2^{<\omega})^{<\omega}$ which projects into the set of all binary trees $U \subset 2^{<\omega}$ such that $[U] \notin I$. As in the previous argument find an uncountable collection $U_\alpha : \alpha \in \omega_1$ of binary trees such that the sets $[U_\alpha]$ are I-positive and mutually disjoint. The following claim is critical:

Claim 3.7.9. *Suppose that $t \in T$ is a node and $S \subset \omega_1$ is a stationary set such that $\{U_\alpha : \alpha \in S\} \subset \mathrm{proj}[T \restriction t]$. Then there are disjoint basic open sets O_0, O_1 and nodes $s_0, s_1 \in T$ such that $T \restriction s_0$ projects into trees which are subsets of O_0, $T \restriction s_1$ projects into trees which are subsets of O_1 and both sets $S_0 = \{\alpha \in S : U_\alpha \in \mathrm{proj}[T \restriction s_0]\}$ and $S_1 = \{\alpha \in S : U_\alpha \in \mathrm{proj}[T \restriction s_1]\}$ are stationary.*

The proof is just a repetition of the argument for the previous claim. As before, construct a perfect collection $t_u : u \in 2^\omega$ of nodes in the tree T such that for incompatible sequences $u, v \in 2^\omega$ there are disjoint basic open sets O_u, O_v such that $T \upharpoonright t_u$ and $T \upharpoonright t_v$ project into trees that are subsets of O_u and O_v respectively and for each sequence $u \in 2^{<\omega}$ the set $\{\alpha \in \omega_1 : U_\alpha \in \text{proj}[T \upharpoonright u]\}$ is stationary. For each infinite binary sequence $v \in 2^\omega$ let U_v be the binary tree into which the path $\bigcup_{u \subset v} t_u \subset T$ projects. Then $U_v : v \in 2^\omega$ is the required collection of I-positive mutually disjoint sets. □

Note that the only property of the ideal I necessary in the above proof was that compact sets are dense in the poset P_I. There is a number of forcings which are not bounding but possess this feature; some of them can be found in Section 4.2.

Proposition 3.7.10. *(CH) Suppose that I is a σ-ideal on a Polish space X such that the forcing P_I is $< \omega_1$-proper and nowhere c.c.c. Then there exists a perfect collection of mutually disjoint Borel I-positive sets.*

Here, a forcing P is $< \omega_1$-proper if for every countable $\in$-tower $\langle M_\beta : \beta \in \alpha \rangle$ of countable elementary submodels of some large structure and every $p \in P \cap M_0$ there is a condition $q \in P, q \leq p$ which is M_β-master condition for every ordinal $\beta \in \alpha$. It can be proved that almost all forcings considered in this book are $< \omega_1$-proper and this property is even absolute throughout forcing extensions. Thus if they fail the c.c.c., an absoluteness argument will show that they must contain a perfect collection of mutually disjoint Borel I-positive sets.

Proof. Let θ be a large enough regular cardinal, and let $y \in H_\theta$. The key step in the argument is to find two countable towers $\vec{M}, \vec{N} \prec H_\theta$ of elementary submodels containing y such that no point of the space X can be simultaneously $\vec{M}$ and $\vec{N}$-generic for the poset P. Once this is done, the argument is a breeze: choose countable towers of elementary submodels $\vec{M}_n, \vec{N}_n : n \in \omega$ such that $n \in m$ implies $\vec{M}_n, \vec{N}_n \in \vec{M}_m, \vec{N}_m$ and no point of the space X can be simultaneously $\vec{M}_n$- and $\vec{N}_n$-generic, and use $< \omega_1$-properness of the forcing P to argue that for every function $f \in 2^\omega$ the Borel set $B_f = \{x \in X : \text{if } f(n) = 0 \text{ then } x \text{ is } \vec{M}_n\text{-generic, and if } f(n) = 1 \text{ then } x \text{ is } \vec{N}_n\text{-generic}\}$ is I-positive – it is the only candidate for a master condition for the tower of models indicated by the function f. Then $\{B_f : f \in 2^\omega\}$ is the desired perfect collection of mutually disjoint Borel I-positive sets.

Now the problem of finding the two towers $\vec{M}, \vec{N}$ as in the previous paragraph is in itself interesting. I do not know if there is an a priori bound on the necessary length of these towers, such as 1 or ω. My argument runs as follows. Consider an $\in$-tower $\langle M_\beta : \beta \in \omega_1 \rangle$ of countable elementary submodels, and a countable submodel N containing this tall tower. Let $\alpha = N \cap \omega_1$ and $\vec{M} = \langle M_\beta : \beta \in \alpha \rangle$.

I claim that no point $x \in X$ can be at the same time $\vec{M}$-generic and N-generic. To see this, note that since the forcing P_I is nowhere c.c.c. and of size $\mathfrak{c} = \aleph_1$, there is a name $\dot{g}$ for a function from ω_1 to ω_1 such that no ground model uncountable set contains only closure points of $\dot{g}$. By elementarity it is possible to find such a name $\dot{g}$ in the model M_0. Now let $C = \{M_\beta \cap \omega_1 : \beta \in \omega_1\}$. If a point $x \in X$ is $\vec{M}$-generic, then all the ordinals in $C \cap \alpha$ are closure points of the function $\dot{g}/x$, and if the point x is N-generic then, since $C \in N$, some of the ordinals in $C \cap \alpha$ must fail to be closure points of the function $\dot{g}/x$. Both cannot be true at the same time!   □

An elaboration of the above proof will be used in Sections 4.5 and 4.3 to show the following related statement.

Fact 3.7.11. *Suppose that ϕ is either an outer regular subadditive stable capacity or a pavement submeasure derived from a countable set of Borel pavers. Then the following are equivalent:*

1. *the forcing P_{I_ϕ} is nowhere c.c.c.;*
2. *every Borel set can be partitioned into perfectly many Borel subsets of the same ϕ-mass as the original set.*

3.8 Π_1^1 on Σ_1^1 ideals

There is an important property of forcings that completely escaped detection by the classical combinatorial methods. I must thank Vladimir Kanovei for turning my attention in the correct direction.

Definition 3.8.1. *A σ-ideal I on a Polish space X is Π_1^1 on Σ_1^1 if for every analytic set $A \subset 2^\omega \times X$ the set $\{y \in 2^\omega : A_y \in I\}$ is coanalytic.*

The reader should consult the textbook [40], 29.E, for several classical theorems: the ideals of countable sets, meager sets and Lebesgue null sets are Π_1^1 on Σ_1^1. It turns out that many σ-ideals encountered in forcing theory are Π_1^1 on Σ_1^1 while others are not. The distinction is crucial for someone who wants to develop the theory of iterated and product forcing without the auxiliary large cardinal assumptions. The attentive reader will have noticed that most ideals from Chapter 4 are Δ_2^1 on Σ_1^1 but this does not help with the ZFC dichotomy treatment of iterations of the resulting forcings. What a difference half a quantifier can make.

At several places in the book I will need the effective version of the above definition.

Definition 3.8.2. *A σ-ideal I on a Polish space X is Π_1^1 on Σ_1^1 with a parameter $z \in \omega^\omega$ if for the universal Σ_1^1 set $A \subset 2^\omega \times 2^\omega \times X$ the set $\{\langle u, v \rangle \in 2^\omega \times 2^\omega : A_{u,v} \in I\}$ is $\Pi_1^1(z)$.*

3.8.1 Motivating applications

In order to motivate the investigation below, I include several ZFC theorems to be proved later.

Fact 3.8.3. *(Theorem 5.1.9) Suppose that I is an iterable Π_1^1 on Σ_1^1 ideal and $\alpha \in \omega_1$ is a countable ordinal. Then the ideal I^α is Π_1^1 on Σ_1^1 as well and the countable support iteration $(P_I)^\alpha$ is isomorphic to the forcing P_{I_α}.*

Fact 3.8.4. *(Theorem 6.3.9) Suppose that I is an iterable Π_1^1 on Σ_1^1 ideal, and J is a σ-ideal generated by an analytic collection of closed sets. If $I \perp J$ then $I^\alpha \perp J$ for all countable ordinals α.*

Fact 3.8.5. *(Theorem 5.2.6) Suppose that $I_n : n \in \omega$ is a collection of Π_1^1 on Σ_1^1 ideals such that the forcings P_{I_n} are proper, bounding, and preserve bases for the meager ideal. Then the product $J = \Pi_n I_n$ is a Π_1^1 on Σ_1^1 σ-ideal, the full support product forcing $\Pi_n P_{I_n}$ is proper, bounding, and preserves bases for the meager ideal, and it is isomorphic to the forcing P_J.*

Thus the considered property propagates through countable support iterations and products. I do not know if it propagates through illfounded iterations and unions. Any theorem in that direction would presumably imply better understanding of the two operations.

3.8.2 Non-c.c.c. examples

I will identify two classes of examples of non-c.c.c. Π_1^1 on Σ_1^1 ideals. Many forcings fall into both of them. The following proposition will be critical in many situations where Π_1^1 on Σ_1^1 ideals occur. It will also make it possible to identify the first large class of such ideals.

Proposition 3.8.6. *Let I be a σ-ideal on a Polish space X. I is Π_1^1 on Σ_1^1 with parameter z if and only if for every real $y \in \mathbb{R}$ the set $U_y = \bigcup(\Sigma_1^1(y, z) \cap I)$ is uniformly $\Pi_1^1(y, z)$.*

The set U_y has the remarkable property that every $\Sigma_1^1(y, z)$ set disjoint from it, if nonempty, is actually I-positive.

Proof. To simplify the notation assume that $z = 0$. The right-to-left direction is almost trivial. Suppose that the set U_y is uniformly $\Pi_1^1(y)$, and let $B \subset \mathbb{R} \times X$ be a $\Sigma_1^1(y)$ set for some real y. Clearly, for every real x the section B_x is $\Sigma_1^1(x, y)$ and therefore it is in the ideal I if and only if it is a subset of the $\Pi_1^1(x, y)$ set $U_{x,y}$, which is a uniformly $\Pi_1^1(x, y)$ condition.

The opposite direction is just a computation of the complexity of the set U_y. First note that if y is a real and $C \subset X$ is a $\Sigma_1^1(y)$ set in I, then it has a $\Delta_1^1(y)$ superset still in the ideal I. To see this, let $<$ be a $\Pi_1^1(y)$ rank on the complement of the set C and consider the set $D = X \setminus \{x \in X : \{u \in X : x \leq u\} \in I\}$. The definability assumptions on I and $<$ show that this is a $\Sigma_1^1(y)$ set, and it is disjoint from the set C. By the effective boundedness theorem, there is an ordinal γ recursive in the real y bounding the ranks of the reals in the set D. The set $\{x \in \mathbb{R} : \text{rank}(x) \geq \gamma\}$ is then the desired $\Delta_1^1(y)$ superset of the set C in the ideal I.

Thus $U_y = \bigcup(\Sigma_1^1(y) \cap I) = \bigcup(\Delta_1^1(y) \cap I)$ which is easily found to be uniformly $\Pi_1^1(y)$ using any standard $\Pi_1^1(y)$ parametrization of $\Delta_1^1(y)$ sets like the one in [40], 35.B. $\square$

Theorem 3.8.7. *[18] If the σ-ideal I obtains from some coanalytic porosity then it is Π_1^1 on Σ_1^1.*

Consult Section 4.2 for definitions and results concerning porosity. Suppose that X is a Polish space and U is a countable collection of its Borel subsets. Any inclusion-preserving function $\text{por} : \mathcal{P}(U) \to \mathcal{B}(X)$ is called an *abstract porosity*. The σ-ideal I σ-generated by sets of the form $\text{por}(a) \setminus \bigcup a$ for $a \subset U$ is the associated *porosity* ideal. The generating sets of the form $\text{por}(a) \setminus \bigcup(a)$ are called *porous*. To say that the porosity por is coanalytic is to say that the set $\{\langle a, r \rangle \in \mathcal{P}(U) \times X : r \in \text{por}(a)\} \subset \mathcal{P}(U) \times X$ is coanalytic.

The class of σ-ideals described in the theorem includes:

- all σ-ideals generated by a coanalytic σ-ideal of closed sets, such as the meager ideal or the ideal of countable sets and many more;
- the metric porosity ideals;
- the σ-continuity ideal.

Proof. To simplify the notation assume that the underlying space X is just the Baire space ω^ω and that the collection U is lightface Δ_1^1, and that the abstract porosity is lightface Π_1^1. By Proposition 3.8.6, for every real u I must prove that $\bigcup(I \cap \Sigma_1^1(u)) \in \Pi_1^1(u)$. For simplicity put $u = 0$. First, a small claim.

Claim 3.8.8. *If $A \subset \omega^\omega \times \omega^\omega$ is Σ_1^1 then the set $\{x \in \omega^\omega : A_x \text{ is porous}\}$ is Π_1^1.*

This is a direct computation. The vertical section A_x is porous if and only if $A_x \subset$ por(a) where $a = \{u \in U : u \cap A_x = 0\}$ by the monotonicity of the abstract porosity. This can be restated as $\forall r \in A_x \forall b \subset U \; a \subset b \to r \in$ por(b) by the monotonicity again, or $\forall r \in A_x \forall b \subset U \; r \in$ por(b) $\vee \exists v \in b \; A_x \cap v = 0$. This is a uniformly $\Pi_1^1(x)$ statement as desired.

The effective version of the First Reflection Theorem [40], 35.10, now shows that every Σ_1^1 porous set has a Δ_1^1 porous superset. A Π_1^1 parametrization of Δ_1^1 sets [40], 35.B, then can be used to show that the set $C = \bigcup(\Sigma_1^1 \cap$porous sets$) = \bigcup(\Delta_1^1 \cap$porous sets$)$ is Π_1^1. It will be enough to show that $C = \bigcup(I \cap \Sigma_1^1)$.

The right-to-left inclusion is clear. For the other, let $A \in \Sigma_1^1$ be a set such that $A \setminus C \neq 0$ and argue that $A \notin I$. Suppose that $\{a_n : n \in \omega\}$ is a countable collection of subsets of U; I must find an element $r \in A \setminus \bigcup_n($por($a_n) \setminus \bigcup a_n)$. To this end, by induction construct recursive trees T_n as well as nodes $t_m^n \in T_m$ for $m \leq n$ so that:

- T_0 is some recursive tree projecting into the Σ_1^1 set $A \setminus C$, $t_0^0 = 0$;
- the nodes $t_n^i \in T_n$ are defined for all $i \geq n$ and form a strictly decreasing sequence in the tree;
- the set $A_n = \bigcap_{m \leq n}$proj$[T_m \restriction t_m^n]$ is nonempty;
- $A_{n+1} \cap$por($a_n) \setminus \bigcup a_n = 0$.

It is clear that in the end the branches through the trees T_n obtained from the nodes t_n^i project into the same real $r \in A$, and the last item of the induction hypothesis will imply that $r \notin \bigcup_n($por($a_n) \setminus \bigcup a_n)$ as desired. To find the tree T_{n+1} and the nodes t_m^{n+1} for $m \leq n+1$, consider the set $b = \{u \in U : A_n \cap u = 0\}$ and the set por(b). A similar complexity computation as in the proof of the claim shows that por(b) is a Π_1^1 set. The set $A_n \setminus$por(b) is then Σ_1^1 and nonempty, because if it were empty, the Σ_1^1 set A_n would be porous, covered by the set por($b) \setminus \bigcup b$ which contradicts the fact that $A_n \cap C = 0$ and the definition of the set C. There are now two cases. Either por($a_n) \subset$ por(b). In this case let T_{n+1} be some recursive tree projecting into the nonempty Σ_1^1 set $A_n \setminus$por(b) and find suitable nodes $t_m^{n+1} : m \leq n+1$ in the trees. Or, por($a_n) \not\subset$ por(b), and this means that $a_n \not\subset b$ by the monotonicity of the abstract porosity. Choose a set $u \in a_n \setminus b$, a recursive tree T_{n+1} projecting into the nonempty Σ_1^1 set $A_n \cap u$, and find suitable nodes $t_m^{n+1} : m \leq n+1$. This concludes the induction step and the proof. □

Theorem 3.8.9. [83] *If the poset P_I is proper and bounding and there is a dense analytic collection of compact sets in P_I and every analytic I-positive set contains a Borel I-positive subset then the ideal I is Π_1^1 on Σ_1^1.*

This includes the following ideals:

- the ideal of capacity zero sets for every subadditive capacity, as long as the resulting forcing is proper, such as the Lebesgue measure or the Newtonian capacity;
- any ideal σ-generated by a σ-compact collection of compact sets;
- the σ-ideals generated by sets of finite Hausdorff measure;
- the E_0-ideal of Section 4.7.1.

In fact this example makes it very difficult to find natural bounding forcings which are not associated with a $\mathbf{\Pi}^1_1$ on $\mathbf{\Sigma}^1_1$ ideal.

Proof. Let X be the underlying Polish space and let $D \subset P_I$ be the analytic dense set of compact sets. Suppose that $A \subset X \times 2^\omega$ is an analytic set, a projection of a closed set $C \subset X \times 2^\omega \times \omega^\omega$. I will show that for a point $y \in 2^\omega$ the set A_y is I-positive if and only if the following statement (*) holds: there is a compact set $K \in D$ and a continuous function $f : K \to \omega^\omega$ such that for every $x \in K$ it is the case that $\langle x, y, f(x) \rangle \in C$. It is not difficult to see that (*) is an analytic condition, which completes the proof of the theorem.

If (*) holds then clearly $K \subset A^y$ which shows that $A^y \notin I$. On the other hand, if $A^y \notin I$ then there is a Borel I-positive set $B \subset A^y$. There is also a name $\dot{z} \in \omega^\omega$ such that $B \Vdash \langle \dot{x}_{gen}, \check{y}, \dot{z} \rangle \in \check{C}$, and since the forcing P_I is bounding, there is a compact set $K \in D$ and a continuous function $f : K \to \omega^\omega$ such that $B \Vdash \dot{z} = \dot{f}(\dot{x}_{gen})$ and in fact for every $x \in K$, $\langle x, y, f(x) \rangle \in C$. The condition (*) follows, completing the proof. $\square$

3.8.3 C.c.c. examples

The situation in the c.c.c. realm is considerably more slippery. It turns out that Shelah and Roslanowski [57] considered a notion that is more or less equivalent to $\mathbf{\Pi}^1_1$ on $\mathbf{\Sigma}^1_1$ for c.c.c. ideals:

Definition 3.8.10. *Suppose $P, \leq$ is a partial order on the reals. Call P Suslin if the sets $P, \leq, \perp$ are analytic. Call P very Suslin if it is c.c.c. and the set $\{A \in P^{\aleph_0} : A$ is a maximal antichain in $P\}$ is analytic as well.*

Proposition 3.8.11. *If P is a c.c.c. very Suslin poset adding a single point in the Baire space then the ideal $I(P)$ is $\mathbf{\Pi}^1_1$ on $\mathbf{\Sigma}^1_1$. Vice versa, if I is a c.c.c. ergodic $\mathbf{\Pi}^1_1$ on $\mathbf{\Sigma}^1_1$ σ-ideal on the real line and P is a Borel collection of closed sets closed under finite positive intersections whose equivalence classes are dense in the factor algebra of P_I, then the poset $\langle P, \subset \rangle$ is very Souslin.*

Proof. Let $B \subset \omega^\omega$ be an analytic set, $B = \text{proj}[T]$ for some tree $T \subset \omega^{<\omega} \times \omega^{<\omega}$. Then $B \notin I(P)$ iff there is a condition $p \in P$ forcing the generic real into $\dot{B}$ iff there is a condition $p \in P$ and a name τ such that p forces τ to be a branch of the tree $\check{T}$ with the generic real as the first coordinate iff there is p and a system of maximal antichains $\{A_n : n \in \omega\}$ below p and a map $f : \bigcup_n A_n \to T$ such that

- every antichain A_{n+1} is a refinement of the antichain A_n;
- if $q \in A_n$ then $f(q)$ is a node in the tree T of length n such that q forces the generic point to start with the first coordinate of $f(q)$;
- f carries comparable conditions to comparable nodes in the tree T.

This is an analytic condition since the notion of a maximal antichain is analytic.

For the second sentence in the proposition, let E be the equivalence relation witnessing the ergodicity, by Proposition 3.7.6 it can be chosen so that E-saturations of I-small sets are I-small. The poset P is Borel. Sets $C, D \in P$ are compatible iff $C \cap D \notin I$ which is an analytic statement by the definability conditions on the ideal. Sets $C, D \in P$ are incompatible iff $C \cap D \in I$ iff the E-saturation of $C \cap D$ is in the ideal I iff (by the ergodicity) the complement of this saturation is not in the ideal I, which is again an analytic statement. And finally, $A \in P^\omega$ is a maximal antichain iff $\omega^\omega \setminus \bigcup A \in I$ iff the E-saturation of the set $\omega^\omega \setminus \bigcup A$ is in I iff (by the ergodicity) its complement is not in I, which is an analytic statement. $\square$

Shelah and Roslanowski worried that the only examples of $\mathbf{\Pi}_1^1$ on $\mathbf{\Sigma}_1^1$ c.c.c. Souslin ideals may be the meager ideal, the Lebesgue null ideal, and the ideals associated with their finite support iterations, and they devised a scheme to generate many other ones [57]. However, it turns out that examples of such σ-ideals are readily at hand:

Proposition 3.8.12. *[83] The σ-ideal associated with the eventually different real forcing is $\mathbf{\Pi}_1^1$ on $\mathbf{\Sigma}_1^1$.*

The *eventually different real forcing* [2], 7.4.8, P is the set of all pairs $\langle t, a \rangle$ where t is a finite sequence of integers and $a \subset \omega^\omega$ is finite, and $\langle s, b \rangle \leq \langle t, a \rangle$ if $t \subset s$, $a \subset b$ and $(s \setminus t) \cap \bigcup a = 0$.

Proof. It is just necessary to verify that the collection of all maximal antichains is a Borel set. Let $A \in P^\omega$; it is enough to show that for every sequence $s \in \omega^{<\omega}$ and every number $m \in \omega$ the statement $\phi(s, m, A) =$ "there is a set $b \subset \omega^\omega$ of size m such that the condition $\langle s, b \rangle$ is incompatible with all elements of the set A" is Borel, because then A is a maximal antichain iff it is an antichain and for every s and m the statement $\phi(s, m, A)$ fails.

I will just show that $\phi(0, 1, A)$ is a Borel statement, the general case is essentially identical with a little more complicated notation. Let $A = \{\langle t_n, a_n \rangle : n \in \omega\}$. $\phi(0, 1, A)$ holds iff there is a function $f \in \omega^\omega$ with nonempty intersection with every t_n (since the set $b = \{f\}$ will witness ϕ) iff the tree $T(0, 1, A) = \{g : g$ is a function from some $k \in \omega$ to $\omega \times \omega$ such that $\forall n \in k$ $g(k) \in t_k$ and $\mathrm{rng}(g)$ is a function$\}$ has an infinite branch (since the range of any infinite branch will give such function f) iff the tree $T(0, 1, A)$ is infinite (since it is finitely branching). This is a Borel statement. $\qquad \square$

Proposition 3.8.13. *[87] Suppose that J is an ideal on ω containing all finite sets, and let I be the σ-ideal on 2^ω associated with the Prikry forcing $P(J)$. The ideal I is Π_1^1 on Σ_1^1 if and only if the ideal J is F_σ.*

Here the Prikry forcing $P(J)$ is the set of all pairs $p = \langle t_p, a_p \rangle \in 2^{<\omega} \times J$ ordered by $q \leq p$ iff $t_p \subset t_q$, $a_p \subset a_q$, and $\{n \in \mathrm{dom}(t_q \setminus t_p) : t_q(n) = 1\} \cap a_p = 0$.

Proof. For the right-to-left direction note that the ideal I is ergodic. Now if I is Π_1^1 on Σ_1^1 then the σ-ideal $I \cap K(2^\omega)$ of compact sets is analytic: a set $C \in K(2^\omega)$ is in I if and only if the complement of the closure of C under finite changes is I-positive, which is an analytic statement. Analytic ideals of compact sets are G_δ by a theorem of Kechris, Louveau, and Woodin [36]. Now it is not difficult to see that for a set $a \subset \omega$, $a \in J$ iff $\{x \in 2^\omega : x(n) = 1 \to n \notin a\} \notin I$ which is an F_σ condition!

For the left-to-right direction, write $K = \{a \subset \omega^{<\omega} \setminus \{0\} : \exists b \in J \forall x \in a$ $x \cap b \neq 0\}$. It is clear that this is an ideal on the set $\omega^{<\omega} \setminus \{0\}$. A useful observation:

Claim 3.8.14. *If J is an F_σ ideal then K is F_σ again.*

Proof. By a theorem of Mazur [50] there is a lower semicontinuous submeasure μ on $\mathcal{P}(\omega)$ such that $J = \{a \subset \omega : \mu(a) < \infty\}$. Let $\mu^{<\omega}$ be a function on $\mathcal{P}(\omega^{<\omega})$ defined by $\mu^{<\omega}(b) = \inf\{\mu(a) : \forall x \in b$ $x \cap a \neq 0\}$. It is not difficult to verify that this is a lower semicontinuous submeasure such that $K = \{b \subset \omega^{<\omega} : \mu^{<\omega} < \infty\}$. The claim follows.

In fact the proof of the proposition shows that if the ideal J is not F_σ then the ideal K is not even analytic. $\qquad \square$

By Proposition 3.8.11, to prove the left-to-right implication of the proposition it is just necessary to show that the collection of countable subsets of $P(J)$ which are maximal antichains is a Borel set. In order to do this, let $A \subset P(J)$ be a countable set. Then A is a maximal antichain if and only if it is an antichain and for every finite set $t \subset \omega$, every condition of the form $\langle t, a \rangle$ is compatible with some element of A. The latter condition is equivalent to: either there is some condition $\langle u, b \rangle \in A$ such that $u \subset t$ and $b \cap t \setminus u = 0$, or the set $a_t = \{x \subset \omega : \exists b \langle t \cup x, b \rangle \in A\}$ is not in the ideal K. By the claim, this is a Borel statement. $\qquad \square$

3.8.4 Nonexamples

Finally the promised class of σ-ideals which are definitely not $\mathbf{\Pi}_1^1$ on $\mathbf{\Sigma}_1^1$, and it is a really important one.

Proposition 3.8.15. *If the poset P_I adds a dominating real then the ideal I is not $\mathbf{\Pi}_1^1$ on $\mathbf{\Sigma}_1^1$.*

Proof. Let X be the underlying Polish space. By Proposition 2.3.1, the assumptions imply that there is a Borel I-positive set $B \subset X$ and a Borel function $f : B \to \omega^\omega$ such that for every function $g \in \omega^\omega$ the set $\{x \in X : f(x)$ does not modulo finite dominate the function $g\}$ is in the ideal I.

Let $\mathbb{T}$ be the set of all trees on ω. The set $U \subset \mathbb{T}$ of all wellfounded trees is well-known to be $\mathbf{\Pi}_1^1$-complete. Let $B \subset \mathbb{T} \times X$ be the set of all pairs $\langle T, x \rangle$ such that $x \in C$ and for all sequences $s \in T$ the tree $T_{xs} = \{t \in T \restriction s : \forall n \in \text{dom}(t) \setminus \text{dom}(s)\ t(n) \in f(x)(n)\}$ is finite. Clearly, if the tree T is wellfounded, the set B_T is equal to C, since the trees T_{xs} are all finitely branching and wellfounded, therefore finite. On the other hand, if the tree T is illfounded as witnessed by some branch $g \in [T]$ then $B_T \subset \{x \in C : f(x)$ does not modulo finite dominate the function $g\} \in I$. Thus the set $\{T \in \mathbb{T} : B_T \in I\}$ is properly $\mathbf{\Sigma}_1^1$ and not $\mathbf{\Pi}_1^1$, proving the proposition. $\qquad\square$

3.8.5 The Suslin number

One may wonder if adding a dominating real is the weakest explicit forcing property that implies that the associated ideal is complex. This is false, as Arnold Miller showed [52]. He produced a Souslin c.c.c. ideal which does not add a dominating real but is still not $\mathbf{\Pi}_1^1$ on $\mathbf{\Sigma}_1^1$. There is an underlying cardinal invariant. Fix a Polish space X and some complete analytic subset $A \subset X$. The exact choice turns out to be irrelevant, and I will use $X =$ the trees on natural numbers and $A =$ the illfounded trees.

Definition 3.8.16. *The Suslin ideal J is the collection of all Borel sets $B \subset X$ such that $A \cap B$ is a relatively Borel subset of B. The Suslin number is the uniformity of this ideal, $= \text{non}(J)$.*

It is not difficult to verify that J in fact is a σ-ideal. Notably the forcing P_J is not proper, and this was one of the basic examples. The preservation of the Suslin number seems to be a forcing property very closely related to the $\mathbf{\Pi}_1^1$ on $\mathbf{\Sigma}_1^1$ property:

Proposition 3.8.17. *If P_I is a $\mathbf{\Pi}_1^1$ on $\mathbf{\Sigma}_1^1$ forcing then $P_I \Vdash 2^\omega \cap V \notin J$; in other words, the forcing P_I does not increase the Suslin number.*

I do not know if under suitable large cardinal assumptions every definable real forcing which preserves the Suslin number must be locally $\mathbf{\Pi}_1^1$ on $\mathbf{\Sigma}_1^1$.

Proof. Consider the space X of all trees on ω and the complete analytic set $A \subset X$ of all illfounded trees. Suppose that I is a $\mathbf{\Pi}_1^1$ on $\mathbf{\Sigma}_1^1$ ideal on some space Y such that the factor forcing P_I is proper, and for contradiction assume that some condition in P_I forces "$\dot{B} \subset \check{X}$ is a Borel set such that $\dot{B} \cap V = \check{A}$." By Proposition 2.3.2, there is an I-positive Borel set $C \subset Y$ and a Borel set $D \subset X \times C$ such that $C \Vdash \dot{B} = \dot{D}^y$. For a given point $x \in X$, $C \Vdash \check{x} \in \dot{B} \leftrightarrow \{y \in C : \langle x, y \rangle \notin D\} \in I$, and the condition on the right hand side of this equivalence is coanalytic since the ideal I is $\mathbf{\Pi}_1^1$ on $\mathbf{\Sigma}_1^1$. However, $C \Vdash \check{x} \in \dot{B} \leftrightarrow x \in A$ by the choice of the name $\dot{B}$, and $A \subset X$ is not coanalytic, contradiction. $\square$

In this light it is perhaps reasonable to attempt to find the position of the Suslin number relative to the standard Cichoń diagram invariants. It turns out that $\mathfrak{b} \leq \mathfrak{sn} \leq \mathrm{non}(\mathrm{meager})$ is provable in ZFC, and $\mathfrak{b} < \mathfrak{sn}$ is consistent by the work of Arnold Miller [52]. I do not know if $\mathfrak{sn} \leq \mathfrak{d}$ is provable in ZFC.

3.9 Dichotomies

A careful review of this book will reveal that it is full of dichotomies and the dichotomies are really the driving force behind most arguments. In this section I will discuss the three most important dichotomy schemes.

3.9.1 The first dichotomy

Definition 3.9.1. *A σ-ideal I on a Polish space satisfies the* first (universally Baire) dichotomy *if every universally Baire set is either covered by a Borel set in I or it contains a Borel I-positive subset.*

All c.c.c. ideals as well as the σ-ideals discussed in Chapter 4 except for Section 4.7 satisfy the first dichotomy under suitable large cardinal assumptions. On the other hand, there are many σ-ideals encountered in this book which do not satisfy this dichotomy. Perhaps surprisingly, the first dichotomy has forcing consequences.

Proposition 3.9.2. *(LC+CH) Suppose that I is a universally Baire σ-ideal on a Polish space X satisfying the first universally Baire dichotomy.*

1. *Suppose that the forcing P_I is $< \omega_1$-proper. If $G \subset P_I$ is a generic filter and $V \subseteq W \subseteq V[G]$ is an intermediate extension, then either W is a c.c.c. extension of V or $W = V[G]$.*

2. *Suppose that the forcing P is ω-proper. If $f \in 2^{\omega_1}$ is a function in the generic extension with all countable initial segments in the ground model, then it is itself in the ground model.*

Here, ω-properness and $< \omega_1$-properness are the strengthenings of properness introduced by Shelah [64]: a forcing P is ω-proper if for every $\in$-increasing ω-tower $\vec{M}$ of countable elementary submodels of a large enough structure and every condition $p \in P$ there is a strengthening $q \leq p$ which is a master condition simultaneously for all models on the sequence. And a forcing P is $< \omega_1$-proper if this statement is true for all towers of models of countable length. All proper forcings considered in this book with the exception of Section 5.5 are in fact $< \omega_1$-proper.

Proposition 3.9.2 can be restated in several ways. The CH assumption is not necessary if the forcing P_I is $< \omega_1$-proper or ω-proper in all forcing extensions, a condition which is invariably satisfied in all applications. In all specific cases discussed in Chapter 4 to which this theorem applies, a careful argument will eliminate the need for the large cardinal assumptions.

Proof. Towards the proof of (1), suppose that P is a poset in which P_I is dense, and $Q \subset P$ is a nowhere c.c.c. regular subposet. A piece of notation: if M is a countable elementary submodel of a large enough structure and $x \in X$ is an M-generic point, let $g(x) \subset M \cap Q$ be the M-generic filter on Q associated with it, more precisely $g(x) = \{q \in Q : \exists B \in M \; B \leq q \wedge x \in B\}$. As in Proposition 3.7.10, if M is a countable elementary submodel of a large enough structure there is a Borel set $B \subset 2^\omega \times X$ whose vertical sections are I-positive sets of M-generic points such that whenever $y_0 \neq y_1 \in 2^\omega$ and $x_0 \in B_{y_0}, x_1 \in B_{y_1}$ then $g(x_0) \neq g(x_1)$. Note the use of CH in Proposition 3.7.10. Now let $A \subset 2^\omega \times X$ be an analytic set enumerating all analytic subsets of the space X, and use the Kondô–Novikov uniformization to find a partial coanalytic function $f : 2^\omega \to X$ uniformizing the set $B \setminus A$. The range $\text{rng}(f) \subset X$ is a Σ^1_2, therefore universally Baire set. It is not covered by any analytic set in the ideal I, since whenever $y \in 2^\omega$ is a point such that $A_y \in I$ then $f(y) \in \text{rng}(f)$ is defined and not in A_y. By the first universally Baire dichotomy, there is a Borel I-positive subset $B \subset \text{rng}(f)$. The function $g \upharpoonright B$ is one-to-one, therefore B forces that the P_I-generic point $\dot{x}_{gen} \in X$ can be recovered from the generic filter $G \subset Q$ as the only point x in the set B such that $G \cap M = g(x)$. Thus $B \Vdash \dot{x}_{gen} \in V[\dot{G}]$ as desired.

For (2), suppose for contradiction that $B \in P_I$ is a condition forcing $\dot{f} : \omega_1 \to 2$ is a function with all initial segments in the ground model which does not itself belong to the ground model. Let M be a countable elementary submodel of a large enough structure. I will show that there is a Borel I-positive set $C \subset \{x \in B : x \text{ is}$

M-generic} such that, writing $g : C \to 2^{M \cap \omega_1}$ for the function defined by $g(x) = \dot{f}/x$, the g-preimages of singletons are in the ideal I. An absoluteness argument the shows that $C \Vdash \dot{f} \upharpoonright M \cap \omega_1 \notin V$, contradicting the initial assumption. The following claim is key.

Claim 3.9.3. *For every countable set $U \subset 2^{M \cap \omega_1}$ the set $\{x \in B : x \text{ is } M\text{-generic}$ and $\dot{f}/x \notin U\}$ is I-positive.*

With the claim in hand, consider the set $W = \{h \in 2^{M \cap \omega_1} : B_h \notin I\}$ where $B_h = \{x \in B : x \text{ is } M\text{-generic and } \dot{f}/x = h\}$. The set W is universally Baire, so it is either countable or it contains a perfect subset. In the former case, the set $C = \{x \in B : x \text{ is } M\text{-generic and } \dot{f}/x \notin W\}$ is Borel, by the claim it is I-positive, and g-preimages of singletons are I-small in C by the definition of C. In the latter case, find a continuous injection $\pi : 2^\omega \to W$, let $A \subset 2^\omega \times X$ be an analytic set enumerating all analytic subsets of the space X, and use the Kondo–Novikov uniformization [40], 36.14, to find a partial coanalytic function $k : 2^\omega \to X$ uniformizing the set $\{\langle y, x \rangle : x \in B_{\pi(y)}, \langle y, x \rangle \notin A\}$. As in the previous proof, the range of k is a Σ^1_2 I-positive set on which g is an injection. Now use the first dichotomy to find a Borel I-positive subset $C \subset \mathrm{rng}(k)$. The set C will be as required.

To prove the claim, first use ω-properness to find a winning strategy $\sigma \in M$ for Player II in the following game. Player I starts with a condition $p_{ini} \in P_I$ and then plays open dense sets $D_n \subset P$, and Player II answers with countable sets $d_n \subset D_n$. Player II wins if the result of the game, the set $p_{ini} \cap \bigcap_n \bigcup d_n$, is I-positive. Let $U \subset 2^{M \cap \omega_1}$ be a countable set, with some enumeration $U = \{h_n : n \in \omega\}$. I will find a play according to the strategy σ whose moves $p_{ini} = B, D_n, d_n : n \in \omega$ will be in the model M, such that for every number n and every condition $q \in d_n$ will be in the n-th open dense subset of P_I in the model M in some fixed enumeration and there will be an ordinal $\alpha \in M \cap \omega_1$ such that q decides $\dot{f} \upharpoonright \alpha$ to be a particular ground model function different from $\check{h}_n \upharpoonright \alpha$. The I-positivity of the result of such a play then confirms the veracity of the claim.

Now suppose that the moves D_n, d_n have been constructed. For every ordinal $\beta \in \omega_1$ let $E_\beta \subset P_I$ be the open dense set of conditions that are in the n-th open dense subset of the forcing P_I in the model M and moreover decide $\dot{f} \upharpoonright \beta$, let $e_\beta \subset E_\beta$ be the answer of the strategy σ to this open dense set, and let $S_\beta = \{k \in 2^\beta : \exists q \in e_\beta \ q \Vdash \dot{f} \upharpoonright \beta = \check{k}\}$. Let $T = \{k \in 2^{<\omega_1} : \text{every initial segment of } k \text{ including } k \text{ is in } \bigcup_\beta S_\beta\}$; thus T is an ω_1-tree. There are two cases to consider:

- h_n is not a branch of the tree $T \cap M$. In such a case, there is an ordinal $\beta \in M \cap \omega_1$ such that the moves $D_{n+1} = E_\beta$, $d_{n+1} = e_\beta$ satisfy the induction demands.
- h_n is a branch of the tree $T \cap M$. In this case first note that the proper definable forcing P_I cannot add a new branch through the ω_1-tree T. (Note that there is a proper forcing Q adding no reals and no new branches through T such that in the resulting extension it is impossible to add new branches through T without collapsing ω_1. On the other hand, the forcing P_I should be proper even in the Q-extension by Corollary 2.2.9.) In particular, $P_I \Vdash \dot{f}$ is not a branch through T. Let D_{n+1} be the open dense set of all conditions q that are in the n-th open dense subset of the poset P_I in the model M and moreover decide some initial segment of $\dot{f}$ which does not belong to the tree T and let d_{n+1} be the answer of the strategy σ.

In each case, the induction step of the construction has been successfully completed.

$\square$

Example 3.9.4. Let I be the E_0-ideal of Section 4.7.1, or the Mathias ideal of Section 4.7.7. Then I does not satisfy the first dichotomy. This is true since the associated forcings – the E_0-forcing and the Mathias forcing – can be decomposed into a σ-closed*c.c.c. iteration and so contain an intermediate σ-closed extension and at the same time, under CH, a function $f \in 2^{\omega_1}$ which has all initial segments in the ground model and itself is not in the ground model.

Example 3.9.5. The ideals associated with products or iterations of non-c.c.c. forcings cannot satisfy the first dichotomy.

Example 3.9.6. Suppose that I is a universally Baire σ-ideal generated by closed sets such that the forcing P_I is bounding or does not add splitting reals. Then the forcing P_I generates a minimal extension. Let me briefly sketch the argument. If there was a proper intermediate extension, it would have to be c.c.c. by the proposition–σ-ideals generated by closed sets satisfy the first dichotomy. The c.c.c. intermediate extension would have to contain a new real – by Corollary 3.5.9 the forcing P_I does not add any branches through Suslin trees. That new real is obtained over the ground model by forcing with a universally Baire c.c.c. σ-ideal which does not make the set of the ground model reals meager, since nonmeagerness is preserved by P_I. Finally, by Corollary 3.5.7 the forcing with this c.c.c. ideal is equivalent to the Cohen forcing, contradicting the bounding property or the no splitting real property. Note that while the forcing P_I is *embeddable* into a σ-closed*c.c.c. iteration it cannot be *equivalent* to such an iteration.

Definition 3.9.7. *Let M be a transitive model of ZF and $I \in M$ be a σ-ideal on a Polish space X. The σ-ideal satisfies the first dichotomy in the model M if every subset of the space X in M has either a Borel I-small superset or a Borel I-positive subset.*

This notion will be only interesting in choiceless models such as $L(\mathbb{R})$, $L(\mathbb{R})[U]$ for a Ramsey ultrafilter U, and notably the choiceless Solovay model. Notably all universally Baire c.c.c. ideals and ideals from Chapter 4 except for Section 4.7 satisfy the first dichotomy in the Solovay model. An absoluteness argument shows that the first dichotomy in the choiceless Solovay extension implies the first universally Baire dichotomy in the ground model. I do not know if the converse is true under any additional assumptions.

3.9.2 The second dichotomy

Definition 3.9.8. *The ideal I satisfies the* second (universally Baire) *dichotomy if every universally Baire set is either in the ideal or contains a Borel I-positive subset.*

Every σ-ideal I such that the forcing P_I is proper can be amended to one satisfying the second dichotomy without changing the forcing using the following operation.

Definition 3.9.9. *Suppose I is a σ-ideal on a Polish space X. The σ-ideal I^* is generated by all universally Baire sets containing no Borel I-positive subset.*

Proposition 3.9.10. *(LC) Suppose that the forcing P_I is proper. A Borel set B is I-positive if and only if it is I^*-positive.*

Proof. Suppose that B is a Borel I-positive set such that $B \subset \bigcup_n A_n$ for some universally Baire sets $A_n : n \in \omega$. A universally Baire absoluteness argument shows that B is still covered by these sets in the P_I-generic extension. There must be a condition $C \in P_I$, $C \subset B$, and a number $n \in \omega$ such that $C \Vdash \dot{x}_{gen} \in \dot{A}_n$. Let M be a countable elementary submodel of a large enough structure and let $D = \{x \in C : x$ is M-generic$\}$; since the forcing P_I is proper, this Borel set is I-positive. By the forcing theorem, for every point $x \in D$ it is the case that $M[x] \models x \in A_n$ and by the universally Baire absoluteness $x \in A_n$. It follows that $D \subset A_n$ as required. $\qquad\square$

Definition 3.9.11. *Let M be a transitive model of ZF and let I be a σ-ideal on a Polish space X, in the model M. The ideal satisfies the second dichotomy in the model M if every subset of the space X in the model M is either in the ideal or it contains a Borel I-positive subset.*

This will be again interesting only in the choiceless models such as $L(\mathbb{R})$, $L(\mathbb{R})[U]$ for a Ramsey ultrafilter U, or the choiceless Solovay model. Given a σ-ideal I there is a natural attempt to amend it to a σ-ideal satisfying the second dichotomy in a given model M:

Definition 3.9.12. *Let I be a σ-ideal on a Polish space. The collection I^{**} consists of all sets without a Borel I-positive subset.*

The key problem of course is that the collection I^{**} may fail to be a σ-ideal.

Proposition 3.9.13. *(LC) Suppose that I is a universally Baire σ-ideal such that the forcing P_I is proper in all forcing extensions. Then in the choiceless Solovay model the collection I^{**} is closed under wellordered unions.*

Proof. Suppose that $A_\alpha : \alpha \in \gamma$ is some sequence of sets in the Solovay model such that their union contains a Borel I-positive set B. I must find an ordinal $\alpha \in \gamma$ such that the set A_α contains an I-positive set. By the usual homogeneity arguments I may assume that the set B is coded in the ground model and the sequence $A_\alpha : \alpha \in \gamma$ is definable from parameters in the ground model. In the ground model, let $C \subset B$ be a condition in the forcing P_I and let $\alpha \in \gamma$ be an ordinal such that $C \Vdash \mathrm{Coll}(\omega, < \kappa) \Vdash \dot{x}_{gen} \in \dot{A}_\alpha$. I will show that the set A_α contains an I-positive Borel set.

Consider the properness game on the partial order P_I – Section 3.10.2. Since the payoff set is universally Baire, Fact 1.4.2 shows that Player II has a winning strategy σ in the game which remains a winning strategy in all forcing extensions, in particular in the Solovay model. There, look at the play of the game against the strategy σ in which Player I starts out with the condition $C \in P_I$ and then enumerates all open dense subsets of the poset P_I in the ground model. The result of the play is a Borel I-positive set $D \subset C$ consisting of V-generic reals only. By the forcing theorem and the usual homogeneity arguments it must be the case that $D \subset A_\alpha$ as required. $\square$

It follows by an absoluteness argument that the collection I^{**} is closed under wellordered unions in all models with absolute definitions such as $L(\mathbb{R})$. It is not true though that the closure of I^{**} under wellordered unions implies properness of the forcing P_I: A decreasing intersection of countably many σ-ideals closed under well-ordered unions is again so closed, but the resulting forcing is not proper by Proposition 2.2.6.

If the ideal I itself satisfies the first dichotomy then $I = I^{**}$, but if it does not it may be interesting to find a description of I^{**} in more informative terms.

Example 3.9.14. Let I be the E_0-ideal on 2^ω of Section 4.7.1. The following holds in the Solovay model as well as in transitive inner models of AD+ containing all the reals. The ideal I^{**} consists of sets which can be decomposed into a wellordered union of partial E_0-transversals.

Example 3.9.15. Let I be an iterable ideal such that $I = I^{**}$, and let $\alpha \in \omega_1$ be a countable ordinal. The results of Section 5.1 show that in the Solovay model as well as in transitive inner models of $AD\mathbb{R}$ containing all the reals it is the case that the iterated Fubini product I^α also satisfies $I^\alpha = (I^\alpha)^{**}$.

Example 3.9.16. Let I be the Mathias ideal. Then both in the Solovay model and in transitive models of AD+ containing all the reals it is the case that $I = I^{**}$.

In the Solovay model a nice characterization of ideals with the second dichotomy is available. To introduce the terminology used in it, if I is an ideal on a Polish space X, write $\hat{I}$ for the ideal on $X \times 2^\omega$ defined by $A \in \hat{I} \leftrightarrow \text{proj}(A) \in I$.

Proposition 3.9.17. *(In the Solovay model) For every σ-ideal I on a Polish space, $\hat{I}$ satisfies the second dichotomy if and only if I is closed under well-ordered unions.*

Note that the proof shows that it is enough to find an *analytic* $\hat{I}$-positive subset of a given $\hat{I}$-positive set for the equivalence to go through.

Proof. The right-to-left implication is easy. In the Solovay model, every subset of a Polish space is a wellordered union of Borel sets. Now if I is a σ-ideal closed under wellordered unions then so is $\hat{I}$ and if B is a $\hat{I}$-positive set, it can be written as a wellordered union of some of its Borel subsets, and one of these must be $\hat{I}$-positive.

For the left-to-right implication note that in the Solovay model, increasing unions of subsets of a Polish space X stabilize in $\aleph_1$ many steps, and so it is enough to prove that closure of a σ-ideal I on X under $\aleph_1$ unions follows from the second dichotomy for $\hat{I}$. For contradiction assume that $\langle B_\alpha : \alpha \in \omega_1 \rangle$ is an $\aleph_1$-collection of I-small sets with I-positive union. Let Z be the Polish space of all trees on ω with a natural topology, let $Y = X \times Z$ and let $A = \{\langle x, T \rangle \in Y : x \in B_\alpha$, the tree T is wellfounded and has rank $\alpha\}$. The projection of the set A is exactly the union $\bigcup_\alpha B_\alpha$, and so $A \notin \hat{I}$. Use the strong dichotomy to find an analytic $\hat{I}$-positive subset $C \subset A$. By the boundedness theorem, there is a countable ordinal β such that whenever $\langle x, T \rangle \in C$ then the rank of the tree T is less than β. Then it must be the case that the projection of the set C to the space X is included in the set $\bigcup_{\alpha \in \beta} B_\alpha \in I$, contradicting the $\hat{I}$-positivity of the set C. $\qquad\square$

In the context of determinacy, one implication of the above equivalence survives.

Proposition 3.9.18. *(ZF+DC+AD+) If $\hat{I}$ satisfies the second dichotomy then I is closed under wellordered unions.*

Proof. Just the same as the previous argument with an additional ingredient, Fact 1.4.6. Assume AD+ and assume that I is a σ-ideal on a Polish space X whose pullback satisfy the second dichotomy. By transfinite induction on $\kappa \in \theta$ argue that whenever $\langle A_\alpha : \alpha \in \kappa \rangle$ is a collection of sets in I then its union is in the ideal I as well. The successor, countable and singular steps in the induction are trivial. So suppose $\kappa \in \theta$ is a regular cardinal and the statement has been verified up to κ. Fix a collection $\langle A_\alpha : \alpha \in \kappa \rangle$ of sets in the ideal I. If $\bigcup_\alpha A_\alpha \notin I$, use Steel's result to find a suitable prewellorder $\leq$ on a set $Y \subset \mathbb{R}$ of length κ and let $B \subset X \times \mathbb{R}$ be the set of all pairs $\langle x, r \rangle$ such that $x \in A_\alpha$ where r is in the α-th $\leq$-equivalence class. The set B is $\hat{I}$-positive and by the second dichotomy it has an analytic positive subset $C \subset B$. The projection of C into the $\mathbb{R}$ coordinate is an analytic subset of Y, and therefore it meets only $< \kappa$ many $\leq$-classes, bounded by some ordinal $\beta \in \kappa$. The projection of C into the X coordinate is an I-positive set, and it is a subset of the set $\bigcup_{\alpha \in \beta} A_\alpha$. However, this contradicts the induction hypothesis, which implies that the set $\bigcup_{\alpha \in \beta} A_\alpha$ is in the ideal I. $\square$

I do not know if the first or the second dichotomy implies that there is a fixed countable ordinal $\alpha \in \omega_1$ such that every positive set contains a positive Σ^0_α set. There are several σ-ideals exposed in this book for which I do not know such a bound.

The previous two propositions have an interesting counterpart in the study of submeasures and capacities. Let X be a Polish space and $\phi : \mathcal{P}(X) \to \mathbb{R}^+$ be a function such that $A \subset B$ implies $\phi(A) \leq \phi(B)$. I will be frequently interested in the continuity of ϕ in increasing wellordered unions, meaning that $\phi(\bigcup_\beta A_\beta) = \sup_\beta \phi(A_\beta)$ if $A_\beta : \beta \in \alpha$ is an inclusion-increasing sequence of sets. The main tool for the verification of this property is the function $\hat{\phi} : \mathcal{P}(X \times 2^\omega) \to \mathbb{R}^+$ defined by $\hat{\phi}(A) = \phi(\mathrm{proj}(A))$, together with the following two propositions.

Proposition 3.9.19. *(In the Solovay model) The function ϕ is continuous in increasing wellordered unions of uncountable cofinality if and only if every subset of $X \times 2^\omega$ has a Borel subset of the same $\hat{\phi}$-mass.*

Proposition 3.9.20. *(ZF+DC+AD+) If every subset of $X \times 2^\omega$ has a Borel subset of the same $\hat{\phi}$-mass, then the function ϕ is continuous in increasing wellordered unions of uncountable cofinality.*

The proofs are essentially identical to the arguments for Propositions 3.9.17 and 3.9.18, and as such are left to the reader.

3.9.3 The third dichotomy

This is the weakest of the dichotomies considered above, their ZFC shadow.

Definition 3.9.21. *A σ-ideal I on a Polish space X satisfies the* third dichotomy *if every I-positive analytic set contains a Borel I-positive subset.*

As before, given a σ-ideal I on a Polish space there is a natural operation to amend it so that it satisfies the third dichotomy. The fact of life is though that all σ-ideals considered in this book satisfy the third dichotomy as they are. It seems to be necessary to supply a different argument every time. The forcing Q_I as in Proposition 2.1.11 is always a useful tool here.

The corresponding dichotomy for coanalytic sets almost always fails in models such as the constructible universe, with exceptions such as the meager ideal or the Lebesgue null ideal. The following proposition is frequently useful in proving the failure of the dichotomy for coanalytic sets in L.

Proposition 3.9.22. *Suppose that I is a σ-ideal on a Polish space X such that provably*

1. *there is a Borel set $B \subset 2^\omega \times X$ whose vertical sections generate the ideal I;*
2. *the forcing P_I is proper;*
3. *there is a Borel function $f : X \to 2^\omega$ such that for every set $C \in I$ there is a set $D \in I$ such that $C \cap D = 0$ and $f''D = 2^\omega$.*

Then in the constructible universe L there is a coanalytic I-positive set without a Borel I-positive subset.

Proof. Work in the constructible universe. Consider the set $A = \{x \in X : \text{for some } y \in 2^\omega \cap L_{\omega_1^{f(x)}}, x \in B_y\}$. I claim that this set works.

First of all, the set $A \subset X$ is coanalytic: $x \in A$ if and only if for every $z \in 2^\omega$, either there is a recursive-in-z wellordering o such that every countable structure $M \models V = L$ whose ordinals are isomorphic to o contains a point y such that $\langle x, y \rangle \in B$, or $z \neq f(x)$. This is a coanalytic statement.

Second, the set $A \subset X$ contains no Borel I-positive subset. Suppose that $C \subset A$ is Borel and I-positive, let M be (the transitive collapse of) a countable submodel of a large enough structure, and let $x \in C$ be a M-generic real for the forcing P_I. Now $\omega_1^M = \omega_1^{M[x]}$ by the properness of the forcing P_I, and so $\omega_1^{f(x)} \in \omega_1^M$, $L_{\omega_1^{f(x)}} \subset M$, the point x falls out of all I-small sets coded by elements of the structure $L_{\omega_1^{f(x)}}$ and therefore does not belong to the set A and $C \not\subset A$.

Finally, the set $A \subset X$ is I-positive. To see this, whenever $C \in I$ is a set, I must produce an element $x \in A \setminus C$. Use the third item of the assumptions to find a set

$D \in I$ such that $C \cap D = 0$ and $f''D = 2^\omega$, let $y \in 2^\omega$ be a point such that $D \subset B_y$, and let $x \in D$ be a point such that $y \in L_{\omega_1^{f(x)}}$. Clearly $x \in A \setminus C$ as desired. □

The coding tool from the statement (3) of this proposition is constructed in later sections for various σ-ideals, giving the following corollaries.

Corollary 3.9.23. *Suppose that ϕ is a pavement submeasure such that the forcing P_{I_ϕ} is nowhere c.c.c. Then in L, there is a coanalytic set of full ϕ-mass without a Borel ϕ-positive subset.*

Corollary 3.9.24. *Suppose that ϕ is an outer regular subadditive stable capacity such that the forcing P_{I_ϕ} is nowhere c.c.c. Then in L, there is a coanalytic subset of full ϕ-mass without a Borel ϕ-positive subset.*

3.10 Games on Boolean algebras

Many forcing properties of partial orders can be characterized via infinitary games on the associated complete Boolean algebras. This line of thinking has been frequently explored in the past [14], [30]. The point I want to make in this section is that in the context of definable forcing, these games are often determined, and the existence of winning strategies for one of the sides can yield surprisingly strong conclusions. The games considered in this section are invariably quite complex – with complicated moves and payoff sets even in the cases of the simplest forcings. This means that the general theorems necessarily use quite strong large cardinal assumptions even in the cases of syntactically very simple partial orders. This being said, the reader should revisit some simple examples to see that in all particular cases the winning strategies are long known and readily at hand. This section contains just the definitions of the games, the determinacy results, and the proof of an equivalence with a forcing property. The applications are scattered throughout the book and the reader is advised to consult the index to find them.

3.10.1 Precipitousness

Let I be a σ-ideal on a Polish space X.

Definition 3.10.1. *The* Borel precipitous game G *is played between Players Empty and Nonempty who alternate to obtain a descending chain $B_0 \supset B_1 \supset \dots$ of conditions in the forcing P_I. Player Nonempty wins if $\bigcap_n B_n \neq 0$.*

Proposition 3.10.2. *Player Nonempty has a winning strategy in the game G.*

Proof. The strategy is the following. On the side, player Nonempty will create an increasing sequence $M_n : n \in \omega$ of countable elementary submodels of some large structure and use some simple bookkeeping tool to make sure that $B_n \in M_n$ and the chain $B_n : n \in \omega$ is M-generic where $M = \bigcup_n M_n$. Now the basic Proposition 2.1.2 applied in the model M shows that $\bigcap_n B_n \neq 0$ as required. It is not difficult to see that player Nonempty can in fact make sure that the intersection is a singleton. $\square$

3.10.2 Properness

There is the classical game theoretical restatement of properness due to Charles Gray.

Definition 3.10.3. *Suppose P is a partial order. Fix a partition* $\omega = \bigcup_k a_k$ *of* ω *into infinite sets such that* $\min(a_k) \geq k$. *The properness game G between Players I and II proceeds as follows. First Player I indicates an initial condition* $p_{ini} \in P$. *After that, the moves alternate, at round n Player I produces an open dense set* $D_n \subset P$ *and Player II responds with a condition* $q_n \in P$ *such that* $q_n \in D_k$ *where k is such that* $n \in a_k$. *Player II wins if the result of the play, the expression*

$$p_{ini} \wedge \bigwedge_k \bigvee_{n \in a_k} q_n$$

denotes a non-zero element in the complete Boolean algebra RO(P).

Theorem 3.10.4. *The following are equivalent:*

1. *Player II has a winning strategy;*
2. *the forcing P is proper.*

and the following are equivalent:

1. *Player I has a winning strategy;*
2. *some condition forces the set* $([P]^{\aleph_0})^V$ *to be nonstationary in* $([P]^{\aleph_0})^{V[G]}$.

Moreover, if suitable large cardinals exist and the forcing P is universally Baire then the game G is determined.

Proof. The first equivalence appeared in the work of Charles Gray and Saharon Shelah. If Player II has a winning strategy σ, M is a countable elementary submodel of a large enough structure and $p \in P \cap M$ is an arbitrary condition, simulate a play against the strategy σ in which Player I indicates $p = p_{ini}$ and then enumerates all the open dense sets in the model M, and Player II follows the strategy $\sigma \in M$. The result of such a play is the required master condition for the model M. On the other hand, if the forcing P is proper, Player II can win the properness game by following some simple bookkeeping tool to make sure that in the end of the play there is a model M which contains all the moves played, and for every dense set $D \in M$ that Player I played, Player II enumerated the whole set $M \cap D$. The result of the game is a condition weaker than the master condition for the model M and therefore nonzero.

For the second equivalence first suppose that there is a winning strategy σ for Player I in the properness game. Let $p_{ini} \in P$ be the initial condition indicated by the strategy, let $G \subset P$ be a generic filter containing the condition p_{ini} and in $V[G]$ let $f : P^{<\omega} \to P$ be the function defined by $f(\vec{p}) = $ some condition in $D \cap G$ if $\vec{p}$ is a legal sequence of Player II's answers to the strategy σ and D is the appropriate set played by the strategy σ, and $f(\vec{p}) = p_{ini}$ otherwise. I claim that no countable set in V is closed under the function f. For contradiction assume that $q \leq p_{ini}$ is a condition and $a \subset P$ is a countable set such that $q \Vdash \check{a}$ is closed under the function $\dot{f}$. Consider the counterplay against the strategy σ in which Player II keeps all of his moves in the set a and to each open dense set $D \subset P$ the strategy σ produces he enumerates the countable set $D \cap a$. A review of the definitions reveals that the condition q is stronger than the result of this play, contradicting the assumption that σ was a winning strategy for Player I.

Now suppose that some condition $p \in P$ forces the set $([P]^{\aleph_0})^V$ to be nonstationary in $([P]^{\aleph_0})^{V[G]}$, as witnessed by some function $f : P^{<\omega} \to P$ in the extension under which no countable subset of P in the ground model is closed. Player I will win by indicating the initial condition $p = p_{ini}$ and then employing some simple bookkeeping tool to play in such a way that in the end there is a countable elementary submodel M of a large enough structure such that all Player II's moves stayed inside M and Player I enumerated all open dense subsets of the poset P in the model M. If the result $q \leq p_{ini}$ of such a play was nonzero then clearly q is a master condition for the model M and it forces $M \cap P$ is closed under $\dot{f}$, contradicting the assumed properties of the name $\dot{f}$. Thus the described strategy is winning for Player II and the second equivalence follows.

For the determinacy of the properness game, fix a universally Baire σ-ideal I on a Polish space X such that $P = P_I$. Note that it is harmless to require Player II to provide codes for the Borel sets he is playing, since the outcome of the game does not depend on these codes. The sequence of the codes is then continuously read off the play, and the determination whether the result of the play is not in the ideal I is a universally Baire operation on the sequence of the codes. Now use Fact 1.4.2 to make the determinacy conclusion. □

In this particular case, there is an interesting consequence of the determinacy of the properness game answering the concerns of Carlos DiPrisco. It does not fit anywhere else in the book, and I include it here.

Proposition 3.10.5. *(LC+CH) Let P be a proper universally Baire forcing. Let κ be a weakly compact cardinal and $V(\mathbb{R})$ be the derived Solovay model. Force with P over $V(\mathbb{R})$ and let $\mathbb{R}^*$ be the set of reals of the resulting extension. Then $V(\mathbb{R}^*)$ is a Solovay model as well.*

Proof. It is enough to show that every element of $\mathbb{R}^*$ is generic over V for a poset of size $< \kappa$ [13], 1.1. To this end, suppose that $p \in P$ and τ are a condition in P and a P-name for a real in the model $V(\mathbb{R})$. By a weak compactness argument, there is an inaccessible cardinal $\lambda < \kappa$ and a V-generic filter $g \subset Coll(\omega, < \lambda)$ in $V(\mathbb{R})$ such that $p \in V[g]$ and $\tau \cap V[g] \in V[g]$ is a name for a real in the forcing $P \cap V[g]$. In the model $V[g]$, the forcing P is proper by Corollary 2.2.9. This means that in that in the properness game $G^{V[g]}$, Player II has a winning strategy σ which remains a winning strategy in every forcing extension of $V[g]$, in particular, in the model $V(\mathbb{R})$. In the model $V(\mathbb{R})$, find a play against the strategy σ in which Player I indicates $p_{ini} = p$ and then enumerates all the open dense subsets of the poset P in the model $V[g]$. Since σ remains a winning strategy in the model $V(\mathbb{R})$, the result of this play is then a condition $q \in P \cap V(\mathbb{R})$. That condition q by its definition forces in P that the intersection $H \cap V[g]$ of the generic filter on P with the model $V[g]$ will be $V[g]$-generic. Since $\tau \cap V[g]$ was a P-name in $V[g]$, this implies that $\tau/H \in V[g][H \cap V[g]]$ is a real in a generic extension of the ground model V by a poset of size $< \kappa$. $\qquad\square$

3.10.3 The bounding condition

Definition 3.10.6. *Suppose P is a partial order. The* bounding game G *between Players I and II proceeds as follows. First Player I indicates an initial condition $p_{ini} \in P$. After that, the moves alternate, at round n Player I produces an open dense set $D_n \subset P$ and Player II responds with a finite set $d_n \subset D_n$. Player II wins if the result of the play, the expression*

$$p_{ini} \wedge \bigwedge_n \bigvee d_n$$

denotes a non-zero element in the complete Boolean algebra $RO(P)$.

Theorem 3.10.7. *Suppose that P is a proper partial order. The following are equivalent:*

- *Player I has a winning strategy in the game G;*
- *some condition forces that P adds an unbounded real.*

Moreover, if suitable large cardinals exist and the forcing P is universally Baire and proper then the game G is determined. If $P = P_I$ for a $\mathbf{\Pi}_1^1$ on $\mathbf{\Sigma}_1^1$ ideal I on a Polish space then the game is determined without the need for the large cardinal assumptions.

Proof. If $p \Vdash \dot{f} \in \omega^\omega$ is an unbounded real then Player I has a winning strategy in the game which completely ignores the moves of the other player. Just put $p = p_{ini}$ and $D_n = \{q \in P : q$ decides the value of $\dot{f}(\check{n})\}$. It is clear that whatever Player II answers, the result of the play would have to bound the function $\dot{f} \in \omega^\omega$ by a ground model function, and therefore the result has to be zero.

On the other hand, if σ is a winning strategy for Player I then a name for an unbounded real can be extracted as follows. Let $p = p_{ini}$, the initial condition dictated by the strategy σ, let M be a countable elementary submodel of a large enough structure, and let $q \leq p$ be an M-master condition. If $\{A_n : n \in \omega\}$ is the enumeration of all maximal antichains of P which are in the model M, the countable antichains $\{M \cap A_n : n \in \omega\}$ are predense below q. I claim though that there is no strengthening $r \leq q$ which is compatible with at most finitely many elements of each antichain $A_n \cap M$. If r was such a strengthening, Player II could win against the strategy σ by choosing a maximal antichain $A_n \subset D_n$, $A_n \in M$ against Player II's move D_n and letting $d_n \subset A_n \cap M$ be the finite set of all conditions compatible with r. It is clear that in this way all moves of the play will be in the model M and the result of the game will be larger than r, therefore nonzero, contradicting the choice of σ as a winning strategy. With this information at hand, it is easy to see that with any enumeration $P \cap M = \{s_m : m \in \omega\}$ the name $\dot{f} \in \omega^\omega$, $\dot{f}(\check{n}) =$ least m such that $s_m \in A_n$ and $\check{s}_m \in \dot{G}$, is forced by the condition q to be an unbounded real.

The argument for the determinacy of this game is literally the same as in the previous section. For the $\mathbf{\Pi}_1^1$ on $\mathbf{\Sigma}_1^1$ version, suppose that the forcing P_I is bounding. I will describe a closed version of the bounding game which is more difficult to play for Player II than the original bounding game, and I will use the determinacy of closed games to find a winning strategy for Player II in the more difficult game.

For the simplicity of notation assume that $X = 2^\omega$. Note that compact sets are dense in the poset P_I, and the set of I-positive compact sets is analytic in $K(X)$ by the definability assumption. Let $T \subset \omega^{<\omega} \times (2^{<\omega})^{<\omega}$ be a tree which projects into the set of all trees $S \subset 2^{<\omega}$ such that $[S] \notin I$. The closed bounding game is then played between Player I and II in the following fashion. The moves of Player I are the same as in the original bounding game. Player II responds to each open set D_n with a finite set d_n of binary trees such that for all $U \in d_n$, $[U] \in D_n$. Moreover he indicates a node $t_n \in T$. Player II wins if the nodes $t_n : n \in \omega$ form a branch through the tree T and its projection S, a binary tree, is covered by $\bigcup d_n$ for every number $n \in \omega$.

The game is closed for Player II and clearly more difficult for him to play than the original game. This closed version of the game is determined and therefore it will be enough to show that Player I has no winning strategy. Suppose σ is a strategy for Player I, M is a countable elementary submodel of a large enough structure, $A_n : n \in \omega$ is a list of all maximal antichains in the model M consisting of compact sets, let p be the initial condition dictated by the strategy σ, and let $q \leq p$ be an M-master condition. The bounding property shows that there is a closed set $r \leq p$ such that r is covered by a finite number of conditions in the antichain A_n, this for all numbers $n \in \omega$. Player II will now win as in the previous proof, producing a path through the tree T which projects into a binary tree S such that $[S] = r$. The argument is complete.   □

3.10.4 Not adding a bounded eventually different real

Definition 3.10.8. *Suppose P is a partial order. The BED game G between Players I and II proceeds as follows. First Player I indicates an initial condition $p_{ini} \in P$. After that, the moves alternate, at round n Player I produces a natural number k_n and a P-name $\dot{m}_n$ for a number smaller than k_n and Player II responds with a number $l_n \in k_n$. Player II wins if the result of the play, the expression*

$$p_{ini} \wedge \bigwedge_n \bigvee_{o > n} |\check{l}_o = \dot{m}_o|$$

denotes a nonzero element in the complete Boolean algebra $RO(P)$.

Theorem 3.10.9. *Let P be a partial order. The following are equivalent:*

- *Player I has a winning strategy in the BED game;*
- *some condition forces that P adds a bounded eventually different real.*

Moreover, if suitable large cardinals exist and the forcing P is universally Baire and proper then the game G is determined.

Proof. First suppose that Player I has a winning strategy σ in the game calling for some initial condition p_{ini}. Then he actually has a winning positional strategy τ, a sequence $k_n, \dot{m}_n$ of names for natural numbers such that he wins playing these partitions no matter what the opponent's answers are. To obtain the positional strategy note that at each round n there are only finitely many names for natural numbers the strategy σ can produce as n-th move for Player I, and choose $\dot{m}_n$ to be a name for a number coding them all. It is clear that τ must be a winning strategy if σ is. Now the name $\dot{g} \in \omega^\omega$, $\dot{g}(n) = \dot{m}_n$, is forced by the condition p_{ini} to be eventually different from every ground model function. For if $q \leq p_{ini}$ forced

$|\dot{g} \cap \check{f}| = \aleph_0$ for some ground model function f, Player II could play $l_n = f(n)$ against the strategy τ, and q would be smaller than the result of the play, showing that Player II wins, a contradiction.

On the other hand, a name $\dot{g} \in \omega^\omega$ for a function bounded by some ground model function $f \in \omega^\omega$ which some condition $p \in P$ forces to be eventually different, provides a winning strategy for Player I. Just let $k_n = f(n)$ and $\dot{m}_n = \dot{g}(\check{n})$.

The determinacy of the game is an issue more sensitive than in the previous cases since the payoff set is not of a form to which Fact 1.4.2 can be immediately applied. I need to consider a new game H which is only slightly harder for Player II than the game G, and which is determined.

The moves of Player I in the game H are the same as in the game G. At round n, Player II answers by choosing a number $l_n \in k_n$ as in the game G, but in addition he plays finite, possibly empty, sets $d_i^n : i \in n$ of conditions in P such that for every $i \in n$ and $q \in d_i^n$, $q \Vdash \dot{m}_i = \check{l}_i$. Player II wins if the expression $\bigwedge_n \bigvee_{o > n} \bigvee_i d_o^i$ denotes a nonzero element in the complete Boolean algebra $RO(P)$.

Fact 1.4.2 shows that the game H is determined whenever the forcing P is universally Baire and suitable large cardinals exist. I will show that the game H is really the same as G for Player I in that Player I has a winning strategy in the game H if and only if some condition forces P to add a bounded eventually different real. The determinacy of the game G then immediately follows: if Player I does not have a winning strategy in the game G, then the forcing P does not add a bounded eventually different real, Player I has no winning strategy in the game H, the determinacy of the game H yields a winning strategy for Player II in it, and if Player II erases from that strategy the finite sets of conditions he obtains a winning strategy for the game G as desired.

Now, on one hand, if the forcing P adds a bounded eventually different real then Player I can win the game H in exactly the same way he won the game G. On the other hand, suppose that P is proper, does not add an eventually different real, and σ is a strategy for Player I in the game H. I must produce a play against the strategy in which Player II wins. Let M be a countable elementary submodel of a large enough structure, let $p_{ini} \in P \cap M$ be the first move dictated by the strategy σ and let $q \le p$ be a master condition. Let $p_n : n \in \omega$ be an enumeration of the set $P \cap M$. Consider the set S of all plays in which Player I uses his strategy and Player II plays just finite sets in the model M such that $d_i^n \subset \{p_0, p_1, \ldots p_n\}$. This is a finitely branching tree of plays. Consider the name $\dot{g} : \omega \to \omega^{<\omega}$ given by the demand that $\dot{g}(n)$ is the collection of all possible names played at round n in the plays in the set S. Since the poset P does not add a bounded eventually different real, there is a condition $r \le q$ and a function $f : \omega \to \omega^{<\omega}$ such that $r \Vdash |\dot{g} \cap \check{f}| = \aleph_0$. Now consider the play in the set S in which Player II at round n plays the number l_n predicted for the name

$\dot{m}_n$ by the function f and uses the sets $d_n^k : k \in \omega$ to slowly exhaust the countable set $\{s \in P \cap M : s \Vdash \dot{m}_n = \check{l}_n\}$. A review of the definitions shows that the result of this play will be larger than the condition r and Player II wins as desired.   □

3.10.5 Laver property

Definition 3.10.10. *[2], 6.3.27. A forcing P has the* Laver *property if for every ground model function $f \in \omega^\omega$ and every ground model nondecreasing function $g \in \omega^\omega$ converging to infinity, for every function $h \in \omega^\omega$ dominated pointwise by f in the extension, there is a ground model function $e : \omega \to [\omega]^{<\aleph_0}$ such that the set $e(n)$ has size $\leq g(n) + 1$ and contains the value $h(n)$.*

A basic definable example of a partial ordering with Laver property is the Mathias forcing. It seems to be difficult to come up with substantially more complex examples. There is a natural game theoretic counterpart to the Laver property.

Definition 3.10.11. *Suppose P is a partial order. The* Laver game G *between Players I and II proceeds as follows. First Player I indicates an initial condition $p_{ini} \in P$. After that, the moves alternate, at round n Player I produces a natural number k_n and a P-name $\dot{m}_n$ for a number smaller than k_n, and a number $g_n \in \omega$ and Player II responds with a set $a_n \subset k_n$ of size at most $g_n + 1$. Player II wins if either the sequence of numbers $g_n : n \in \omega$ was not nondecreasing and diverging to infinity or else the result of the play, the expression*

$$p_{ini} \wedge \bigwedge_n |\dot{m}_n \in \check{a}_n|$$

denotes a non-zero element in the complete Boolean algebra $RO(P)$.

Theorem 3.10.12. *Let P be a partial order. The following are equivalent:*

- *Player I has a winning strategy in the Laver game;*
- *the forcing P does not have the Laver property.*

Moreover, if suitable large cardinals exist and the forcing P is universally Baire and proper then the game G is determined.

Proof. Suppose first that Player I has a winning strategy σ in the Laver game. It is then not difficult to see that he has a *positional* winning strategy τ, that is, a nondecreasing function $g \in \omega^\omega$ and names $\dot{m}_n$ such that he wins playing these objects regardless of the oponnent's moves. To see this, note that at every move there are only finitely many options for Player II and so there are only finitely many possible answers the strategy σ can supply. Let $\dot{m}_n$ be a name for a number coding all the finitely many names the strategy σ can supply at round n. Finally, use a compactness argument to find an increasing sequence $\{n_i : i \in \omega\}$ of natural numbers such that the strategy σ asks for at least i many pieces of the partition at each round after round n_i, no matter what Player II's moves. Then define the function g by $g(n) = i$ if $n_m \leq n < n_{i+1}$. It is not difficult to check that the positional winning strategy τ given by g and $\{\dot{m}_n : n \in \omega\}$ is winning since it is a better strategy than σ. But then, if $p_{ini} \in P$ is the initial condition dictated by the strategy, $f \in \omega^\omega$ is a function defined by $f(n) = k_n$, and $\dot{h}$ is a name for a function in ω^ω defined by $\dot{h}(n) = \dot{m}_n$ it is immediate that these objects witness the failure of the Laver property of the poset P_J.

On the other hand, if Player I has no winning strategy in the Laver game then the Laver property is rather easy to check. Suppose $p \in P$ is a condition, $f \in \omega^\omega$ a function, $g \in \omega^\omega$ a nondecreasing function diverging to infinity, and $\dot{h}$ a name for a function in ω^ω dominated by f. The condition $p = p_{ini}$ together with the names $\dot{m}_n = \dot{h}(\check{n})$ and the function g does not constitute a positional winning strategy for Player I, and there must be a winning counterplay for Player II, with moves $B_n = \bigcup_{k \in e(n)} C_n^k$ for some function $e : \omega \to [\omega]^{<\aleph_0}$ such that $|e(n)| \leq g(n) + 1$. Any condition $q \leq$ the result of the play then clearly forces $\forall n \in \omega \; \dot{h}(n) \in \check{e}(n)$ as desired.

The determinacy of the game is handled in the same way as in Section 3.10.4.

$\square$

3.10.6 Strong preservation of submeasures

Definition 3.10.13. *Suppose that ϕ is a universally Baire submeasure on a Polish space X. A forcing P strongly preserves ϕ if for every real number $\varepsilon > 0$ and every set $A \subset X$ in the extension with $\phi(A) < \varepsilon$ there is a Borel set $B \subset X$ coded in the ground model such that $A \subset B$ and $\phi(B) < \varepsilon$.*

This is a commonly studied forcing property; in the case of outer Lebesgue measure, it coincides with the preservation of basis of the Lebesgue null ideal. In other cases, it apparently has a life of its own. In this book it appears in Theorems 4.3.13 and 4.4.17. I will always use it with outer regular submeasures. It may make sense to study it for functions which are not necessarily subadditive, such as general outer regular capacities.

For several classes of submeasures the strong preservation has a suitable game-theoretic restatement. I will restrict to the case of strongly subadditive capacities.

Definition 3.10.14. *Suppose that P is a forcing and ϕ is an outer strongly subadditive capacity on the Cantor space 2^ω. The strong preservation game G*

is played by Players I and II in the following fashion. First, Player I indi-cates real numbers $0 < \varepsilon_{ini} < \delta_{ini}$, a clopen set $U_{ini} \subset 2^\omega$, and a condition $p_{ini} \in P$. Then he plays one by one P-names $\dot{U}_n : n \in \omega$ for clopen sets so that p_{ini} forces $\dot{U}_{ini} \subset \dot{U}_0 \subset \dot{U}_1 \subset \ldots$ and for every pair of natural numbers $n \in m$, $\phi(\dot{U}_n) - \phi(\dot{U}_{ini}) < \varepsilon_{ini}$ and $\phi(\dot{U}_m) - \phi(\dot{U}_n) < 2^{-n}\varepsilon_{ini}$. Player II plays a sequence $V_n : n \in \omega$ of clopen sets such that $U_{ini} \subset V_0 \subset V_1 \subset \ldots$ and for every $n \in m \in \omega$, $\phi(V_n) - \phi(U_{ini}) < \delta_{ini}$ and $\phi(V_m) - \phi(V_n) < 2^{-n}\delta_{ini}$. Player II is allowed to wait for arbitrary finite number of steps before placing the next set on his sequence. Player II wins if the result of the play, the expression

$$p_{ini} \wedge \bigwedge_n \bigvee_m |\dot{U}_n \subset \dot{V}_m|$$

denotes a nonzero element in the complete algebra $RO(P)$.

Note that as the sets $\dot{U}_n$ and V_n are (forced to be) clopen in the compact space 2^ω, the result of the game is equal to $p_{ini} \wedge |\bigcup_n \dot{U}_n \subset \bigcup_n \dot{V}_n|$.

Theorem 3.10.15. *Suppose that P is a forcing and ϕ is a strongly subadditive capacity on the Cantor space 2^ω. The following are equivalent:*

1. *Player I has a winning strategy in the strong preservation game;*
2. *P does not strongly preserve the capacity ϕ.*

If suitable large cardinals exist and P is proper and $P = P_I$ for some universally Baire σ-ideal I on a Polish space, then the game is determined. If the ideal I is in addition Π^1_1 on Σ^1_1 and the forcing P_I is bounding, then the large cardinal assumptions are not necessary.

Proof. The (2)→(1) direction is easy. If P does not strongly preserve ϕ the there is a condition $p \in P$, real numbers $0 < \varepsilon < \delta$ and a P-name for an open set $\dot{O} \subset 2^\omega$ such that $p \Vdash \phi(\dot{O}) < \varepsilon$ and $\dot{O}$ cannot be covered by a ground model open set of mass $< \delta$. Then Player I can win by indicating $p_{ini} = p$, $\varepsilon = \varepsilon_{ini}$, $\delta = \delta_{ini}$, and $U_{ini} = 0$, and then playing so that in the end $p \Vdash \dot{O} = \bigcup_n \dot{U}_n$.

The (1)→(2) direction is harder. Suppose that Player I has a winning strategy σ in the game, with initial choices $p = p_{ini}$, $U = U_{ini}$, $\varepsilon = \varepsilon_{ini}$, $\delta = \delta_{ini}$. I will find a P-name $\dot{O} \subset X$ for an open set such that $p_{ini} \Vdash \dot{U} \subset \dot{O} \wedge \phi(\dot{O}) - \phi(\dot{U}) < \frac{\varepsilon+\delta}{2}$ such that every legal sequence $\langle V_n : n \in \omega \rangle$ of nontrivial Player II's moves can be played against the strategy σ in such a way that the strategy σ answers with a sequence of names $\langle \dot{U}_n : n \in \omega \rangle$ such that $p \Vdash \bigcup_n \dot{U}_n \subset \dot{O}$. It immediately follows that the set $\dot{O}$ cannot be covered by a ground model set of ϕ-mass $< \delta$: if $R \subset 2^\omega$ is a ground model open set with $\phi(R) < \delta$, $q \leq p$ a condition, and $q \Vdash \dot{O} \subset \dot{R}$, Player II could produce a counterplay against the strategy σ such that in it, $\bigcup_n V_n = R$

and $p_{ini} \Vdash \bigcup_n \dot{U}_n \subset \dot{O}$. The result of such a play would have to be bigger than q, contradicting the assumption that σ was a winning strategy for Player I.

To produce the name $\dot{O}$, consider the tree T of all finite legal sequences of nontrivila moves of Player II against the strategy σ. To each node $t \in T$, $t \neq 0$, assign a positive number m_t such that $\Sigma_t 2^{-m_t} \varepsilon < \frac{\delta - \varepsilon}{2}$ and $s \subset t \to m_s \in m_t$. To each node $t \in T$ assign the unique finite partial play τ_t against the strategy σ such that the sequence of nontrivial moves of Player II in it is exactly t, and for every $n \in \mathrm{dom}(t)$, $t(n)$ is played exactly at round $m_{t \restriction n+1}$. The play τ_t ends with the last move on the sequence t. Let $\dot{O}$ be the name for the union of all the names for basic open sets that the strategy σ produces against the plays $\tau_t : t \in T$. I claim this name works.

It is clearly enough to prove that $p \Vdash \phi(\dot{O}) - \phi(\dot{U}_{ini}) < \frac{\varepsilon + \delta}{2}$. Just enumerate the nodes of the tree T as $t_n : n \in \omega$, respecting the extension order. For every number $n \in \omega$ $t \in T$, let $\dot{O}_n$ be the name for the union of all the basic open sets the strategy σ produces in the infinite extension of the play τ_{t_n} in which Player II adds no nontrivial moves past τ_t. It will be enough to prove by induction on $n \in \omega$ that $p \Vdash \phi(\bigcup_{k \in n} \dot{O}_k) - \phi(\dot{U}) < \varepsilon + \Sigma_{k \in n} 2^{-m_{t_k}} \varepsilon$. This is clear for $n = 0$. If it is known for some $n \in \omega$, find $l \in n$ such that t_l is an immediate predecessor of t_n, and observe that $p \Vdash \phi(\dot{O}_n) < \phi(\dot{O}_n \cap \dot{O}_l) + 2^{-m_{t_n}} \varepsilon$ by the definition of the game. The strong subadditivity now kicks in to give $p \Vdash \phi(\bigcup_{k \in n} \dot{O}_k \cup \dot{O}_n) < \phi(\bigcup_{k \in n} \dot{O}_k) + 2^{-m_{t_n}} \varepsilon$ and the induction step follows.

Now suppose that $P = P_I$ is a proper forcing with some universally Baire σ-ideal I. The determinacy of the game is a nontrivial statement, since the payoff set of the game is not of the form to which Fact 1.4.2 can be applied. Instead, I have to define an auxiliary game $\bar{G}$ which is determined under suitable large cardinal assumptions, and show that the game $\bar{G}$ is more difficult for Player II than the game G, but still Player I has a winning strategy in one of the games iff he has a winning strategy in the other. The game $\bar{G}$ is obtained from G by having Player II additionally indicate conditions $p_k \in P, k \in \omega$, and stipulating that Player II wins if the result of the game, the expression

$$p_{ini} \wedge \bigwedge_n \bigvee_m \bigvee \{p_k : p_k \Vdash \dot{U}_n \subset \check{V}_m\}$$

is nonzero in the complete algebra $RO(P)$. It is not difficult to use Fact 1.4.2 to verify that if $P = P_I$ for a universally Baire σ-ideal I and suitable large cardinals exist, then the game $\bar{G}$ is determined. (The set $\{\langle n, m, k \rangle \in \omega^3 : p_k \Vdash \dot{U}_n \subset \check{V}_m\}$ as well as the sequence of (codes for) the conditions $p_k : k \in \omega$ are obtained continuously from the play of the game, and then checking if the result of the play is in the ideal I is a universally Baire procedure on these two objects and the ideal I.) Clearly, the game $\bar{G}$ is more difficult for Player II than the original game. I must prove that if Player I has a winning strategy in the auxiliary game, then he has one in the original game. The properness of the forcing P is a necessary ingredient here.

Suppose that σ is a winning strategy for Player I in the auxiliary game, with initial choices p, U, ε, δ. Let M be a countable elementary submodel of a large enough structure, and let $p_k : k \in \omega$ be an enumeration of the set $P \cap M$. Now consider only the plays in which Player II indicates this sequence of conditions. Just as before, it is possible to find a name $\dot{O}$ for an open set such that $p \Vdash \phi(\dot{O}) - \phi(U) < \frac{\varepsilon+\delta}{2}$ and every legal sequence of clopen sets can be played against the strategy σ such that the resulting name for an open set the strategy σ produces is forced to be a subset of $\dot{O}$. I will show that, writing $q \leq p$ for any M-master condition, $q \Vdash \dot{O}$ cannot be covered by any open set from the ground model of mass $\phi(U)$. This of course shows that Player I has a winning strategy in the original game.

Now if $R \subset 2^\omega$ was an open set with $U \subset R$ and $\phi(R) - \phi(U) < \delta$, and $r \leq q$ was a condition forcing $\dot{O} \subset \check{R}$, then stratify $R = \bigcup_n V_n$ as a union of a legal sequence of nontrivial moves for Player II's moves and find a way to play them so that the resulting set the strategy σ produces is forced by p to be a subset of $\dot{O}$. Now observe that the moves $\dot{U}_n$ the strategy σ produced in such a play are all elements of the model M. Since $r \in P$ is a master condition, this means that $r \Vdash \bigwedge_n \bigvee_m \dot{U}_n \subset \dot{V}_m$ is equivalent to $r \leq \bigwedge_n \bigvee_m \bigvee \{s \in M \cap P : s \Vdash \dot{U}_n \subset \dot{V}_m\}$. Thus the condition r witnesses that Player II won against the strategy σ, a contradiction.

The ZFC case of bounding forcing associated with a Π^1_1 on Σ^1_1 ideal is left to the interested readers. It is just necessary to further adjust the auxiliary game by requiring Player II to produce a witness to the fact that the result of the game is not in the ideal I. □

3.10.7 Analytic P-ideals

Given a suitably definable ideal J on ω and a forcing P, one may ask whether every set in J in the extension is covered by a set in J in the ground model. In general, this is a difficult question. In this section I offer a game-theoretic reformulation of this preservation property for analytic P-ideals. I will use the characterization of P-ideals due to Solecki: if J is an analytic P-ideal on ω then there is a lower semicontinuous submeasure ϕ on $\mathcal{P}(\omega)$ such that $J = J_\phi = \{a \subset \omega : \lim_n \phi(a \setminus n) = 0\}$.

Definition 3.10.16. *Suppose that ϕ is a lower semicontinuous submeasure on $\mathcal{P}(\omega)$ and P is a forcing. The J_ϕ-preservation game is played between Players I and II in the following fashion. First, Player I indicates real numbers $0 < \varepsilon_{ini} < \delta_{ini}$, a condition $p_{ini} \in P$, as well as a finite set $u_{ini} \subset \omega$. After that, at round $n \in \omega$ Player I indicates a P-name $\dot{u}_n$ for a finite subset of ω such that p_{ini} forces $\check{u}_{ini} \subset \dot{u}_0 \subset \dot{u}_1 \subset \ldots$ and $\phi(\dot{u}_n \setminus \check{u}) < \varepsilon$ and $\phi(\dot{u}_n \setminus \dot{u}_m) < 2^{-m}\varepsilon_{ini}$. Player II answers* *with a sequence $v_0, v_1, v_2 \ldots$ of finite subsets of ω such that $u_{ini} \subset v_0 \subset v_1 \subset \ldots$, $\phi(v_n \setminus u_{ini}) < \delta_{ini}$ and $\phi(v_n \setminus v_m) < 2^{-m}\delta_{ini}$. Player II can postpone putting another set on his sequence for any number of rounds. Player II wins if the result of the play, the expression*

$$p_{ini} \wedge |\bigcup_n \check{u}_n \subset \bigcup_n \check{v}_n|$$

is a nonzero element in the algebra $RO(P)$.

Theorem 3.10.17. *Suppose that P is a forcing and ϕ is a lower semicontinuous submeasure on $\mathcal{P}(\omega)$. The following are equivalent:*

1. *P forces that every element of the ideal J_ϕ in the extension is covered by a set in the ideal J_ϕ in the ground model;*
2. *Player I does not have a winning strategy in the game.*

Moreover, if suitable large cardinals exist and $P = P_I$ for a universally Baire ideal I, and P_I is proper, the game is determined. If in addition the ideal I is $\mathbf{\Pi}_1^1$ on $\mathbf{\Sigma}_1^1$ and the forcing P_I is bounding, then large cardinal assumptions are not necessary.

Proof. Obviously, if item (1) fails and $p \Vdash \dot{a} \in J_\phi$ is a set that cannot be covered by a ground model element of J_ϕ, then Player I can win the game by indicating $p_{ini} = p$ and then arranging $\dot{a} = \bigcup_n \dot{u}_n$. On the other hand, suppose that (1) holds and σ is a strategy for Player I. I must find a counterplay winning for Player II. Let $p, \varepsilon, \delta, u$ be the initial choices of the strategy σ. I will find a set $b \in J_\phi$ and a condition $q \leq p$ such that $u \subset b$, $\phi(b \setminus u) < \delta$ and for every legal sequence $v_n : n \in \omega$ of nontrivial moves for Player II there is a way to play it so that $q \Vdash \bigcup_n \dot{u}_n \subset b$, where $\dot{u}_n : n \in \omega$ is the sequence of names the strategy σ produces. Then it is easy to stratify the set b into an increasing union $b = \bigcup_n v_n$ so that these sets form a legal sequence of nontrivial moves for Player II, and there must be a way to play them so that in the end, Player II wins as witnessed by the condition q.

To find the set b, consider the tree of all finite sequences of legal answers by Player II, and for each $t \in T$ fix a number n_t such that $s \subset t \to n_s \in n_t$ and $\Sigma_t 2^{-n_t} < \frac{\delta - \varepsilon}{\varepsilon}$. For every $t \in T$ let τ_t be the shortest finite play in which Player II produces exactly the sequence t as his nontrivial moves, and he plays $t(k)$ at round $n_{t \restriction k+1}$. Let $\dot{a}$ be the name for the union of all sets the strategy σ indicates in the course of all these plays $\tau_t : t \in T$. I claim that $p_{ini} \Vdash \dot{a} \in J_\phi$ and $\phi(\dot{a}) < \delta$. Then use the preservation properties of the forcing P to find a condition $q \leq p$ and a ground model set $b \in J_\phi$ of ϕ-mass $< \delta$ such that $q \Vdash \dot{a} \subset b$; this will conclude the proof.

To verify the properties of the name $\dot{a}$, for a node $t \in T$ write $\dot{a}_t$ for the name for the union of all names the strategy σ produces in the infinite play that starts with τ_t and has no nontrivial moves by Player II past t. Clearly $\dot{a} = \bigcup_t \dot{a}_t$. It follows from the definition of the game that p forces these sets to be in the ideal J and $\phi(\dot{a}_t \setminus \dot{a}_s) < 2^{-n_s}\varepsilon$ whenever $s \in T$ is an immediate predecessor of the node $t \in T$. Thus, $p \Vdash \phi(\dot{a}) \leq \phi(\dot{a}_0) + \Sigma\{\phi(\dot{a}_t \setminus \dot{a}_s) : t$ ius an immediate succesor of $s\} \leq \varepsilon + \Sigma_t 2^{-n_t}\varepsilon < \delta$. Moreover, if $\gamma > 0$ is a real number, find a finite set $S \subset T$ closed under initial segment so that $\Sigma_{t \in T \setminus S} 2^{-n_t} < \gamma\varepsilon/2$. Since $p \Vdash \dot{a}_t \in J_\phi$ for every node $t \in T$, it is also true that p forces $\bigcup_{t \in S} a_t \in J_\phi$ and there must be a number $m \in \omega$ such that $\phi(\bigcup_{t \in S} \dot{a}_t \setminus m) < \gamma/2$. For such a number $m \in \omega$, $\phi(\dot{a} \setminus m) < \gamma$, and therefore $\dot{a} \in J_\phi$.

The determinacy of the game is handled in a way parallel to the previous cases, and I omit the argument. $\qquad\square$

3.10.8 Continuous reading of names

Definition 3.10.18. *Suppose I is a σ-ideal on a Polish space X and $\mathcal{O}$ is a countable topology basis for X closed under finite unions. Suppose $\omega = \bigcup\{a_{i,j} : i, j \in \omega\}$ is a partition of ω into an array of infinite pairwise disjoint sets. The* continuous reading of names *game G between Players I and II proceeds as follows. First Player I indicates and initial condition $p_{ini} \in P_I$. After that, the moves alternate, Player I producing a name τ_n for a natural number and Player II answering with a basic open set $O_n \in \mathcal{O}$. Player II wins if the result of the play, the expression*

$$p_{ini} \wedge \bigwedge_i \bigwedge_j |\dot{x}_{gen} \in \bigcup_{n \in a_{i,j}} O_n \leftrightarrow \tau_i = j|$$

denotes a nonzero element in the complete Boolean algebra $RO(P_I)$.

Theorem 3.10.19. *Suppose that the forcing P_I is proper. The following are equivalent:*

1. *P_I fails to have the continuous reading of names;*
2. *Player I has a winning strategy in the game G.*

Moreover, if suitable large cardinals exist and the ideal I is universally Baire then the game G is determined.

Proof. On one hand, if P_I fails to have the continuous reading of names then there is an I-positive Borel set $B \subset X$ and a Borel function $f : B \to \omega^\omega$ such that f is not continuous on any Borel I-positive subset of B. Player I will then win by setting $p_{ini} = B$ and $\tau_i = f(\dot{x}_{gen})(i)$ and ignoring the moves of Player II altogether. If the result of such a play was positive, with an I-positive Borel set $C \subset B$ below it, it is clear that for every $i, j \in \omega$ it is the case that for all but I-many points $x \in C$,

$f(x)(i) = j \leftrightarrow x \in \bigcup_{n \in a_{i,j}} O_n$. Throwing out these countably many *I*-small sets from *C*, the remainder is an *I*-positive Borel subset $C' \subset C$ such that the *f*-preimages of basic open sets in ω^ω are relatively open in C'.

On the other hand, suppose that Player I has a winning strategy σ in the game *G* starting with a condition $p_{ini} \in P_I$. Let *M* be a countable elementary submodel of a large enough structure containing p_{ini}, *I* and σ and let *B* be the *I*-positive Borel set of the *M*-generic points meeting the condition p_{ini}. Enumerate all names for natural numbers in the model *M* by $\tau_i : i \in \omega$, and define $f : B \to \omega^\omega$ by $f(x)(i) = \tau_i/x$. This is a Borel function, and I claim that *f* is not continuous on any Borel *I*-positive subset of *B*. If $C \subset B$ was a Borel *I*-positive set such that $f \upharpoonright C$ was continuous, then Player II could defeat the strategy σ in the following way. For the *n*-th move Player I will make, find a number i_n such that this move is τ_{i_n}, for every number $j \in \omega$ find an open set $O_{i_n,j} \subset X$ such that $\forall x \in C\ f(x)(i_n) = j \leftrightarrow x \in O_{i_n,j}$, and use the moves with indices in the set $a_{i_n,j} \subset \omega$ to cover the open set $O_{i_n,j} \subset X$ with its basic open subsets. It is clear than in the end the condition *C* is below the result of the play, and therefore Player I lost.

The determinacy of the game is handled as in the previous cases and I omit the argument. □

3.10.9 Preservation of Baire category

Definition 3.10.20. *Suppose P is a partial order. The* Baire category game *G between Players I and II proceeds as follows. First Player II indicates an initial condition $p_{ini} \in P$. After that, the moves alternate, Player I producing a condition p_n and Player II playing its strengthening $q_n \leq p_n$. Player I wins if the result of the play, the expression*

$$p_{ini} \wedge \bigwedge_m \bigvee_{n > m} q_n$$

denotes a nonzero element in the complete Boolean algebra RO(P).

Theorem 3.10.21. *Suppose that the forcing P is proper. The following are equivalent:*

1. P below some condition makes the set of the ground model reals meager;

2. Player II has a winning strategy in the game G.

Moreover, if suitable large cardinals exist and the forcing P is universally Baire and proper then the game G is determined.

Proof. (1)→(2) is easier. Suppose that $p \Vdash \dot{O}_n : n \in \omega$ are open dense subsets of the Cantor space such that $2^\omega \cap V \cap \bigcap_n \dot{O}_n = 0$. Then Player II will win by first indicating $p = p_{ini}$ and then playing conditions q_n in such a manner that there are finite binary sequences t_n such that

- $t_0 \subset t_1 \subset \ldots$
- $q_n \Vdash \dot{O}_{t_n} \subset \bigcap_{m \in n} \dot{O}_m$.

This is a winning strategy, since if the result of the play was nonzero, it would be a condition forcing the ground model real $\bigcup_n t_n$ into the comeager set $\bigcap_n \dot{O}_n$, contradicting the assumptions.

For the opposite direction, fix a winning strategy σ for Player II in the game and let M be a countable elementary submodel of a large enough structure. Let q be an M-master condition below the inital condition $p_{ini} \in P \cap M$ indicated by the strategy σ. Let T be the tree of all finite plays of the game T with moves in the model M in which Player II follows his strategy σ. Let $\dot{G}$ be the P-name for the generic filter and let $\dot{A} = \{\tau \in [T] : \exists^\infty n \; q_n(\tau) \in \dot{G}\} \subset [T]$, where $q_n(\tau)$ is the n-th condition indicated by Player II in the play τ. I will show that $q \Vdash \dot{A} \subset [T]$ is a comeager set which contains no ground model elements, and this will complete the proof.

To see that $\dot{A}$ is forced to be comeager it is enough to produce for each play $t \in T$ ending with Player II's move a two move extension in the tree T whose last move by Player II belongs to the generic filter. Note that $D_t = \{r \in P : \exists s \in P \; t^\frown s^\frown r$ is a play according to the strategy $\sigma\} \subset P$ is a dense set in the model M. Therefore the condition q forces $D_t \cap \dot{G} \cap M \neq 0$, yielding the desired extension of the play t. To see that no ground model infinite play $\tau \in [T]$ can belong to the set $\dot{A}$, note that if $r \leq q$ is a condition such that $\check{\tau} \in \dot{A}$ then r must be a condition stronger than the result of the game τ. This is impossible since the strategy σ is winning and therefore the result of the game τ is zero.

The determinacy of the game follows immediately from Fact 1.4.2. $\qquad\square$

3.10.10 Preservation of category basis

Definition 3.10.22. *Suppose P is a partial order. Say that P preserves category basis if every meager set in the extension is contained in a meager set coded in the ground model.*

Essentially by a theorem of Fremlin [2], 2.2.11, the preservation of category basis is equivalent to the conjunction of ω^ω-bounding and the preservation of nonmeagerness.

Definition 3.10.23. *Suppose P is a partial order. The category basis game G between players I and II proceeds as follows. First Player II indicates an initial* *condition $p_{ini} \in P$. After that, the moves alternate, Player I producing a condition p_n and Player II playing its strengthening $q_n \leq p_n$, after which Player I may or may not raise a flag. Let $n_0 < n_1 < \ldots$ be the rounds at which Player I raised the flag; we agree that $n_0 = 0$. Player I wins if he raised the flag infinitely many times and the result of the play, the expression*

$$p_{ini} \wedge \bigwedge_i \bigvee_{n_i \leq n < n_{i+1}} q_n$$

denotes a non-zero element in the algebra $RO(P)$.

Theorem 3.10.24. *Suppose that the forcing P is proper. The following are equivalent:*

1. P does not preserve category basis;
2. Player II has a winning strategy in the category basis game.

Moreover, if suitable large cardinals exist and the forcing P is universally Baire and proper then the game G is determined. If $P = P_I$ for some $\mathbf{\Pi}_1^1$ on $\mathbf{\Sigma}_1^1$ σ-ideal I then the game is determined without the use of the large cardinal assumptions.

Proof. First suppose that P does not preserve category basis, and find a name $\dot{O}_n : n \in \omega$ for a sequence of open dense subsets of the Cantor space 2^ω such that the set $\bigcap_n \dot{O}_n$ contains no comeager subset coded in the ground model. I will assume that each $\dot{O}_n$ is in fact a name for a dense open subset of $2^{<\omega}$. Player II will win in the following way. He enumerates the binary tree $2^{<\omega}$ by $t_m : m \in \omega$ and in each round $m \in \omega$ he finds a condition $q_m \leq p_m$ and a binary sequence $s_m \supset t_m$ so that $q_m \Vdash \check{s}_m \in \bigcap_{n \in m} \dot{O}_n$. Suppose for contradiction that this strategy is not winning and find a play according to it whose result is some condition $r \in P$. Consider the set $B = \{x \in 2^\omega : \exists^\infty m\ s_m \subset x\}$. A review of the definitions reveals that $B \subset 2^\omega$ is a comeager set and $r \Vdash \dot{B} \subset \bigcap_n \dot{O}_n$, contradicting the assumption on the open dense sets $\dot{O}_n : n \in \omega$.

Now suppose that Player II has a winning strategy σ. First I will argue that there must be also a winning strategy for Player II which disregards the flag.

Now let $p_{ini} \in P$ be the initial condition indicated by the strategy σ for contradiction assume that there is a condition $p \leq p_{ini}$ such that the forcing $P \upharpoonright p$ preserves category basis. Let M be a countable elementary submodel of a large enough structure containing the strategy τ. Let T be the countable tree of all plays according to the strategy σ with all moves in the model M. Let $q \leq p$ be an M-master condition. As in the previous proof, writing $\dot{A} = \{\tau \in [T] : \exists^\infty n\ q_n(\tau) \in \dot{G}\} \subset [T]$ it is the case that $q \Vdash \dot{A} \subset [T]$ is comeager. Since the forcing $P \upharpoonright p$ preserves category, there is a strengthening $q' \leq q$ and a play $\tau \in [T] \cap V$ such that $q' \Vdash \check{\tau} \in \dot{A}$. Let

$\dot{b} = \{n \in \omega : \tau(n) \in \dot{G}\} \subset \omega$. This is forced by q' to be an infinite subset of ω, and since the forcing $P \restriction p$ is bounding, there is a ground model infinite set $c \subset \omega$ and a condition $q'' \leq q$ such that $q'' \Vdash$ there is an element of $\dot{b}$ between any two distinct elements of $\check{c}$. Consider the play τ' which proceeds just like τ except that Player I raises flags at rounds indicated by the numbers in the set c. Since the strategy σ disregards the flag entirely, this is a play which observes the strategy σ. A review of the definitions shows that $q'' \in P$ is a condition smaller than the result of the play τ', thus Player I won, contradicting the assumption that σ was a winning strategy.

The determinacy of the game is clear in the context of large cardinals and universally Baire forcing P from Fact 1.4.2. The determinacy of the game in the $\mathbf{\Pi}^1_1$ on $\mathbf{\Sigma}^1_1$ case in ZFC is handled similarly to Theorem 3.10.7. I will produce a Borel variation of the game which in the case of the category preserving forcing is more difficult for Player I than the original game, show that if the forcing P_I preserves category basis then Player II has no winning strategy in the Borel game, and use a determinacy argument to find a winning strategy for Player I.

For the simplicity of notation assume that $X = 2^\omega$. Since the ideal I is $\mathbf{\Pi}^1_1$ on $\mathbf{\Sigma}^1_1$ the collection of compact I-positive subsets of X is analytic in $K(X)$. Let $T \subset \omega^{<\omega} \times (2^{<\omega})^{<\omega}$ be a tree that projects into the set of all trees $S \subset 2^{<\omega}$ such that $[S] \notin I$. The Borel category basis game is played in the following fashion. Player I indicates compact sets $p_n \in P_I$ and Player II answers with compact sets $q_n \subset p_n$. Moreover, Player II raises the flag after some rounds indexed by $0 = n_0 \in n_1 \in n_2 \in \ldots$, and at n-th round also indicates a node $t_n \in T$ such that all numbers used in t_n are smaller than n and $|t_n| \leq n$. Player I wins if he raised the flag infinitely many times and the nodes $t_n : n \in \omega$ form a branch through the tree T and its projection S, a binary tree, is covered by each of the sets $\bigcup\{q_n : n_k \leq n < n_{k+1}\}$ for every $k \in \omega$.

The payoff of this game is clearly Borel. Moreover, if the forcing P_I is bounding then compact sets are dense in it, and since smaller move is a better move for Player II, the Borel game is in such a case harder for Player I than the original game. I will show that if the forcing P_I preserves category basis then Player II does not have a winning strategy in the Borel game. A Borel determinacy argument the gives a winning strategy for Player I in the Borel game as well as in the original game.

Suppose that the forcing P_I preserves category basis and for contradiction assume that Player II has a winning strategy. Similarly to the work above first find a winning strategy σ for Player II which does not depend on the flag and the nodes of the tree T Player I is producing along the play. This is possible since there are only finitely many possibilities for the node t_n at each round n and, as always, a smaller move is a better move for Player II. The rest of the argument is essentially identical to the proof above. □

3.10.11 The Sacks property

The paper [86] proves a theorem similar to the previous section for the Lebesgue measure in place of category. I will state the result without proof.

Definition 3.10.25. *A forcing P has* Sacks property *if for every ground model nondecreasing function $f \in \omega^\omega$ diverging to infinity in the ground model and every function $g \in \omega^\omega$ in the extension there is a function $h : \omega \to [\omega]^{<\aleph_0}$ in the ground model such that $|h(n)| \leq f(n)$ and $g(n) \in h(n)$.*

[2] shows that a forcing has the Sacks property if and only if it strongly preserves outer Lebesgue measure: for every real number $\varepsilon > 0$, every open set of measure $< \varepsilon$ in the extension is covered by such an open set from the ground model.

Definition 3.10.26. *Suppose P is a partial order. The* measure basis game *G between Players I and II proceeds as follows. First Player II indicates an initial condition $p_{ini} \in P$. After that, the moves alternate, Player I producing a condition p_n and Player II playing its strengthening $q_n \leq p_n$, after which Player II may or may not raise a flag. Let $n_0 < n_1 < \ldots$ be the rounds at which Player I raised the flag; we agree that $n_0 = 0$. Player I wins if either the numbers $n_{i+1} - n_i : i \in \omega$ do not form a nondecreasing sequence diverging to infinity or else the result of the game, the expression*

$$p_{ini} \wedge \bigwedge_i \bigvee_{n_i \leq n < n_{i+1}} q_n$$

denotes a non-zero element in the algebra RO(P).

Theorem 3.10.27. *Suppose that the forcing P is proper. The following are equivalent:*

1. *P does not have the Sacks property;*
2. *Player II has a winning strategy in the measure basis game.*

Moreover, if suitable large cardinals exist and the forcing P is universally Baire and proper then the game G is determined. If $P = P_I$ for some $\mathbf{\Pi}^1_1$ on $\mathbf{\Sigma}^1_1$ σ-ideal I then the game is determined without the use of the large cardinal assumptions.

3.11 Ramsey properties

The Ramsey theory is a rich subject. In this section I define several definable Ramsey theoretic properties that may hold for various σ-ideals. Their verification is frequently very difficult and typically involves forcing fusion arguments.

3.11.1 Rectangular Ramsey properties

The rectangular Ramsey properties are directly related to side-by-side product forcing.

Definition 3.11.1. *σ-ideals I, J on respective Polish spaces X, Y are said to satisfy the rectangular Ramsey property $\mathrm{MRR}(I, J)$ if for every partition $B \times C = \bigcup_n D_n$ of a rectangle with Borel I-positive side $B \subset X$ and a Borel J-positive side $C \subset Y$ into countably many Borel sets, one of the sets D_n contains a rectangle $B' \times C'$ with a Borel I-positive side $B' \subset B$ and a Borel J-positive side $C' \subset C$. The Ramsey notation used is $(I^+, J^+) \to_B (I^+, J^+)^1_{\aleph_0}$. A similar definition is applied in the case of a product of finitely or countably many σ-ideals.*

It is immediately clear that if I, J are σ-ideals on Polish spaces X, Y satisfying the rectangular Ramsey property then the side-by-side product $P_I \times P_J$ is naturally isomorphic to a dense subset of $P_{I \times J}$ where $I \times J$ is the ideal on the space $X \times Y$ generated by Borel sets that do not contain a rectangle with Borel I, J-positive sides; and similarly for products of arbitrary countable collection of forcings. This feature greatly simplifies the understanding of the product forcing, as will become clear in Section 5.2. If in this case the forcing $P_I \times P_J$ is proper then the Ramsey property holds for partitions in which the sets $D_n : n \in \omega$ come from larger definability classes: analytic in ZFC, and universally Baire with a suitable large cardinal assumption. This follows from Proposition 3.9.10 applied to the forcing $P_{I \times J}$.

Before I indulge in a list of examples and nonexamples, it is worth mentioning that the typical method for proving the negation of the rectangular Ramsey property is to demonstrate a strong negative partition property:

Definition 3.11.2. *$(I^+, J^+) \not\to_B [I^+, J^+]^1_c$ is the statement that there is a rectangle $B \times C$ with Borel I-positive and J-positive sides respectively and a Borel function $f : B \times C \to 2^\omega$ such that the f-image of every subrectangle $B' \times C'$ with Borel I-positive and J-positive sides respectively is the full set 2^ω. Similarly for a product of finitely or countably many σ-ideals.*

Note that if a Borel function f witnesses $(I^+, J^+) \not\to_B [I^+, J^+]$ and $2^\omega = \bigcup_n E_n$ is a partition of the Cantor space into countably many sets then the sets $D_n \subset B \times C : n \in \omega$ given by $\langle x, y \rangle \in D_n \leftrightarrow f(x, y) \in C_n$ witness the relation $(I^+, J^+) \not\to_B (I^+, J^+)$.

Many examples of the rectangular Ramsey properties are proved in Section 5.2. They are surprisingly closely related to a part of the Cichoń diagram:

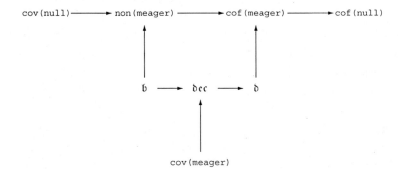

The upper right hand corner (meaning the forcings which do not increase the invariants in the upper right hand corner) is well-behaved, as exemplified in

Theorem 3.11.3. *(LC) (Theorem 5.2.6) If $I_n : n \in \omega$ is a sequence of universally Baire σ-ideals such that the forcings P_{I_n} are all proper and preserve category bases then* $\mathrm{MRR}(I_n : n \in \omega)$. *In the case of* $\mathbf{\Pi}^1_1$ *on* $\mathbf{\Sigma}^1_1$ *ideals this is provable in ZFC.*

The search for rectangular Ramsey theorems not implied by this theorem leads to the following train of thought. The invariant cof(meager) is well-known to be the maximum of non(meager) and $\mathfrak{d}$. Thus the forcings which increase it must increase either non(meager) or $\mathfrak{d}$.

For the forcings which increase $\mathfrak{d}$ the positive results are

Example 3.11.4. [31] Let I be the σ-ideal generated by the compact subsets of the Baire space; thus P_I is the Miller forcing, the optimal way to increase $\mathfrak{d}$. $\mathrm{MRR}(I, I)$ holds.

Example 3.11.5. (Spinas, personal communication) Let I be the σ-ideal generated by the sets on which the Pawlikowski function is continuous. The forcing P_I is the Steprāns forcing of Section 4.2.3, the optimal way to increase $\mathfrak{d}_{cc}$. $\mathrm{MRR}(I, I)$ holds. The proof is close to that of [31]. Added in proof: this is not so clear.

The negative results are the following. They are all established later using a much stronger negative square bracket Ramsey property.

Example 3.11.6. [77] Let I be the σ-ideal generated by the compact subsets of the Baire space. Then $\neg\mathrm{MRR}(I, I, I)$ and in fact $[I^+, I^+, I^+] \not\rightarrow_B [I^+, I^+, I^+]^1_c$. It immediately follows that if J is a σ-ideal on a Polish space X such that the forcing

P_J adds an unbounded real then $\neg$MRR(J, J, J). To see this, fix a Borel J-positive set $B \subset X$ and a Borel function $f : B \to \omega^\omega$ such that f-preimages of compact sets are I-small. Let $\pi : (\omega^\omega)^3 \to \omega$ be a function witnessing the negative square bracket relation with I. A trivial diagram chasing argument shows that the function $\pi \circ f$ witnesses the same relation for the ideal J.

Example 3.11.7. Let I be the Laver ideal. The MRR(I, I) fails and in fact the negative square bracket relation holds. As in the previous example this means that whenever J is a σ-ideal such that the forcing P_J is proper and adds a dominating real then MRR(J, J) fails. This follows from Example 3.11.13 below.

Example 3.11.8. Suppose that I is the ideal of meager sets. Then MRR(I, I) and in fact $(I^+, I^+) \nrightarrow_B [I^+, I^+]^1_c$ holds. The proof closely follows the case for Lebesgue measure below.

The forcings that increase the cardinal non(meager) are not as well understood from the Ramsey point of view.

Example 3.11.9. Let I be the Lebesgue null ideal on the unit interval. Then MRR(I, I) fails and in fact $(I^+, I^+) \nrightarrow_B [I^+, I^+]^1_c$ holds. To see this, let $r_x : x \in 2^\omega$ be a perfect set of reals which are pairwise not a rational multiple of each other. Let $f : [0, 1] \times [0, 1] \to \mathbb{R}$ be defined by $f(s, t) = s - t$. By an old result of Steinhaus [2], 3.2.10, for Borel I-positive sets B, C the set $B - C$ of all differences contains an interval. Now let $a_x : x \in 2^\omega$ be a Borel collection of pairwise disjoint dense subsets of $\mathbb{R}$ and define the map $g : [0, 1] \times [0, 1] \to 2^\omega$ by $g(s, t) = x$ if $s - t \in a_x$. Clearly the function g witnesses the negative square bracket partition relation.

The subject of asymmetric rectangular Ramsey theorems (those where $I \neq J$) offers much more variation.

Example 3.11.10. (LC) If I is a universally Baire σ-ideal on a Polish space X such that P_I is proper and preserves Lebesgue measure basis, and if J is a polar ideal then MRR(I, J). In the case of a Π^1_1 on Σ^1_1 ideal I this is provable in ZFC.

Example 3.11.11. (LC) (Theorem 5.2.8) If I is a universally Baire σ-ideal on a Polish space X such that P_I is proper and preserves category basis, and if J is an ideal such that P_J is proper and preserves category then MRR(I, J).

Example 3.11.12. [16] If X, Y are Polish spaces equipped with respective continuous submeasures ϕ and ψ then $[I^+_\phi, I^+_\psi] \nrightarrow_B [I^+_\phi, I^+_\psi]$.

Example 3.11.13. (LC) If I is a σ-ideal on a Polish space X such that P_I is proper then the following are equivalent:

1. P_I is bounding and does not add splitting reals;
2. $MRR(I, J)$ where J is the Mathias ideal;
3. $MRR(I, K)$ where K is the Laver ideal;
4. $[I^+, K^+] \to_B [I^+, K^+]^1_c$.

Let me just show here that if P_I adds a splitting real then the square bracket relation with the Laver ideal fails. Suppose first that P_I adds a splitting real and let $B \in P_I$ be a condition and $f : B \to 2^\omega$ be a Borel function such that for every bit $b \in 2$ and every infinite set $a \subset \omega$ the set $\{x \in B : \forall n \in a \ f(x)(n) = b\}$ is in the ideal I. Let $g : \omega^\omega \times B \to 2^\omega$ be the function defined by $g(y, x) = f(x) \circ y$. I will show that the g-image of every rectangle $D \times E \subset \omega^\omega \times B$ with Borel Laver-positive and I-positive Borel sides contains a nonempty open set. The same trick as in Example 3.11.9 then finishes the argument. It is in fact clear from the proof that Laver$\times P_I$ adds a Cohen real and a similar construction will work for every poset P_J adding a dominating real in place of the Laver forcing.

Suppose that $D \times E$ is a rectangle with Borel positive sides. Let $T \subset \omega^{<\omega}$ be a Laver tree such that $[T] \subset D$ and let $t_0 \in T$ be its trunk. Let $E_0 \subset E$ be some I-positive Borel set such that $f(x) \circ t_0$ is the same binary sequence $u \in 2^{<\omega}$ for all $x \in E_0$. I will show that $O_u \subset g''(D \times E)$. Let $z \in 2^{<\omega}$ be some infinite sequence extending u; I will find points $x \in E_0$ and $y \in [T]$ such that $g(y, x) = z$. Choose a winning strategy σ for the Nonempty player in the precipitous game with the ideal I as in Section 3.10.1. By induction on $n \in \omega$ build sequences $t_n \in T$ and sets $E_n, E'_n \in P_I$ such that

- $t_0 \subset t_1 \subset \ldots$
- for every point $x \in E_n$ the function $f(x) \circ t_n$ equals $z \restriction dom(t_n)$;
- $E_0, E'_1, E_1, E'_2, E_2, \ldots$ is a play of the precipitous game in which the Nonempty player; E'_0 is not defined follows his strategy σ.

Suppose that t_n, E_n have been found. Let $E'_{n+1} \in P_I$ be the move dictated to the Nonempty player at this stage. The set $a_n = \{m \in \omega : t_n \frown m \in T$ and for both $b = 0, 1$ the set $\{x \in E'_{n+1} : f(x)(m) = b\}$ is I-positive$\}$ must be infinite by the assumption on the function f. Choose any number $m \in a_n$ and let $t_{n+1} = t_n \frown m$ and $E_{n+1} = \{x \in E'_{n+1} : f(x)(m) = z(|t_n|)\}$. The induction hypotheses are satisfied.

In the end, let $y = \bigcup_n t_n \in D$ and let $x \in E$ be some point in the intersection $\bigcap_n E_n$. Clearly, $g(y, x) = z$.

The ultimate delimitative negative result in this subject is

Example 3.11.14. There is a c.c.c. ideal I such that letting J be the ideal of countable sets, $[I^+, J^+] \not\to [I^+, J^+]^1_c$. Let P be the eventually different real forcing

[2], 7.4.B, and let I be its associated σ-ideal on ω^ω, so that P is in the forcing sense equivalent to P_I. Let J be the ideal of countable sets. $MRR(I, J)$ fails. The easiest way to see that is to choose a perfect subset $X \subset \omega^\omega$ consisting of mutually eventually different functions, and define a function $f: \omega^\omega \times X \to 2^\omega$ by letting $f(y, x)$ to be the sequence of parities of elements of the set $\{n \in \omega : x(n) = y(n)\}$. It turns out that the f-image of every rectangle with Borel I-positive and J-positive sides respectively, contains a nonempty open subset of 2^ω. Just like in the previous arguments this means that $[I^+, K^+] \to [I^+, K^+]^1_c$ fails for every σ-ideal K containing all singletons.

3.11.2 Square Ramsey properties

Quite a different keg of fish are the square Ramsey properties.

Definition 3.11.15. *Let I be a σ-ideal on a Polish space X. Let n, k, l be natural numbers. The shorthand $I^+ \to_B (I^+)^n_{k,l}$ denotes the following statement: for every Borel I-positive set B and every Borel function $f: [B]^n \to k$ there is a Borel I-positive set $C \subset B$ such that $|f''[C]^n| \leq l$.*

The shorthand $I^+ \to_B [I^+]^n_c$ is the statement that for every Borel I-positive set $B \subset X$ and every Borel function $f: [B]^n \to 2^\omega$ there is a Borel I-positive set $C \subset B$ such that $f''[C]^n \neq 2^\omega$.

Example 3.11.16. The archetype of square Ramsey theorems is the theorem of Blass dealing with the ideal I of countable sets on 2^ω. Blass proved that $I^+ \to_B (I^+)^2_{k,1}$ holds for every natural number k, and for every natural number $n \in \omega$ there is $l(n) \in \omega$ such that $I^+ \to_B (I^+)^n_{k,l(n)}$ holds for every natural number k.

Yuan-Chyuan Sheu proved a parallel theorem for the c_{min} ideal and computed the numbers $l(n)$. Let me include one negative result.

Example 3.11.17. (Blass, personal communication) Let I be the E_0 ideal on 2^ω of Section 4.7.1. Then $I^+ \not\to_B [I^+]^3_c$. For the proof, suppose $x_0, x_1, x_2 \in 2^\omega$ are distinct infinite binary sequences. Define $f(x_0, x_1, x_2) \in 2^\omega$ in the following way. By induction on the number $n \in \omega$ construct numbers $m_n \in \omega$ and bits $f(x_0, x_1, x_2)(n)$. Let $m_0 =$ the least number k such that $x(k), y(k), z(k)$ are not all equal, let $i_0 \in 3$ be such that the bit $x_{i_0}(m_0)$ is different from the other two, and let $f(x_0, x_1, x_2)(0) = x_{i_0}(m_0)$. Suppose the numbers m_n and i_n are known. Then $m_{n+1} =$ the least number $k > m_n$ such that the two bits $\{x_i(k) : i \in 3, i \neq i_n\}$ are distinct, let $f(x_0, x_1, x_2)(n+1) = x_{i_n}(m_{n+1})$, and let $i_{n+1} \in 3$ be the number j for which $x_{i_n}(m_{n+1}) \neq x_j(m_{n+1})$. This completes the definition of the function $f: [2^\omega]^3 \to 2^\omega$.

It is easy to see that the function f is Borel. In order to prove that it witnesses the negative partition relation, I must show that for every E_0 tree $T \subset 2^{<\omega}$ and every point $y \in 2^\omega$ there are branches x_0, x_1, x_2 through the tree T such that $f(x_0, x_1, x_2) = y$. This is left to the reader.

3.11.3 Canonical Ramsey theorems

Still different issues arise in the subject of canonical Ramsey theorems, both rectangular and square.

Definition 3.11.18. *Suppose that I, J are σ-ideals on Polish spaces X, Y respectively, and let F be a set of Borel equivalence relations on $X \times Y$. The shorthand $(I^+, J^+) \to^c_B (I^+, J^+)_F$ stands for the statement that for every Borel I and J positive sets B and C respectively and every Borel equivalence relation e on $B \times C$ there are Borel I and J positive sets $B' \subset B$ and $C' \subset C$ respectively and an equivalence relation $f \in F$ such that $e \restriction B' \times C' = f \restriction B' \times C'$. Similar notation will be used for the square canonization properties.*

As always in canonical Ramsey theory it is the nature of the set F that makes all the difference. The rectangular canonical theorems have a tight connection with the degree structure of the product forcing extension. It is for example well-known that $(I^+, I^+) \to^c_B (I^+, I^+)_F$ holds for $I =$ the ideal of countable sets on the Cantor space and $F = \{E_0, E_1, E_2, E_3\}$, where E_0 is the identity equivalence relation, E_1 is the identity on the first coordinate, E_2 is the identity on the second coordinate, and E_3 is the equivalence relation with a single class. This result implies that in the product Sacks forcing extension, there are four V-degrees of reals: the trivial degree, the degree of the first generic, the degree of the second generic, and the degree of a real coding the two generics. The canonical Ramsey theorem is apparently properly stronger than the classification of degrees in the product forcing extension though.

As the last remark, note that the relation $(I^+, J^+) \not\to_B [I^+, J^+]$ implies the failure of the canonical Ramsey relation for any reasonable family F of equivalence relations. If $f : B \times C \to 2^\omega$ is a Borel function witnessing the negative square bracket relation, then for every Borel equivalence E on 2^ω define a Borel equivalence E' on $B \times C$ by $(x_0, y_0)E'(x_1, y_1)$ if $f(x_0, y_0)Ef(x_1, y_1)$, and note that the complexity of the relation E' remains the same at every rectangle with Borel positive sides.

3.12 Pure decision property

The pure decision is a property frequently used in certain types of forcing arguments. This section offers its topological restatement.

Definition 3.12.1. *[87] Let I be a σ-ideal on some Polish space X with a fixed metric d. We say that the poset P_I has the* pure decision property *or* contractive reading of names *(with respect to the metric d) if for every I-positive Borel set $B \subset X$ and every Borel map $f : (B, d) \to (Y, e)$ into a compact metric space there is a Borel I-positive set $C \subset B$ on which the map f is a contraction.*

Example 3.12.2. The Laver forcing has the pure decision property in the standard representation, with respect to the metric of least difference on ω^ω: $d(x, y) = 2^{-n}$ where n is the smallest number where the functions $x, y \in \omega^\omega$ differ.

Proof. Let I be the Laver σ-ideal on the Baire space. Let B be Borel I-positive set and $f : (B, d) \to (Y, e)$ be a Borel map into a compact metric space. Thinning out the set B if necessary we may assume that $B = [T]$ for some Laver tree $T \subset \omega^{<\omega}$. To simplify the notation assume that T has an empty trunk.

Before proceeding recall the well known fact that for every Laver tree S and Borel partition $[S] = \bigcup_{i \in n} A_i$ into finitely many pieces there is a Laver tree $U \subset S$ with the same trunk such that the set $[U]$ is included in one of the pieces of the partition.

Now for every n find a finite 2^{-n-1}-network $y_n \subset Y$, that is, a set such that every point of the space Y is 2^{-n-1}-close to one of its elements. By induction on $n \in \omega$ build a fusion sequence of Laver trees T_n so that $T_0 = T$, T_{n+1} agrees with T_n on sequences of length $n + 1$ and for every such a sequence $t \in T_n$ there is an element $x_t \in y_n$ such that for every path r through T_{n+1} extending the sequence t, the element $f(r) \in Y$ is 2^{-n-1}-close to x_t. This is possible by the observation in the previous paragraph. Note that by the triangle inequality this means that for two such paths r_0, r_1 the elements $f(r_0), f(r_1) \in Y$ will have e-distance $\leq 2^{-n}$. Let S be the fusion of the sequence of trees T_n. It is not difficult to see that the set $C = [S]$ has the required properties. $\qquad\qquad\square$

4

Examples

4.1 Ideals σ-generated by closed sets

This is the historically first class of ideals discovered to generate proper forcings.

Definition 4.1.1. *A σ-ideal I is σ-generated by closed sets if every set in I is contained in an F_σ set in I.*

Several results of this section are in fact special cases of the results of Section 4.2 concerning the abstract porosity ideals. Nevertheless I decided to treat them separately here because the abstract porosity ideals form a much more complex family.

One feature, which sets this class of ideals apart from the other classes considered in this chapter, is that it is not invariant under different presentations. Every forcing of the form P_I where I is an ideal on a Polish space X σ-generated by closed sets has a presentation P_J on the Baire space ω^ω such that J is σ-generated by closed sets. To see this note that the set X is a one-to-one continuous image of a closed subset C of the Baire space, $X = f''C$. The ideal J on ω^ω generated by $\omega^\omega \setminus C$ and the preimages of I-small sets is σ-generated by closed sets since the open set $\omega^\omega \setminus C$ is F_σ and f-preimages of closed sets are closed. Clearly P_I is naturally isomorphic to P_J.

4.1.1 General results

Theorem 4.1.2. *[82] Suppose that I is a σ-ideal on a Polish space X generated by closed sets. The forcing P_I is proper. The forcing P_I preserves the Baire category and has the continuous reading of names.*

Proof. Fix a countable basis $\mathcal{O}$ of the space X. Suppose that M is a countable elementary submodel of a large enough structure and $B \in P_I \cap M$ is a condition. I must show that the set $C = \{x \in B : x \text{ is } M\text{-generic}\}$ is I-positive. To this end, fix a countable collection $F_n : n \in \omega$ of closed sets in the ideal I and find a point $x \in C \setminus \bigcup_n F_n$. To produce the point x, enumerate the open dense subsets of the forcing P_I in the model M by $D_n : n \in \omega$ and by induction build conditions $B_n : n \in \omega$ so that

- $B = B_0 \supset B_1 \supset \ldots$
- $B_{n+1} \in D_n \cap M$
- $B_{n+1} \cap F_n = 0$

This is easy to do. Suppose that $B_n \in M$ has been obtained. There must be a basic open set $O \in \mathcal{O}$ such that $B_n \cap O$ is disjoint from F_n and I-positive, otherwise the set $B_n = \bigcup \{B_n \cap O : O \cap F_n = 0, O \in \mathcal{O}\} \cup F_n$ would be in the ideal I. The set $B_n \cap O \in M$ is a condition, by an elementarity argument there must be a set $B_{n+1} \subset B_n \cap O$ in the set D_{n+1}.

In the end, the filter generated by the conditions $B_n : n \in \omega$ is M-generic, and by an application of Proposition 2.1.2 in the model M the intersection $\bigcap_n B_n$ contains a single point x. Clearly $x \in C \setminus \bigcup_n F_n$ and the forcing P_I is proper.

For the continuous reading of names suppose that $B \notin I$ is a Borel set and $f : B \to 2^\omega$ is a Borel function. By a result of Solecki [69], there is an I-positive G_δ subset $B_0 \subset B$; throwing out all sets $O \cap B_0$ such that $O \in \mathcal{O}$ and $O \cap B_0 \in I$ I may assume that for every open set $O \subset X$ the intersection $O \cap B_0$ is either I-positive or empty. Then every I-small closed set $F \subset X$ is nowhere dense in B_0 and so the ideal I is included in the ideal of meager subsets of B_0. The set $C \subset X$ is G_δ and therefore Polish in the inherited topology. The function f is continuous on some comeager subset $B_1 \subset B_0$–[40], 8.38. Clearly, $B_1 \notin I$ and the function $f \restriction B_1$ is continuous as desired.

The category preservation follows similar lines. Suppose that $B \in P_I$ is a Borel set, $C \subset 2^\omega$ is a nonmeager Borel set, and $D \subset B \times C$ is a set with meager vertical sections. I must find a horizontal section of the complement of D which is not in the ideal I. As before, I can assume that the set B is in fact G_δ and the ideal I below B is contained in the ideal of meager subsets of B. Since the set B is a Polish space in the inherited topology, the Kuratowski–Ulam theorem [40], 8.41(iii), says that there is a horizontal section of the complement of the set D which is nonmeager in the set B and therefore I-positive as desired. □

Theorem 4.1.3. *Suppose that I is a σ-ideal generated by closed sets on a Polish space X.*

1. *I satisfies the third dichotomy, in fact every analytic I-positive set contains a G_δ I-positive subset.*
2. *In the choiceless Solovay model I satisfies the first dichotomy and it is closed under wellordered unions.*
3. *(ZF+DC+AD+) I satisfies the first dichotomy and it is closed under wellordered unions.*

Proof. The first item is a result of Solecki [69]. The second item is a result of DiPrisco and Todorcevic [13]; a generalization is proved in Section 4.2. For the third item, fix a countable topology basis $\mathcal{O}$ for the space X and let $A \subset X \times 2^\omega$ be a set. I will show that if $\text{proj}(A) \notin I$ then there is an analytic set $B \subset A$ such that $\text{proj}(B) \notin I$ and then use Proposition 3.9.18. Fix a Borel bijection $\pi : 2^\omega \to X \times 2^\omega$ and consider the game $H(I, A)$ in which Player I gradually produces open sets $O_n : n \in \omega$ and Player II produces a sequence $z \in 2^{\leq\omega}$; if z is infinite let $\pi_1(z) \in X$ be the first coordinate of the point $\pi(z) \in X \times 2^\omega$. At round n Player I must decide which of the first n basic open sets in $\mathcal{O}$ are a subset of $O_m : m \in n$, however Player II is allowed to postpone placing new bits on his sequence arbitrarily. The winning condition for Player II is (if the sequence z is infinite then $\pi(z) \in A$) and (if all sets $X \setminus O_n : n \in \omega$ are in the ideal I then z is infinite and $\pi_1(z) \in \bigcap_n O_n$).

Claim 4.1.4. *Player I has a winning strategy if and only if* $\text{proj}(A) \in I$.

Proof. If $\text{proj}(A) \in I$ then Player I will win simply by producing some open sets $O_n : n \in \omega$ such that $X \setminus O_n \in I$ and $\text{proj}(A) \subset \bigcup_n(X \setminus O_n)$. On the other hand suppose that $\text{proj}(A) \notin I$ and σ is a strategy for Player I. I must produce a winning counterplay against the strategy σ.

For each finite sequence τ that is a position of the game $H(I, A)$ in which Player I followed the strategy σ, let $O_n(\tau) : n \in \omega$ be the n-th open set the strategy σ will produce if Player II places no new bits on his sequence z after τ as the play extends to infinity. If there is a number $n \in \omega$ such that $X \setminus O_n(\tau) \notin I$, then Player II won this infinite extension of τ and we are done. Suppose this does not happen for any play τ, and use the assumption on the set A to find a sequence $z \in 2^\omega$ such that $\pi(z) \in A$ and $\pi_1(z) \in \bigcap_{\tau,n} O_n(\tau)$. I will produce a play against the strategy σ in which Player II produces the sequence $z \in 2^\omega$ and wins.

By induction on n build the initial segments τ_n of the play so that the point $\pi_1(z) \in X$ belongs to one of the open sets the strategy σ decided to put in O_n in the play τ_{n+1}. This is easily possible: Let $\tau_0 = 0$ and once τ_n has been constructed, Player II just waits long enough after τ_n, adding no bits on his sequence, until the strategy σ puts a basic open set into O_n which contains the point $\pi_1(z)$. After that,

Player II adds another bit on the sequence z, and that will conclude the construction of the play τ_{n+1}. In this way, Player II clearly wins. □

Now suppose that $\text{proj}(A) \notin I$. To produce an analytic set $B \subset A$ such that $\text{proj}(B) \notin I$, just apply the previous claim with a determinacy argument to find a winning strategy σ for Player II in the game $H(I, A)$ and let $B = \{\pi(z) \in A :$ for some counterplay by Player II the strategy σ produces the sequence $z\}$. Clearly, this is an analytic subset of A by its definition and by the winning condition for Player II. Moreover, $\text{proj}(B) \notin I$ since the strategy σ remains winning in the game $H(I, B)$ and the claim can be used for the set B. □

Theorem 4.1.5. *If I is a σ-ideal generated by closed sets, then the forcing P_I is embeddable into σ-closed*c.c.c. iteration.*

This is proved in Section 4.2.

Theorem 4.1.6. *Suppose that I is a σ-ideal on a Polish space X σ-generated by closed sets. Then P_I forces that every function $\dot{f} : \omega_1 \to 2$ has an infinite subfunction in the ground model.*

Proof. Suppose $B \Vdash \dot{f} : \omega_1 \to 2$ is a function. Let σ be a winning strategy for the Nonempty Player In the Borel precipitous game of Section 3.10.1. Choose a large enough structure and let $\vec{M}$ be a continuous $\in$-tower of countable elementary submodels of this structure of length ω_1 such that $I, \sigma, B, \dot{f} \in \vec{M}(0)$. For a Borel set $C \subset B$ and an ordinal $\alpha \in \omega_1$ let $C(\alpha) = \{x \in C : x \text{ is } \vec{M}(\beta)\text{-generic for all } \beta \leq \alpha\}$, and for a bit $b \in 2$ let $C(\alpha + 1, b) = \{x \in C(\alpha + 1) : \dot{f}(\vec{M}(\alpha) \cap \omega_1)/x = b\}$. A bootstrapping argument based on the proof of Theorem 4.1.2 shows that if $C \in P_I \cap \vec{M}(0)$ then $C(\alpha) \notin I$, and so $C(\alpha) \in \vec{M}(\alpha + 1)$. A review of the definitions reveals that the Borel set $C(\alpha + 1, b)$, if I-positive, is a condition forcing $\dot{f}(\vec{M}(\alpha) \cap \omega_1) = b$.

There are two separate cases:

- there is an ordinal $\beta \in \omega_1$ and a condition $C \subset B(\beta)$, $C \in \vec{M}(\beta + 1)$ such that for uncountably many ordinals $\alpha \in \omega_1$ the set $C(\alpha + 1, 1)$ is in the ideal I;
- otherwise.

In the first case, fix C, let $\alpha_n : n \in \omega$ be infinitely many ordinals as in the description of the first item, let $\alpha = \bigcup_n \alpha_n$ and let $D = C(\alpha) \setminus \bigcup_n C(\alpha_n + 1, 1)$. The set D is a difference of an I-positive and I-small set, therefore a condition in the poset P_I and it forces $\dot{f} \upharpoonright \{\vec{M}(\alpha_n) \cap \omega_1 : n \in \omega\}$ is constantly zero.

In the second case, use the failure of the first case to find an uncountable set $A \subset \omega_1$ such that for every $\alpha \in A$, every ordinal $\beta \in \alpha$ and every condition $C \subset B(\beta)$ in $P_I \cap \vec{M}(\beta + 1)$ the set $C(\alpha + 1, 1)$ is I-positive. Let $\alpha_n : n \in \omega$ enumerate the first

ω many elements of the set A. I will show that $D = \bigcap_n B(\alpha_n + 1, 1) \notin I$ which will complete the argument for the second case since $D \Vdash \dot{f} \upharpoonright \{\breve{M}(\alpha_n) \cap \omega_1 : n \in \omega\}$ is constantly equal to one. Let $F_n : n \in \omega$ be a countable collection of closed sets in the σ-ideal I. To find a element $x \in D \setminus \bigcup_n F_n$, by induction on $n \in \omega$ build conditions $p_n : n \in \omega$ so that

- $B = p_0 \supset p_1 \supset \ldots$, and there are conditions $q_0, q_1 \ldots$ such that the sequence $p_0, q_0, p_1, q_1, \ldots$ is a play of the Borel precipitous game in which the Nonempty player follows his strategy σ; in this way $\bigcap_n p_n \neq 0$ and the point $x \in X$ will be an arbitrary point in the intersection;
- $p_{n+1} \in \breve{M}(\alpha_n + 1)$ and $p_{n+1} \subset B(\alpha_n + 1, 1) \setminus F_n$.

Suppose that the condition p_n has been constructed, and let $q_n \subset p_n$ be the answer dictated by the strategy σ to the Nonempty player. By an elementarity argument, $q_n \in \breve{M}(\alpha_{n-1})$. Since F_n is a closed set in the ideal I there must be a basic open set O such that $q_n \cap O \notin I$ and $F_n \cap O = 0$. Now let $p_{n+1} = (q_n \cap O)(\alpha_n + 1, 1) = q_n \cap O \cap B(\alpha_n + 1, 1)$ which is still I-positive by the choice of the ordinal α_n. This completes the inductive construction. In the end, any element in the nonempty intersection $\bigcap_n p_n$ shows that the set D is not covered by the sets $F_n : n \in \omega$ as desired.   □

Note that the proof really gives a bit more. If $\mathrm{add}(\mathrm{meager}) > \aleph_1$ then either there is an infinite ground model set $b \subset \omega_1$ such that $f \upharpoonright b$ is constantly zero, or there is an uncountable set $b \subset \omega_1$ such that $f \upharpoonright b$ is constantly one.

Theorem 4.1.7. *(LC) Suppose that I is a universally Baire σ-ideal on a Polish space X σ-generated by closed sets. Suppose that $V \subseteq V[H] \subseteq V[G]$ are the ground model, the P_I-extension, and an intermediate extension. Then $V[H]$ is either equal to V, or it is an extension of V by a single Cohen real, or it is equal to $V[G]$.*

Proof. The first step is to show that the intermediate extension is c.c.c. This immediately follows from Proposition 3.9.2 and Theorem 4.1.2. These two use large cardinal assumptions, however a manual ZFC construction will give the same conclusion in the case that the ideal I is generated by an analytic collection of closed sets.

Suppose now that $Q \subset RO(P_I)$ is a complete c.c.c. algebra. The second step in the proof is to show that Q is countably generated. This uses no large cardinal or definability assumptions. Suppose for contradiction that Q is not countably generated, and by induction on $\alpha \in \omega_1$ build an increasing chain of countably generated complete subalgebras $Q_\alpha \subset Q$ and elements $b_\alpha \in Q_{\alpha+1}$ such that b_α is not decided by any element of Q_α. Let $f : \omega_1 \to 2$ be a function in the generic extension given by $f(\alpha) = 1$ if b_α is in the generic filter. An elaboration of the argument in the previous theorem shows that there are three possibilities:

- there is an uncountable set $A \subset \omega_1$ and a descending chain of conditions $B_\alpha : \alpha \in \omega_1$ in P_I such that $B_\alpha \Vdash \forall \beta \in A \cap \alpha \, f(\alpha) = 1$;
- there is an uncountable set $A \subset \omega_1$ and a descending chain of conditions $B_\alpha : \alpha \in \omega_1$ in P_I such that $B_\alpha \Vdash \forall \beta \in A \cap \alpha \, f(\alpha) = 0$;
- there is a countable set $\{\alpha_n : n \in \omega\}$ of countable ordinals such that for every function $g \in 2^\omega$ there is a condition $B \in P_I$ forcing $\dot{f}(\alpha_n) = 1 \leftrightarrow g(n) = 1$.

I will derive a contradiction in all the three cases. In the first and second case, note that $Q \Vdash \dot{f} \upharpoonright A$ is not constant. To see this let $R = \bigcup_\alpha Q_\alpha$. By the c.c.c. this is a complete subalgebra of Q, $\dot{f}$ is an R-name and no condition $r \in R$ decides uncountably many values of $\dot{f}$ since for some ordinal α, $r \in Q_\alpha$ and then r fails to decide all the values past α. Now by the c.c.c. again there is a fixed ordinal $\alpha \in \omega_1$ such that R (or Q or P_I) forces $\dot{f} \upharpoonright \check{A} \cap \check{\alpha}$ is not constant. However, this contradicts the fact that the condition B_α forces $\dot{f} \upharpoonright \check{A} \cap \check{\alpha}$ to be constant. The third case uses c.c.c. again–since $\dot{f} \upharpoonright \{\alpha_n : n \in \omega\}$ is a Q-name and Q is c.c.c. there are at most countably many ground model functions that can be forced equal to it, however the third item ascertains that there are continuum many of them.

Thus the algebra Q is countably generated, and since it is proper, its generic extension is given by a single binary sequence $y \in 2^\omega$. The last step is to show that $V[y]$ is a Cohen extension of V. Consider the σ-ideal $J = \{B \subset 2^\omega : Q \Vdash \dot{y} \notin \dot{B}\}$. Since Q is c.c.c., so is the forcing P_J and so the real y must be P_J-generic. Now P_J is a definable c.c.c. forcing which preserves Baire category since P_I does, so by Corollary 3.5.7, P_J must be equivalent to the Cohen forcing. The large cardinal and definability assumption used in that corollary can be in this case eliminated by manual construction of the winning strategies necessary–work which is implicitly present in the proof of Theorem 4.1.2. □

There is a large subclass of σ-ideals which have much stronger preservation properties.

Theorem 4.1.8. *Suppose that the σ-ideal I on a compact metric space X is generated by a σ-compact collection of compact sets. Then the forcing P_I is bounding and does not add splitting reals.*

This class of forcings includes, among others, the partial orders of the scheme called $\mathbb{Q}_0$ in [58], Section 1.3. Let me include a very brief exposition. Suppose that T is a finitely branching tree, for every node $t \in T$ let a_t be the set of its immediate successors, and let $\phi_t : \mathcal{P}(a_t) \to \mathbb{R}^+$ be monotone functions such that the numbers

$\phi_t(a_t)$ tend to ∞. Let P be the partial order of all trees $S \subset T$ such that for every number $n \in \omega$ and every node $t \in S$ there is an extension $u \in S$ such that writing $a_u^S \subset a_u$ for the set of all of its immediate successors in the tree S, it is the case that $\phi_u(a_u^S) \geq n$. Now let I be the ideal on the space $[T]$ generated by all closed sets $[S]$ where $S \subset T$ is a tree such that all the set $\{a_u^S : u \in S\} \subset \mathbb{R}^+$ is bounded. It is not difficult to see that I is generated by a σ-compact collection of compact sets and the determinacy argument below shows that an analytic set is I-positive if and only if it contains all branches of some tree $S \in P$. Thus the forcings P_I, P are in the forcing sense equivalent.

Proof. For the simplicity of notation I will deal with the case $X = 2^\omega$. The σ-ideal I is generated by the union of compact families $F_n \subset K(X) : n \in \omega$ of closed sets. Taking finite unions I may and will assume that the sets $F_n : n \in \omega$ are increasing. There are two different proofs of the theorem offering different extra information.

The first proof has a familiar dichotomy at its heart. Call a binary tree T *large* if for every node $t \in T$ and every number $n \in \omega$ there is a number m such that no binary tree S such that $[S] \in F_n$ contains all the nodes of T_n below t. The following claim has been repeatedly rediscovered.

Claim 4.1.9. *Suppose that $A \subset X$ is an analytic set. Then either $A \in I$ or A contains all branches of an I-large tree. In the presence of AD this extends to all subsets of X.*

Proof. Let $A \subset X$ be an arbitrary set and consider a game $G(A)$ between Players I and II in which Player I produces binary trees $S_n : n \in \omega$ such that $[S_n] \in F_n$. Player II produces a point $x \in X$. Player II wins if $x \in A \setminus \bigcup_n C_n$. In order to complete the description of the game, I must specify the precise schedule for both players. At round n, Player I must indicate the n-th level of the trees $S_m : m \in n$. On the other hand, Player II can wait for an arbitrary finite number of rounds before placing another bit on his sequence x.

The same argument as in Theorem 4.1.3 now shows that $A \in I \leftrightarrow$ Player I has a winning strategy in the game $G(A)$. Suppose now that $A \notin I$ is a Borel set. Then Borel determinacy implies that Player II has a winning strategy σ in the game $G(A)$. The key point now is that the collection Y of all counterplays by Player I forms a compact set as the sets $F_n \subset K(X)$ are closed. Let $B = \sigma''Y$. This is a compact set since it is a continuous image of a compact set, it is a subset of the set A since the strategy σ is winning, and it is I-positive since the strategy σ remains a winning strategy in the game $G(B)$ for trivial reasons. Thus every Borel I-positive set $A \subset X$ has an I-positive compact subset $B \subset A$, $B = [T]$ for some binary tree T. Removing all the nodes $t \in T$ such that $[T \restriction t] \in I$ if necessary I arrive at an I-large tree as desired in the claim. The extension of this argument to analytic sets can be arranged through Solecki's result 4.1.3(1), and the extension to all sets under AD is immediate.   □

Thus the poset P_I has a dense subset naturally isomorphic to the forcing of all I-large trees ordered by inclusion. The theorem is now proved by familiar fusion arguments. In fact, one can prove that the forcing P_I preserves P-points. Suppose that U is a P-point ultrafilter on ω, T is an I-large tree, and $\dot{a}$ is a name for a subset of ω. Thinning out the tree T if necessary I may assume that there is a continuous function $f : [T] \to \mathcal{P}(\omega)$ so that $T \Vdash \dot{a} = \dot{f}(\dot{x}_{gen})$. There are two distinct cases:

- either there is an I-large tree $T' \subset T$ such that for all $x \in [T']$ the set $f(x)$ belongs to U;
- otherwise.

In the first case, replace T with T', the name $\dot{a}$ with the name for its complement, and proceed as in the second case. In the second case, I will produce an I-large tree $S \subset T$ and a set $b \in U$ such that $\forall x \in [S]$ $b \subset f(x)$. This will conclude the proof of preservation of P-points. Recall the P-point game between Players I and II: Player I plays sets $b_n : n \in \omega$ in the ultrafilter, and Player II answers with finite subsets $c_n \subset b_n$. Player I wins if $\bigcup_n c_n \notin U$. Since U is a P-point, Player I has no winning strategy [2], 4.4.4. I will describe a strategy σ for Player I such that after each move c_n he writes on the side a tree S_n such that these trees form a fusion sequence, and $S_n \Vdash \check{c}_n \subset \dot{a}$. Find a counterplay against this strategy winning for Player II, let $b = \bigcup_n c_n$ and $S = \bigcap_n S_n$ with the sets c_n and trees S_n obtained in this counterplay, and observe that S, b are as required.

In order to describe the strategy σ, let $S_{-1} = T$ and $m_{-1} = 0$. Once the tree S_{n-1} and the number m_{n-1} have been obtained, find a natural number m_n such that every node $s \in S_{n-1}$ of length m_{n-1} and for every compact set $C \in F_n$ there is and extension $t \in S_{n-1}$ of length m_n such that $O_t \cap C = 0$. For every node $t \in S_{n-1}$ of length m_{n-1} find a path $x_t \in [S_{n-1}]$ such that $t \subset x_t$ and $f(x_t) \in U$. The strategy σ then makes the move $b_n = \bigcap_t f(x_t)$. Once Player II responds with a finite set $c_n \subset b_n$, just use the continuity of the function f to find a number $k \in \omega$ such that for every node $t \in S_{n-1} \cap 2^{m_{n-1}}$ and every path $x \in [S_{n-1}]$ which agrees with x_t at the first k entries, $c_n \subset f(x)$. Let $S_n = \{s \in S_{n-1} : \exists t \in S_{n-1} \cap 2^{m_{n-1}} \ s$ is compatible with $x_t \upharpoonright k\}$. It is not difficult to see that the strategy σ has the required properties.

The second proof uses infinitary games on Boolean algebras, and it uses the following fact of independent interest.

Claim 4.1.10. *(LC) Suppose that J is a universally Baire σ-ideal such that P_J is proper. The following are equivalent:*

1. *P_J does not add a bounded eventually different real;*
2. *for every σ-ideal I σ-generated by a σ-compact collection of compact sets, $I \not\perp J$.*

The theorem follows easily from the claim. Consider the ideal J of sets nowhere dense in the algebra $\mathcal{P}(\omega)$ mod fin so that P_J is in the forcing sense equivalent to the Mathias forcing of Section 4.7.7 and it does not add a bounded eventually different real. The claim then says that $I \not\perp J$ which by the results of Section 3.4 is equivalent to the forcing P_I not adding unbounded or splitting reals. Of course it is the implication (1)→(2) which is most interesting from the forcing preservation point of view.

The implication (2)→(1) is easier. If P_J adds a bounded eventually different real then there is a σ-ideal I generated by an F_σ collection of compact sets such that $I \perp J$. Namely, suppose that $B \in P_J$ is a condition forcing that $\dot{g} \in \omega^\omega$ is a function pointwise dominated by some $\check{f} \in \omega^\omega$, yet eventually different from any ground model function. Thinning out the condition B if necessary we may assume that there is a Borel function $G : B \to \prod_n f(n)$ such that $B \Vdash \dot{g} = \dot{G}(\dot{x}_{gen})$. Now let I be the σ-ideal on $\prod_n f(n)$ which is σ-generated by sets $C_{g,n} = \{h \in \prod_n f_n : h$ is different from g at every input $\geq n\}$. This is an F_σ collection of compact sets. It turns out that $I \perp J$ as witnessed by the set $C \subset \prod_n f(n) \times B$, where $\langle h, x \rangle \in C$ if $|h \cap G(x)| = \aleph_0$. Clearly, the vertical sections of the set C are J-small since B forces the real $\dot{g}$ to be eventually different from any given ground model function h. The horizontal sections of the complement of the set C are I-small since P_I forces the generic real to have infinite intersection with any function $G(x)$ for x in the ground model.

Now the (1)→(2) implication. For definiteness assume that the σ-ideal I is on the Cantor space 2^ω, and fix compact sets $K_n : n \in \omega$ of compact subsets of 2^ω, the generators for the ideal I. I may and will assume that $K_0 \subset K_1 \subset \ldots$, and for brevity, I will identify the elements of the sets K_n with their respective trees on $2^{<\omega}$.

Suppose for contradiction that $I \perp J$, as witnessed by some sets $B \in P_I$ and $C \in P_J$ and a Borel set $D \subset B \times C$ such that its complement has I-small horizontal sections. I must find a J-positive vertical section of the set D. Thinning out the set C if necessary it is possible to find Borel functions $f_n : C \to K_n$ such that $((B \times C) \setminus D)^y \subset \bigcup_n f_n(y)$ – Proposition 2.3.4.

Use the determinacy of the BED game on the poset P_J of Section 3.10.4 to find Player II's winning strategy σ in it. Find a winning strategy τ for the Nonempty Player In the Borel precipitous game associated with the poset P_I. I will construct a play p of the Borel precipitous game respecting the strategy τ with moves denoted by $B_{ini} = B, B_0, B_1, B_2, \ldots$ I will also construct a play q of the BED game respecting the strategy σ, in which Player I plays certain partitions $P_0, P_1, \ldots$ of the set C into finitely many Borel sets, and Player II chooses their elements $C_0, C_1, \ldots$ I will proceed in such a way that for every point $y \in C_n$ the sets $\bigcup_{m \in n} f_m(y)$ and B_{2n+1} are disjoint.

In the end, let $C_\infty = \bigcap_n \bigcup_{m>n} C_m \subset C \notin J$ be the result of the play q against the strategy σ, and let $x \in B$ be the real such that $\{x\} = \bigcap_n B_n$. The choice of the functions f_n implies that for every point $y \in C_\infty$ it is the case that $\langle x, y \rangle \in D$, and the proof will be complete.

In order to perform the inductive construction, suppose that the sets C_n and B_{2n+1} have been obtained. For every number $i \in \omega$ consider the equivalence relation E_n^i on the set C given by $y \, E_n^i z$ if and only if $f_m(y) \upharpoonright i = f_m(z) \upharpoonright i$ for every number $m \le n$. The equivalences induce partitions P_n^i of the set C into finitely many Borel equivalence classes. The next move on the play against the strategy σ will be one of the partitions P_n^i, it is just necessary to decide which one:

Let C_n^i be the answer the strategy σ gives if the partition P_n^i is played. Let $T_m^i : m \le n$ be the uniform values of $f_m(y) \upharpoonright i : m \le n$ for every point $y \in C_n^i$. So each T_m^i is a binary tree of height i. Let U be a nonprincipal ultrafilter on ω and define infinite binary trees $T_m : m \le n$ by setting $t \in T_m$ iff $\{i \in \omega : t \in T_m^i\} \in U$. Since the sets $K_m : m \le n$ are closed, it follows immediately that $T_m \in K_m$ for all $m \le n$. Since the set B_{2n+1} is I-positive and the closed sets $[T_m] : m \le n$ are in the ideal I, there must be a finite binary sequence $t \notin \bigcup_{m \le n} T_m$ such that $B_{2n+2} = B_{2n+1} \cap O_t \notin I$. Let B_{2n+3} be the condition dictated to the Nonempty player by the strategy τ in the Borel precipitous game after the move B_{2n+2} is played. By the definitions, there must be a number i larger than the length of the sequence t such that $t \notin T_m^i$, for all $m \le n$. Then P_n^i will be the next Player I's move on the play q against the strategy σ. $\qquad \square$

4.1.2 Sacks forcing

Set theorists get born and die, move to distant countries, get married, bear children, go bankrupt, grow old and sick, and Sacks forcing is still with us, working just as well as the day Sacks invented it. It is the partial order of perfect binary trees ordered by inclusion.

Let I be the σ-ideal of countable subsets of 2^ω. The following fact shows that the forcing P_I has a dense subset naturally isomorphic to Sacks forcing.

Proposition 4.1.11. *[40], 29.1. Every analytic set $A \subset 2^\omega$ is either countable or it contains all branches of some perfect binary tree.*

Proof. Just an outline of the integer game proof. Suppose that $C \subset 2^\omega \times \omega^\omega$ is a set. Consider the game $G(C)$ between Players I and II in which at n-th round Player I plays a finite binary sequence $s_n \in 2^{<\omega}$ and a natural number $m_n \in \omega$. Player II answers with a bit $b_n \in 2$. In the end let $x \in 2^\omega$ be the concatenation

$s_0 \frown b_0 \frown s_1 \frown b_1 \frown \ldots$ and let $y = m_0 m_1 m_2 \cdots \in \omega^\omega$ Player I wins if the pair $\langle x, y \rangle$ belongs to the set C. The following claim is key.

Claim 4.1.12. *Player II has a winning strategy if and only if* proj(C) *is countable. If Player I has a winning strategy then the set* proj(C) *contains a perfect subset.*

Now if $A \subset 2^\omega$ is an uncountable analytic set, find a closed set $C \subset 2^\omega \times \omega^\omega$ projecting into it, note that the game $G(C)$ is closed for Player I and therefore determined, and use the claim to inscribe a perfect subset into A. □

Corollary 4.1.13. *The ideal I is homogeneous.*

Proof. If $B \notin I$ is a Borel set then it contains all branches of some perfect binary tree T. The natural continuous bijection $\pi : 2^\omega \to [T]$ has the required property that π-preimages of countable sets are countable. □

4.1.3 Miller forcing

Arnold Miller introduced the *Miller forcing*: the conditions are superperfect trees ordered by inclusion. Here, a tree $T \subset \omega^{<\omega}$ is *superperfect* if for every node $t \in T$ there is an extension $s \in T$ with infinitely many immediate successors.

Let I be the σ-ideal generated by compact subsets of ω^ω. The following dichotomy proved by Kechris shows that Miller forcing is naturally isomorphic to a dense subset of P_I.

Proposition 4.1.14. *[39] Suppose that $A \subset \omega^\omega$ is an analytic set. Either $A \in I$ or A contains all branches of some superperfect tree.*

Proof. Suppose that $C \subset \omega^\omega \times \omega^\omega$ is a set. Consider the game $G(C)$ between Players I and II in which at n-th round Player I plays a finite sequence $s_n \in \omega^{<\omega}$ and a number $m_n \in \omega$ and Player II answers with a number $k_n \in \omega$. The sequence s_{n+1} must then begin with a number larger than m_n. In the end let $x \in 2^\omega$ be the concatenation $s_0 \frown s_1 \frown \ldots$ and let $y = m_0 m_1 m_2 \cdots \in \omega^\omega$ Player I wins if the pair $\langle x, y \rangle$ belongs to the set C.

Claim 4.1.15. *Player II has a winning strategy if and only if* proj(C) $\in I$. *If Player I has a winning strategy then the set* proj(C) *contains all branches of a superperfect tree.*

Now if $A \subset 2^\omega$ is an uncountable analytic set, find a closed set $C \subset \omega^\omega \times \omega^\omega$ projecting into it, note that the game $G(C)$ is closed for Player I and therefore determined, and use the claim to inscribe a superperfect subset into A. □

Corollary 4.1.16. *The ideal I is homogeneous.*

Proof. If $B \notin I$ is a Borel set then it contains all branches of some superperfect tree T. The natural continuous bijection $\pi : \omega^\omega \to [T]$ has the required property that preimages of compact sets are compact.   □

There is a still different presentation of the Miller forcing. Let X be a Polish space and $B \subset X$ be a Borel subset which is not F_σ. Let J be the σ-ideal generated by the closed sets which are subsets of B. The forcing P_J below B is then in the forcing sense isomorphic to the Miller forcing. To see this, suppose that $C \subset B$ is a condition in P_J. This means that the set C cannot be separated from $X \setminus B$ by an F_σ-set. By a theorem of Kechris, Louveau, and Woodin [40], 21.22, there is a Cantor set $D \subset X$ such that $D \setminus B$ is countable dense and $D \cap B \subset C$. This means that the set $C \cap D$ is hoemorphic to the Baire space ω^ω and it will be enough to prove that the ideal J below the set $C \cap D$ is σ-generated by compact sets. To do this, note that if $E \subset C \cap D$ is compact then $E \subset B$ is closed and therefore in the ideal J. On the other hand, if $E \subset B$ is a closed set, then $E \cap D$ is a compact set, and it is equal to $E \cap C \cap D$ by the choice of the set D.

4.1.4 Cohen forcing and its iteration

Paul Cohen introduced the *Cohen forcing* as the poset of finite binary sequences ordered by inclusion. Consider the σ-ideal I of meager sets on 2^ω. This is a σ-ideal generated by closed sets. The Baire category theorem shows that Cohen forcing is in the forcing sense isomorphic to P_I.

Fact 4.1.17. *[40], 21.6. If $A \subset 2^\omega$ is an analytic set then either $A \in I$ or there is a basic open set $O \subset 2^\omega$ such that $O \setminus A \in I$. In the context of AD this extends to all subsets of 2^ω.*

Corollary 4.1.18. *The ideal I is homogeneous.*

Proof. Suppose that $B \notin I$ is a Borel set, find a nonempty basic open set O such that $O \setminus B \in I$ and let $\pi_0 : 2^\omega \to O$ be the natural continuous bijection. This function has the property that the preimages of meager sets are meager, however its range is not entirely included in the set B. To fix this, let $E = \{x \in 2^\omega : \pi_0(x) \notin B\}$, note that the set E is meager, and let $\pi : 2^\omega \to B$ be defined by $\pi(x) = \pi_0(x)$ if $x \notin E$ and $\pi(x) = $ arbitrary element of the set B if $x \in E$. This function has the desired properties in the definition of homogeneity.   □

Cohen forcing is usually iterated with finite support. It turns out that the σ-ideal $J = I^\omega$ associated with its countable support iteration of length ω is also generated by closed sets in its natural presentation. Note that the ideals associated with longer iterations cannot be generated by closed sets since these iterations contain the nowhere c.c.c. iteration of length ω as an intermediate extension, which would contradict Theorem 4.1.7.

Let J be the σ-ideal on $X = (2^{\omega})^{\omega}$ generated by the sets $C \subset X$ such that for some number n, the projection of the set C into $(2^{\omega})^n$ is a closed nowhere dense set. With a nod to the presentation of the countable support iteration in Section 5.1, I will state the following without proof.

Fact 4.1.19. *If $A \subset X$ is an analytic set then either $A \in J$ or A contains all branches of some $1, \omega$-tree.*

So indeed the poset P_J contains a dense subset naturally isomorphic to the countable support iteration of length ω of the Cohen forcing. Compare this result with Theorem 4.1.7. The ideal J is σ-generated by closed sets, and therefore every intermediate extension of the P_J extension must be given by a Cohen real. There are many such extensions in the P_J extension; however, none of them is the largest.

4.1.5 The c_{min} ideal

Kojman [22], [23] and many others considered the partition $c_{min} : [2^{\omega}]^2 \to 2$ given by $c_{min}(x, y) = \Delta(x, y) \mod 2$ where $\Delta(x, y) = \min\{n : x(n) \neq y(n)\}$. It turns out that the Cantor space cannot be covered by countably many c_{min}-homogeneous set and the partition is in fact minimal in this respect.

Fact 4.1.20. *[22] If $d : [X]^2$ is a continuous partition such that the underlying Polish space X cannot be covered by countably many d-homogeneous sets then there is a continuous function $\pi : 2^{\omega} \to X$ such that the following diagram commutes.*

$$
\begin{array}{ccc}
X & \xrightarrow{\ d\ } & 2 \\
{\scriptstyle \pi}\big\uparrow & & \big\uparrow{\scriptstyle =} \\
2^{\omega} & \xrightarrow{\ c_{min}\ } & 2
\end{array}
$$

A very closely related object is the simplest function which cannot be decomposed into countably many monotonic functions. Let $f : 2^{\omega} \to 2^{\omega}$ be the function defined by $f(x)(n) = (x(n) + n) \mod 2$. With the usual lexicographic ordering on 2^{ω}, this function cannot be decomposed into countably many monotonic functions. It turns out that it is the simplest such a function.

Fact 4.1.21. *[83] If X is a Polish space with a separable Borel ordering and $g : X \to 2^{\omega}$ is an analytic function then*

1. *either g can be decomposed into countably many monotonic subfunctions;*
2. *or there are monotonic embeddings π and ψ such that the following diagram commutes.*

$$
\begin{array}{ccc}
X & \xrightarrow{\ g\ } & 2^\omega \\
\uparrow{\scriptstyle \pi} & & \uparrow{\scriptstyle \psi} \\
2^\omega & \xrightarrow{\ f\ } & 2^\omega
\end{array}
$$

Under AD this extends to all functions g.

It follows directly from the definitions that a set $A \subset 2^\omega$ is c_{min}-homogeneous if and only if the function f is monotonic on A, and these properties are inherited by the closure of the set A. Consider the σ-ideal I generated by c_{min}-homogeneous sets. Since the partition is continuous, the closure of a homogeneous set is itself homogeneous, therefore the ideal I is generated by closed sets. In fact, the collection of closed homogeneous sets is compact in the hyperspace $K(2^\omega)$ and therefore the forcing P_I is bounding and adds no splitting reals by Theorem 4.1.8. To find a combinatorial presentation of the forcing P_I the following concept is useful. A tree $T \subset 2^{<\omega}$ is a c_{min}-tree if for every node $t \in T$ and every bit $b \in 2$ there is a splitnode $s \supset t, s \in T$ whose length has parity b. The following is proved using the usual determinacy arguments.

Proposition 4.1.22. *[83] An analytic set is I-positive if and only if it contains all branches of some c_{min}-tree.*

Proof. Let $C \subset 2^\omega \times \omega^\omega$ be a set. Consider a game $G(C)$ between Players I and II in which at round n Player I chooses a finite binary sequence $t_n \in 2^{<\omega}$ of even length and a number m_n. Player II responds with a bit $b_n \in 2$. In the end, let $x = t_0^\frown b_0^\frown t_1^\frown b_1 \ldots$ and $y = m_0 m_1 m_2 \ldots$ Player I wins if $\langle x, y \rangle \in C$. The following claim is key.

Claim 4.1.23. *Player II has a winning strategy if and only if $\mathrm{proj}(C) \in I$. If Player I has a winning strategy then the set $\mathrm{proj}(C)$ contains all branches of some c_{min} tree.*

Now if $A \subset 2^\omega$ is an analytic I-positive set, choose a closed subset $C \subset 2^\omega \times \omega^\omega$ projecting into it, note that the game $G(C)$ is closed for Player I and therefore determined, and use the claim to inscribe a c_{min} tree into the set A. $\square$

Corollary 4.1.24. *The σ-ideal I is homogeneous.*

Proof. If $B \notin I$ is a Borel set then it contains all branches of some c_{min} tree T; thinning out the tree if necessary, I may assume that the first branching node is at an even level, the next two branching nodes come at odd levels, the next four come at even levels and so on. The natural continuous bijection $\pi : 2^\omega \to [T]$ preserves the partition c_{min} and thus has the required property that preimages of c_{min}-homogeneous sets are again c_{min}-homogeneous.   □

Sheu investigated the Ramsey theoretic properties of the σ-ideal I. He proved that the countable support product of an arbitrary number of copies of P_I does not add a splitting real.

The previous fact shows the c_{min} forcing is the simplest forcing in which the *monotonic reading of names* fails. Here the monotonic reading of names for a σ-ideal J is the statement that every Borel function with J-positive domain has a monotonic restriction with J-positive domain. The notion of monotonicity uses a separable Borel ordering on the domain and the range of the function f and it turns out that the choice of the ordering is immaterial. Some forcings including the Sacks forcing do have the monotonic reading of names, others do not. The whole situation should be compared with Steprāns forcing of Section 4.2.3, which is the simplest forcing without the continuous reading of names.

4.1.6 Packing measures

Suppose that $\langle X, d \rangle$ is a compact metric space and $h : \mathbb{R}^+ \to \mathbb{R}^+$ is a *gauge function*, a continuous nondecreasing function with $h(0) = 0$. There are several definitions of various packing measures. I will define the packing premeasure ϕ_p by $\phi_p(A) = \inf_\delta \sup\{\Sigma_n h(r_n) :$ there are points $x_n \in A$ such that the closed balls around x_n with radius $r_n < \delta$ are pairwise disjoint$\}$ and the packing measure as $\phi(A) = \inf\{\Sigma_n \phi_p(A_n) : A \subset \bigcup_n A_n\}$. It turns out that ϕ is a metric measure and typically it is not σ-finite. Let I be the σ-ideal generated by sets of finite packing measure.

Proposition 4.1.25. *The ideal I is σ-generated by a σ-compact collection of compact sets.*

Proof. For natural numbers $k, l \in \omega$ consider the set $K_{kl} \subset K(X)$ consisting of those compact sets $F \subset X$ such that $\phi_p(F, k) = \sup\{\Sigma_i h(r_i) :$ there are points $x_i \in F$ such that the closed balls around x_i of radius $r_i < 2^{-k}$ are pairwise disjoint$\} \leq l$.

First observe that every set $K_{kl} \subset K(X)$ is compact. Suppose $F_n : n \in \omega$ are sets in K_{kl} converging to another compact set F. If $\phi_p(F, k) > l$ as witnessed by some finite set of points x_i and radii r_i, there will be a large number n such that the set F_n contains points y_i sufficiently close to x_i such that the balls around y_i of radius r_i are disjoint, and then $\phi_p(F_n, k) > l$, a contradiction. Thus $\phi_p(F, k) \leq l$ and $F \in K_{kl}$ as required.

Second, observe that every set $A \subset X$ of finite ϕ_p mass is a subset of a set in one of the collections K_{kl}. Just choose k, l large enough so that $\phi_p(A, k) < l$ and note that $\phi_p(A, k) = \phi_p(\bar{A}, k)$, thus the closure $\bar{A}$ is in K_{kl} as required.

It is now immediately clear that the ideal I is σ-generated by the sets $K_{kl} : k, l \in \omega$.

$\square$

Consider the following measure-theoretic result:

Fact 4.1.26. *[32] Every Borel ϕ-positive set has a Borel subset of positive, finite ϕ-mass.*

Caution – this result is sensitive to the particular definition of packing measure in question, for a discussion see [32]. It now follows from 4.1.8 and 3.6.4 that the forcing P_I is bounding, preserves Baire category, preserves outer Lebesgue measure, and does not add splitting reals. I do not know how the forcing depends on the metric space and gauge function in question.

Example 4.1.27. To construct a thematic forcing which increases the invariant cof(null) and keeps all other Cichoń invariants unchanged choose a decreasing sequence of numbers $0 < d_n < 1 : n \in \omega$ converging to zero such that $d_n > n^2 d_{n+1}$, and let $k_n : n \in \omega$ be positive natural numbers such that the sequence $k_n \cdot d_n : n \in \omega$ diverges to infinity. Consider the metric space $X = \Pi_n k_n$ with the least difference metric $d(x, y) = d_{\Delta(x,y)}$, and consider the one-dimensional packing measure μ associated with it. The numbers k_n were chosen so that the space X is not a countable union of sets of finite μ-measure, and the numbers d_n were chosen so that any tunnel $T = \Pi_n a_n$ where $a_n \subset k_n$ is a set of size $\leq n^2$, has finite packing measure. Let I be the σ-ideal on the space X generated by the sets of finite packing measure. The forcing adds an element of the space X which cannot be enclosed by an n^2 tunnel from the ground model, and therefore it increases the invariant cof(null) by [2], 2.3.9. The forcing is bounding and does not add splitting reals by Theorem 4.1.8, it preserves Baire category by Theorem 4.1.2, and it preserves outer Lebesgue measure by Theorem 3.6.2 and the previous Fact.

4.1.7 The splitting real forcing

Shelah [62] considered a bounding forcing for adding a splitting real. The conditions are trees $T \subset 2^{<\omega}$ such that for every node $t \in T$ there is a number $n \in \omega$ such that for every $m > n$ and every bit $b \in 2$ there is an extension $s \supset t, s \in T$ such that $s(n) = b$. The ordering is that of inclusion.

Consider the σ-ideal I on 2^ω generated by sets $A_a = \{x \in 2^\omega : \forall n \in a\ x(n) = 1\}$ and $B_a = \{x \in 2^\omega : \forall n \in a\ x(n) = 0\}$, as a varies over all infinite subsets of ω. The generating sets are all closed and they form a G_δ-set in $K(2^\omega)$. The following proposition shows that the forcing P_I contains a dense subset isomorphic to Shelah's forcing. Compare this to Theorem 4.1.8

Fact 4.1.28. *[72] An analytic set $A \subset 2^\omega$ is I-positive if and only if it contains all branches of some splitting tree.*

Regarding the finer forcing properties of this partial order I will make the following observation.

Proposition 4.1.29. $I \perp \text{null}$.

Thus under some condition the forcing P_I collapses the outer Lebesgue measure. I do not know if under some other condition the forcing P_I preserves it.

Proof. To facilitate the underlying intuition consider the ideal J for adding a σ-splitting subset of the set z which is a disjoint union of $2^n : n \in \omega$. Thus the domain of the σ-ideal J is $X = 2^z$ and J is clearly just another presentation of the ideal I, and it will be enough to show $J \perp \text{null}$. Consider the Borel set $B \subset X$ of the characteristic functions of graphs of all functions f such that $\forall n\ f(n) \in 2^n$. It is not difficult to see that the set B is J-positive. Now consider the Borel set $D \subset B \times 2^\omega$, $\langle x, y \rangle \in D \leftrightarrow \exists^\infty n\ x(\langle n, y \restriction n \rangle) = 1$. A brief review of the definitions will reveal that the vertical sections of the set D are Lebesgue null while the horizontal sections of its complement are in the ideal J as desired. $\square$

4.1.8 Sets of extended uniqueness

Let $\mathbb{T}$ be the unit circle in the complex plane. For a measure μ on $\mathbb{T}$ let $\hat{\mu}(n)$ be the n-th Fourier coefficient of the measure μ, that is, $\hat{\mu}(n) = \int_{\mathbb{T}} e^{-inx} d\mu(x)$. The measure μ is a *Rajchman* measure if its Fourier coefficients converge to 0. Let $I = \{A \subset \mathbb{T} : \mu(A) = 0$ for every Rajchman measure $\mu\}$. In Fourier analysis, the sets in the ideal I are known as *sets of extended uniqueness*.

Fact 4.1.30. *(Debs, Saint Raymond) [37], VIII.3, Theorem 1. The ideal I is generated by closed sets.*

Note that by the virtue of its definition the ideal I is polar and therefore the forcing P_I is bounding and preserves the outer Lebesgue measure. Other forcing properties of the forcing P_I are unknown.

4.1.9 Destroying a partition into closed sets

Arnold Miller [51] considered a forcing P adding an infinite binary sequence $r \in 2^\omega$ which destroys a partition $2^\omega = \bigcup_{i \in K} C_i$ of the Cantor space into a collection of pairwise disjoint compact sets in the ground model indexed by some uncountable index set K, that is, $P \Vdash \forall i \; \dot{r} \notin \dot{C}_i$. The forcing P consists of binary trees T such that $C_i \cap [T]$ is a nowhere dense subset of the set $[T]$ for every index i.

Consider the σ-ideal I on 2^ω generated by the sets C_i. The following proposition shows that the forcing P_I contains a dense subset isomorphic to the forcing P.

Proposition 4.1.31. *A set $A \subset 2^\omega$ is I-positive if and only if it contains all branches of some tree in the forcing P.*

Proof. Let $B \subset 2^\omega$ be a Borel I-positive set; I must produce a tree $T \subset 2^{<\omega}$ such that $[T] \subset B$ and $T \in P$. The key idea is to find the perfect tree T in such a way that for every node $t \in T$ there is an index i_t such that $|C_{i_t} \cap [T \restriction t]| = 1$. Such a tree must be in the poset P: if i is an index and $t \in T$ is a node, then there are two cases: either $i = i_t$, in which case certainly there is a node $s \in T$ extending t such that $C_i \cap [T \restriction s] = 0$; or $i \neq i_t$, in which case again there is an initial segment $s \in T$ of the unique element of $C_{i_t} \cap [T \restriction t]$ such that $O_s \cap C_i = 0$ since the sets C_i and C_{i_s} are closed and disjoint.

Fix a tree $S \subset (\omega \times 2)^{<\omega}$ projecting into the set $B \subset 2^\omega$. Removing some bad nodes of the tree S and thinning out the set B if necessary I may assume that for all nodes $s \in S$, $p[S \restriction s] \notin I$. Call a node $s \in S$ *incompatible* with an index $i \in K$ if the binary sequence forming the second coordinate of the node s is not on the tree determining the compact set $C_i \subset 2^\omega$ Now by induction on $n \in \omega$ build trees $T_n \subset S$ and functions α_n, β_n such that:

- T_{n+1} end-extends T_n.
- Two distinct endnodes $t = \langle a_t, b_t \rangle$ and $s = \langle a_s, b_s \rangle$ of the tree T_n have $\subset$-incomparable second coordinates.
- $\alpha_n : T_n \to K$ is an injection and $\beta_n : T_n \to [S]$ is a function such that $\beta_n(t)$ is a cofinal branch of the tree S which passes through t as well as through some terminal node of the tree T_n, and its second coordinate is a binary sequence in the compact set $C_{\alpha_n(t)}$. The function π_n returns distinct values of i on distinct nodes $t \in T_n$.
- $\alpha_n, \beta_n \subset \alpha_{n+1}, \beta_{n+1}$.
- Whenever $t \in T_n$ and then the only terminal node of the tree T_{n+1} compatible with $\alpha_n(t)$ is the one on the branch $\beta_n(t) \subset S$.

This is not difficult to do. Suppose that T_n, π_n have been constructed. There are just finitely many pairwise disjoint compact sets indicated by the function α_n

and so there will be a number $m \in \omega$ such that for every $t \in T_n$ among the nodes $\beta_n(s) \upharpoonright m : s \in T_n$ the only one compatible with $\alpha_n(t)$ will be $\beta_n(t) \upharpoonright m$. Let T_{n+1} be the closure of the set $\{\beta_n(s) \upharpoonright m : s \in T_n\}$ under initial segment. For every node $t \in T_{n+1} \setminus T_n$ find an index i_t and a branch $b_t \subset S$ such that the branch b_t passes through t as well as some terminal node of T_{n+1}, its second coordinate is in the set C_{i_t}, and the indices i_t are mutually distinct and outside of the set $\mathrm{rng}(\alpha_n)$. This is easily possible since for every terminal node $s \in T_{n+1}$ the projection $p[S \upharpoonright s]$ meets infinitely many sets among the $C_i : i \in K$. Finally, let $\alpha_{n+1} = \alpha_n \cup \{\langle t, i_t \rangle : t \in T_{n+1} \setminus T_n\}$ and $\beta_{n+1} = \beta_n \cup \{\langle t, b_t \rangle : t \in T_{n+1} \setminus T_n\}$. The induction hypotheses continue to hold.

In the end, let $T_\omega = \bigcup_n T_n \subset S$ and let $T \subset 2^{<\omega}$ be the projection of T_ω into $2^{<\omega}$. The second item of the induction hypothesis implies that any branch through the tree T is a projection of a unique branch through the tree T_ω and so $[T] \subset B$. Also, if $n \in \omega$, $s \in T_n$ is a terminal node and $t \in T$ its projection then $[T \upharpoonright t] \cap C_{\alpha_n(s)}$ is a set containing a single element, the projection of the branch $\beta_n(s)$. This concludes the proof. $\qquad\qquad\square$

4.2 Porosity ideals

The subject of the present section grew out of an attempt to generalize the proof of properness of the forcing associated with the σ-ideal generated by porous subsets of a Polish metric space. It turns out that there is a suitable generalization of the metric porosity.

Definition 4.2.1. *[18] Let X be a Polish space and U a countable collection of its Borel subsets. An* abstract porosity *is an inclusion preserving map* por $: \mathcal{P}(U) \to \mathcal{B}(X)$*, that is $a \subset b \to \mathrm{por}(a) \subset \mathrm{por}(b)$. The porosity ideal I associated with the porosity* por *is σ-generated by sets* $\mathrm{por}(a) \setminus \bigcup a$*, as a runs through all subsets of U. Such sets (and their subsets) are called* porous.

Note that by extending the topology on the space X without changing the Borel structure and the underlying forcing P_I it is possible to assume that the sets in U are clopen, without changing the Borel structure or the poset P_I. However, this is not always a natural step to make; cf. Subsection 4.2.3. Another remark in this vein is that every presentation of a porosity ideal is a porosity ideal; this is in contradistinction to the class of σ-ideals generated by closed sets which is not invariant under various presentations.

In the way of examples, note that every σ-ideal J generated by closed sets is a porosity ideal. Just let U be the collection of all basic open sets and $\mathrm{por}(a) = X$ whenever $X \setminus \bigcup a \in J$, and $\mathrm{por}(a) = 0$ otherwise. It is clear that a set is porous if and only if it is covered by a closed set in the ideal J. Somewhat more substantial example is the following representation of the meager ideal as a porosity ideal. Let U be again the collection of basic open sets, and let $\mathrm{por}(a)$ be the closure of $\bigcup a$. It is not difficult to see that a set is porous if and only if it is covered by a closed nowhere dense set. The real examples are covered by the subsections below. It is nevertheless the case that the extent of the family of porosity ideals is unclear. Question 7.3.1 is one related unsolved problem. There indeed are porosity ideals which cannot be presented as ideals generated by closed sets, such as the metric porosity ideal of Section 4.2.2.

4.2.1 General results

Theorem 4.2.2. *[18] Suppose that I is a porosity σ-ideal on a Polish space X. Then the forcing P_I is proper and preserves Baire category.*

Proof. Suppose that $B \in P_I$ is a Borel set, and $\dot{O}_n : n \in \omega$ is a name for a sequence of open dense subsets of $2^{<\omega}$. Suppose M is a countable elementary submodel of a large enough structure and $y \in 2^\omega$ is a point which meets all open dense subsets of the Cantor space in the model M. I must show that the set $C = \{x \in B : x$ is M-generic for P_I and $y \in \bigcap_n \dot{O}_n/x\}$ is I-positive. Let $a_n \subset U : n \in \omega$ be sets; I must produce a point $x \in C \setminus \bigcup \{\mathrm{por}(a_n) \setminus \bigcup a_n : n \in \omega\}$. In order to do this, enumerate the open dense subsets of P_I in the model M by $D_n : n \in \omega$ and by induction on n build conditions $B_n : n \in \omega$ so that

- $B = B_0 \supset B_1 \supset \ldots$ are all elements of the model M;
- $B_{n+1} \in D_n$;
- $B_{n+1} \cap \mathrm{por}(a_n) \setminus \bigcup a_n = 0$;
- for some initial segment $t_n \subset y$, $B_{n+1} \Vdash \check{t}_n \in \dot{O}_n$.

Once this is done, the filter $g \subset P_I \cap M$ generated by the sets B_n is M-generic by the second item and its intersection is a singleton containing a point $x \in X$. By the third item $x \in B \setminus \bigcup \{\mathrm{por}(a_n) \setminus \bigcup a_n : n \in \omega\}$. Finally by the last item $y \in \bigcap_n \dot{O}_n/x$ as required, completing the proof.

The inductive construction is easy to perform. Suppose that $B_n \in M$ has been found. Let $a = \{u \in U : B_n \cap u \in I\}$ and let $B_n^0 \subset B$ be defined by $B_n^0 = (B \setminus \bigcup a) \setminus \mathrm{por}(a)$. Note that both of the set differences in the previous expression removed only a Borel I-small set from B, so $B_n^0 \in P_I \cap M$. Consider the set $a_n \subset M$. I claim that there is a condition $B_n^1 \subset B_n^0$ in the model M such that $B_n^1 \cap \mathrm{por}(a_n) \setminus \bigcup a_n = 0$. There are two possible cases. Either $a_n \subset a$ in which case $\mathrm{por}(a_n) \subset \mathrm{por}(a)$ by the monotonicity of the porosity function and $B_n^1 = B_n^0$ will work. Or there is a set $u \in a_n \setminus a$ in which case the set $B_n^1 = B_n^0 \cap u \in P_I \cap M$ will work. Now use the elementarity to find a condition $B_n^2 \subset B_n^1$ in the model M and in the open dense set D_n. The set $\{t \in 2^{<\omega} : \exists A \subset B_n^2 \; A \Vdash \check{t} \in \dot{O}_n\} \subset 2^{<\omega}$ is open dense and an element of the model M, and by the choice of the point $y \in 2^\omega$ it is possible to find a set $B_n^3 \subset B_n^2$ in the model M and an initial segment $t_n \subset y$ such that $B_n^3 \Vdash \check{t}_n \in \dot{O}_n$. Finally, let $B_{n+1} = B_n^3$ and the induction step is complete. $\qquad \square$

Theorem 4.2.3. *Suppose that I is a porosity σ-ideal on a Polish space X.*

1. *I satisfies the third dichotomy.*
2. *(In the choiceless Solovay model) I satisfies the first dichotomy and is closed under wellordered unions.*
3. *(ZF+DC+AD+) I satisfies the first dichotomy and it is closed under wellordered unions.*

Proof. For the first item use the forcing Q_I of all analytic I-positive sets ordered by inclusion. The ideal I is generated by Borel sets and so Proposition 2.1.11 applies to show that the forcing adds a point $\dot{x}_{gen}$ which belongs to all sets in the generic filter.

Let $A \notin I$ be an analytic set. Let M be a countable elementary submodel of a large enough structure containing the set A and let $B = \{x \in A : x$ is M-generic for the forcing $Q_I\}$. This is a Borel set by Fact 1.4.8; it will be enough to show that $B \notin I$. The proof follows exactly the lines of the previous proof. The only relevant point is that if $C \notin I$ is an analytic set and $a = \{u \in U : C \cap u \in I\}$ then the set $(C \setminus \bigcup a) \setminus \mathrm{por}(a)$ is analytic again, since the range of the porosity function consists of Borel sets.

The second item is similar. Work in the choiceless Solovay model derived from an inaccessible cardinal κ. Suppose that $A \subset X \times 2^\omega$ is an arbitrary set such that $\mathrm{proj}(A) \notin I$. I will produce a Borel subset $B \subset A$ such that $\mathrm{proj}(B) \notin I$. An application of Proposition 3.9.17 then completes the proof of the second item. By the usual homogeneity arguments I may assume that the set A is definable from parameters in the ground model. Since $\mathrm{proj}(A) \notin I$ there must be in V a forcing P of size $< \kappa$ and a P-name $\langle \dot{x}, \dot{y} \rangle$ such that $P \Vdash \forall a \subset U \; a \in V \to \dot{x} \notin \mathrm{por}(a) \setminus \bigcup a$ and $P \Vdash \mathrm{Coll}(\omega, < \kappa) \Vdash \langle \dot{x}, \dot{y} \rangle \in \dot{A}$. Back again in the Solovay model, let $B = \{\langle x, y \rangle \in X \times 2^\omega : \exists g \subset P \; g$ is a V-generic filter and $\langle x, y \rangle = \langle \dot{x}, \dot{y} \rangle / g\}$. The set B is Borel by Fact 1.4.8 and the usual homogeneity arguments show that $B \subset A$. The proof will be complete once I show that $\mathrm{proj}(B) \notin I$.

The proof follows closely the argument for properness again. Let $a_n : n \in \omega$ be subsets of U; I must produce a V-generic filter $g \subset P$ such that $\dot{x}/g \notin \bigcup_n (\mathrm{por}(a_n) \setminus \bigcup a_n)$. Enumerate the dense open subsets of the poset P in V by $D_n : n \in \omega$ and by induction on $n \in \omega$ build a descending chain of conditions $1_P = p_0 \geq p_1 \geq p_2 \geq \cdots$ so that

- $p_{n+1} \in D_n$;
- for every V-generic filter $h \subset P$ containing the condition p_{n+1}, $\dot{x}/h \notin \mathrm{por}(a_n) \setminus \bigcup a_n$.

If this construction succeeds then certainly the filter g generated by the conditions $p_n : n \in \omega$ is as desired. To perform the inductive step, suppose that the condition $p_n \in P$ has been constructed, and let $a = \{u \in U : V \models p_n \Vdash \dot{x} \notin \dot{a}\}$. The set $a \subset U$ is in the ground model, and there are two cases. Either $a_n \subset a$; then, since $p_n \Vdash \dot{x} \notin \mathrm{por}(a) \setminus \bigcup a$, it is also the case that $p_n \Vdash \dot{x} \notin \mathrm{por}(a)$ and by the monotonicity of the porosity function no V-generic filter h containing the condition p_n has $\dot{x}/h \in \mathrm{por}(a_n)$. In this case any condition $p_{n+1} \leq p_n$ in the open dense set D_n will satisfy the induction hypotheses. Or there is a set $u \in a_n \setminus a$ and consequently a condition $q \leq p_n$ such that $q \Vdash \dot{x} \in \dot{u}$. Clearly, any V-generic filter $h \subset P$ containing the condition $q \in P$ will have $\dot{x}/h \in u$ and so $\dot{x}/h \notin \mathrm{por}(a_n) \setminus \bigcup a_n$. Thus in this case any condition $p_{n+1} \leq p$ in the open dense set D_n will satisfy the induction hypotheses.

The last item is similar again. Suppose for simplicity that the underlying Polish space is 2^ω. Suppose that $A \subset 2^\omega \times 2^\omega$ is a set such that $\mathrm{proj}(A) \notin I$; I must produce an analytic set $B \subset A$ such that $\mathrm{proj}(B) \notin I$. The argument is then concluded by a reference to Proposition 3.9.18.

Fix the countable collection U of Borel sets defining the porosity. Consider a game $G(A)$ between Players I and II in which Player I produces a sequence $a_n : n \in \omega$ of subsets of U and Player II produces a point $\langle x, y \rangle \in 2^\omega \times 2^\omega$. Player II wins if $\langle x, y \rangle \in A$ and $x \notin \bigcup_n(\mathrm{por}(a_n) \setminus \bigcup a_n)$. In order to complete the description of the game, just enumerate the set U in some way. At round $n \in \omega$, Player I decides which among the first n elements of U belong to which sets $a_m : m \in \omega$. Player II answers with a pair of sequences $s_n, t_n \in 2^n$. Player II wins if the sequences s_n converge (the limit is denoted by x), the sequences t_n converge (the limit is denoted by y), $\langle x, y \rangle \in A$ and $x \notin \bigcup_n(\mathrm{por}(a_n) \setminus \bigcup a_n)$.

I will prove that Player I has a winning strategy in the game $G(A)$ if and only if $\mathrm{proj}(A) \in I$. Then, if $\mathrm{proj}(A) \notin I$, Player II must have a winning strategy σ by a determinacy argument. Let $B \subset A$ be the set of all points $\langle x, y \rangle$ the strategy σ can produce against a counterplay by Player II. The set B is analytic by its definition – it is the image of the compact collection of all possible plays by Player I under a total Baire class one function – it is a subset of the set A as σ is a winning strategy, and its projection is I-positive as σ remains a winning strategy in the game $G(B)$.

First of all, if $\mathrm{proj}(A) \in I$ then Player I can easily win ignoring the moves of the adversary altogether. On the other hand, suppose that σ is a strategy for Player I. I will find a set $C \subset 2^\omega$ in the ideal I such that for all $x \notin C$ and all $y \in 2^\omega$ there is a counterplay against the strategy σ which produces the pair $\langle x, y \rangle$ such that x does not belong to the small set obtained by the strategy. This means that either proj$(A) \subset C$ and proj$(A) \in I$, or else σ is not a winning strategy, completing the proof.

For every finite partial play τ of the game respecting the strategy σ and all natural numbers n, m let $a_{\tau,n,m} \subset U$ be the collection of all those sets $u \in U$ that the strategy σ can put into a_n in some play extending τ in which the additional sequences played by Player II all agreed with the last move of τ on the first m bits. Let $C = \bigcup_{\tau,n,m}(\text{por}(a_{\tau,n,m}) \setminus \bigcup a_{\tau,n,m})$. Clearly, $C \in I$. I claim that the set C works.

For any point $x \in 2^\omega \setminus C$ and $y \in 2^\omega$ build by induction on $i \in \omega$ partial plays τ_i against the strategy σ so that

- $0 = \tau_0 \subset \tau_1 \subset \ldots$
- all moves of Player II on $\tau_{i+1} \setminus \tau_i$ agree with x or y respectively on the first i bits;
- for every $i \in \omega$, either $x \notin \bigcup a_{\tau_i,i,i}$ or at some point during the play τ_{i+1} the strategy σ put a set $u \in U$ into a_i such that $x \in u$.

The induction is entirely straightforward given the definition of the set $a_{\tau_i,i,i}$. Consider the infinite play $\bigcup_n \tau_n$. In it, the moves of Player II converge to the points x and y as guaranteed by the second item. Now fix a number $i \in \omega$ and let $a \subset U$ be the i-th set obtained by the strategy σ. I must show that $x \notin \text{por}(a) \setminus \bigcup a$. For this, consider the set $b = a_{\tau_i,i,i}$, use the second item to see that $a \subset b$, and use the monotonicity of the porosity function to see that $\bigcup a \cup \text{por}(a) \subset \bigcup b \cup \text{por}(b)$. There are now two possibilities according to the third item above.

- $x \notin \bigcup b$. In this case, since $x \notin \text{por}(b) \setminus \bigcup b$ by the definition of the set C, it follows that $x \notin \bigcup b \cup \text{por}(b)$ and consequently $x \notin \text{por}(a) \setminus \bigcup a \subset \bigcup b \cup \text{por}(b)$ as desired.
- There is a set $u \in a$ such that $x \in u$. Then certainly $x \notin \text{por}(a) \setminus \bigcup(a)$ as well.

In both cases, the argument came to a successful conclusion.

□

Theorem 4.2.4. *If I is a porosity σ-ideal then the forcing P_I is embeddable into σ-closed*c.c.c. iteration.*

Proof. For simplicity assume that the underlying Polish space is the Cantor space 2^ω. Let P be the partial order of pairs $p = \langle a_p, b_p \rangle$ where:

- a_p is a countable set of trees $T \subset (\omega \times 2)^{<\omega}$ such that proj$[T] \notin I$; for every I-positive element $u \in U$ the set a_p contains a tree which projects to u;
- The set a_p is closed under I-positive intersections: if $T_0, T_1 \in a_p$ and proj$[T_0] \cap$ proj$[T_1] \notin I$ then there is a tree $S \in a_p$ such that proj$[S] = $ proj$[T_0] \cap$ proj$[T_1]$;

- The set a_p is closed under restriction: if $t \in a_p$ and $t \in T$ is a node such that $\mathrm{proj}[T \upharpoonright t] \notin I$ then $T \upharpoonright t \in a_p$;
- The set a_p is closed under the following operation: if $T \in a_p$ is a tree and $a \subset U$ is the set $\{u \in U : \mathrm{proj}[T] \cap u \in I\}$ then there is a tree $S \in a_p$ which projects into the set $(\mathrm{proj}[T] \setminus \bigcup a) \setminus \mathrm{por}(a)$. Note that $\mathrm{proj}[T] \setminus \mathrm{proj}[S] \in I$.
- b_p is a countable collection of sets $z \subset a_p$ such that for every tree $T \in a_p$ there is $S \in z$ such that $\mathrm{proj}[T] \cap \mathrm{proj}[S] \notin I$.

The ordering is that of coordinatewise reverse inclusion. This is clearly a σ-closed forcing. If $G \subset P$ is a generic filter then let $Q \in V[G]$ be the set $\{\mathrm{proj}[T] : \exists p \in G \ T \in a_p\}$ ordered by inclusion. The proof will be complete once I show that Q is c.c.c. and if $D \subset P_I$ is a dense set in the ground model then $D \cap Q \subset Q$ is dense. For then, the poset Q adds a V-generic filter on P_I and so P_I embeds into the iteration $P * Q$. The following claim will be instrumental.

Claim 4.2.5. *Suppose* $p = \langle a_p, b_p \rangle \in P$ *is a condition and consider the set* $B = \bigcap_{z \in b_p} \bigcup \{\mathrm{proj}[T] : T \in z\}$. *Then for every tree* $S \in a_p$, $B \cap \mathrm{proj}[S] \notin I$ *and for every* I-*positive analytic set* $C \subset B$ *there is a condition* $q = \langle a_q, b_p \rangle \leq p$ *such that the set* a_q *includes some tree* T *such that* $C = \mathrm{proj}[T]$.

The claim immediately implies that if $D \subset P_I$ is an open dense set then $P \Vdash \check{D} \cap Q$ is dense. To see this, for every condition $p \in P$ and every tree $S \in a_p$ consider the set $B \subset X$ as in the claim and look at the intersection $B \cap \mathrm{proj}[S]$. It is an I-positive analytic set and therefore contains an I-positive Borel subset and even one from the open dense set D. Choose such a set $C \subset B$ and use the claim to find a condition $q \leq p$ which includes some tree T such that $\mathrm{proj}[T] = B \cap \mathrm{proj}[S]$. The condition q forces that there is an element of the open dense set D below the condition $\mathrm{proj}[S] \in Q$. Finally use a genericity argument to see that $D \cap Q$ is forced to be dense.

For the c.c.c. of the forcing Q in the P-extension, suppose that $p \Vdash \dot{D} \subset \dot{Q}$ is an open dense set. Let M be a countable elementary submodel of a large enough structure, let $q \leq p$ be a the coordinatewise union of some M-generic filter $g \subset P \cap M$ and let $z \subset a_q$ be the set $\{\mathrm{proj}[S] : S \in a_q, q \Vdash \mathrm{proj}[S] \in \dot{D}\}$. A genericity argument concerning the filter g shows that for every tree $T \in a_q$ there is $S \in z$ such that $\mathrm{proj}[S] \cap \mathrm{proj}[T] \notin I$. Ergo, the pair $r = \langle a_q, b_q \cup \{z\} \rangle \leq q$ is a condition in the poset P. It is immediate that $r \Vdash \{\mathrm{proj}[S] : S \in z\} \subset \dot{Q}$ is a countable predense set, proving the c.c.c. of the forcing Q.

All that remains to do is to prove the claim. Fix the condition $p = \langle a_p, b_p \rangle$. First I will show that $B \notin I$. Suppose that $a_n : n \in \omega$ are subsets of the set U defining the porosity ideal I. For every tree $S \in a_p$ I must produce a point $x \in 2^\omega$ such that

$x \in B \cap \text{proj}[S] \setminus \bigcup_n (\text{por}(a_n) \setminus \bigcup a_n)$. Enumerate the set b_p as $\{z_n : n \in \omega\}$ and by induction on $n \in \omega$ build a sequence $S_n : n \in \omega$ of trees in a_p so that

- $S = S_0$, $\text{proj}[S_0] \supset \text{proj}[S_1] \supset \dots$
- for every number n there is a tree $T \in z_n$ such that $\text{proj}[S_{n+1}] \subset \text{proj}[T]$;
- for every number n the set $\text{proj}[S_{n+1}] \cap \text{por}(a_n) \setminus \bigcup a_n$ is empty;
- there are nodes $t_m^n \in S_m$ of length n for every number $m \in n$ such that $t_m^{m+1} \subset t_m^{m+2} \subset \dots$ and $\text{proj}[S_n] \subset \text{proj}[S_m \restriction t_m^n]$ for every $m \in n$.

The fourth item is designed to imply that $\bigcap_n \text{proj}[S_n] \neq 0$. Any point $x \in 2^\omega$ in this intersection is the required element of the set $\text{proj}[S] \cap B \setminus \bigcup_n (\text{por}(a_n) \setminus \bigcup a_n)$ as is clear from the second and third item.

The induction step closely follows the argument from the properness proof. Suppose that S_n has been constructed. Letting $a = \{u \in U : u \cap \text{proj}[S_n] \in I\} \subset U$, the set $(\text{proj}[S_n] \setminus \bigcup a) \setminus \text{por}(a)$ is I-positive and there is a tree $S_n^0 \in a_p$ projecting into it. The next step is to produce a tree $S_n^1 \in a_p$ such that $\text{proj}[S_n^1] \subset \text{proj}[S_n^0]$ and $\text{proj}[S_n^1] \cap \text{por}(a_n) \setminus \bigcup(a_n) = 0$. There are two cases: either $a_n \subset a$, in which case the tree $S_n^1 = S_n^0$ will work; or there is some set $u \in a_n \setminus a$, in which case the analytic set $\text{proj}[S_n^0] \cap u$ is I-positive and any tree $S_n^1 \in a_p$ projecting into it will work. The next step is to find a tree $T \in z_n$ such that the set $\text{proj}[T] \cap \text{proj}[S_n^1]$ is I-positive and let $S_n^2 \in a_p$ be any tree that projects into this set. Finally, by the countable additivity of the σ-ideal I there are nodes $t_m^{n+1} \in S_m$ of length $n+1$ for every number $m \in n+1$ extending the nodes t_m^n such that the set $\text{proj}[S_n^2] \cap \bigcap_{m \in n+1} \text{proj}[S_m \restriction t_m^{n+1}]$ is I-positive and let $S_{n+1} \in a_p$ be any tree projecting into it. This completes the inductive step and shows that the set $B \subset X$ is I-positive.

Now suppose that $C \subset B$ is any I-positive analytic set. Find a tree $T \subset (\omega \times 2)^{<\omega}$ which projects into the set C and close the set $a_p \cup \{T\}$ so that it satisfies the first four items from the definition of the poset p, getting a set a_q. Since the set $a_q \setminus a_p$ contains only trees that project into subsets of the set C, the definition of the set B shows that the pair $q = \langle a_q, b_p \rangle$ satisfies the last item in the definition of the forcing P, and it is the condition sought in the claim. $\qquad \square$

The following two theorems have proofs almost literally identical to the corresponding results in Section 4.1.

Theorem 4.2.6. *If I is a porosity σ-ideal then P_I forces that every function $f : \omega_1 \to 2$ in the extension has an infinite ground model subfunction.*

Theorem 4.2.7. *(ZFC+LC) Suppose that I is a σ-ideal generated by a universally Baire porosity and $V \subseteq V[H] \subseteq V[G]$ are a ground model, a P_I-extension, and an intermediate extension. Then either $V[H]$ is given by a single Cohen real or it is equal to $V[G]$.*

4.2.2 Metric porosity

The study of σ-ideals associated with metric porosities was the original motivation for developing the concept of general porosity.

Definition 4.2.8. *[81] Let X, d be a compact metric space, let $A \subset X$ be a set and $x \in X$ a point. The* metric porosity *of the set A at point x is defined as*

$$\limsup_{\delta > 0} \frac{r(A, x, \delta)}{\delta}$$

where $r(A, x, \delta)$ is the supremum of the radii of open balls which are subsets of the open ball around x with radious δ and are disjoint from the set A. A set is metrically porous *if it has nonzero porosity at all of its points. A set is* metrically σ-porous *if it can be decomposed into countably many metrically porous sets.*

It is not difficult to check that the σ-ideal of σ-porous sets can be generated by an abstract porosity in the sense of Definition 4.2.1. Just let the set U be the collection of open balls with rational radii and centers in some fixed countable dense subset of the space X, and for $a \subset U$ let $x \in \mathrm{por}(a)$ if $\limsup_{\delta > 0} \frac{s(a,x,\delta)}{\delta} > 0$ where $s(a, x, \delta)$ is the supremum of the radii of the balls in a which are subsets of the ball of radius δ around the point x.

The most important result regarding the associated partial orders has been obtained by Zajíček and Zelený as a corollary of their results in [80].

Proposition 4.2.9. *[80] Let X, d be a compact metric space and I the σ-ideal of metrically σ-porous sets. Then the forcing P_I is bounding.*

There is an alternative game theoretic proof of Diego Rojas showing that at least in the case of a zero-dimensional space X compact sets are dense in the forcing P_I. This proof generalites to such variations of metric porosity as strong porosity or symmetric porosity.

What are the finer forcing properties of the resulting partial orders? In general, the σ-porous ideal I on a metric space X, d may not preserve outer Lebesgue measure. Humke and Preiss [27] showed that the ideal associated with the metric porosity on the real numbers is not polar; I do not know if the resulting forcing preserves outer Lebesgue measure. The metric porosity on the space constructed in the following example certainly does collapse outer Lebesgue measure.

Example 4.2.10. Let $0 < p_n < 1 : n \in \omega$ be real numbers such that $\Pi_n p_n \neq 0$, and let $k_n : n \in \omega$ be positive natural numbers such that $\Pi_n p_n^{k_n} = 0$. Consider the space $X = \Pi_n k_n$ with the least difference metric $d(x, y) = 2^{-\Delta(x,y)}$. I claim that the associated forcing P_I makes the ground model reals null. It will be enough to find a Polish probability measure space Y, μ and a Borel set $D \subset X \times Y$ such that the vertical sections of D have μ-mass one and its horizontal sections are σ-porous.

Let T be the tree of all finite sequences t such that $t(n) \in k_n$ for all numbers $n \in \text{dom}(t)$; thus $X = [T]$. Consider the space $Y = 2^T$ equipped with the probability measure μ defined by the demand $\mu(\{y \in Y : y(t) = 0\}) = p_{|t|}$. The set $C = \{y \in Y : \text{all but finitely many nodes } t \in T \text{ have an immediate successor } s \text{ such that } y(s) = 1\} \subset Y$ has μ-mass one. Consider the Borel set $D = \{\langle x, y \rangle \in X \times C : \forall^\infty n \; y(x \restriction n) = 0\} \subset X \times C$. Its vertical sections have μ-mass one, while for every point $y \in Y$, the horizontal section D^y is the union of the sets $A_n : n \in \omega$, where $A_n = \{x \in X : \forall m > n y(x \restriction m) = 1\}$ is a set of porosity $1/2$ at each of its point by the definition of the set C. Thus $I \perp I_\mu$ as required.

4.2.3 σ-continuity

Lusin considered the question whether there is a Borel function which cannot be decomposed into countably many continuous functions (is not σ-*continuous*). The answer turns out to be positive, and Pawlikowski [8] provided a particularly simple example. The function $f : (\omega + 1)^\omega \to \omega^\omega$ is defined by $f(x)(n) = x(n) + 1$ if $x(n) \in \omega$ and $f(x)(n) = 0$ otherwise. Here the space $\omega + 1$ is equipped with the order topology and the space $(\omega + 1)^\omega$ then with the product topology. It turns out that the function f is not σ-continuous and it is the simplest such an example.

Fact 4.2.11. *Suppose that $g : X \to Y$ is a Borel function and $A \subset X$ is an analytic set. Then*

1. *either $g \restriction A$ is σ-continuous;*
2. *or there are topological embeddings π and ψ such that the following diagram commutes.*

$$
\begin{array}{ccc}
A & \xrightarrow{\;g\;} & Y \\
\uparrow{\scriptstyle \pi} & & \uparrow{\scriptstyle \psi} \\
(\omega + 1)^\omega & \xrightarrow{\;f\;} & \omega^\omega
\end{array}
$$

Under AD this is true for all functions and all sets.

This has been proved originally by Solecki [70] for Baire class one functions and later extended [83] to all Borel functions.

There is a natural associated forcing. Suppose that $f : X \to Y$ is a Borel non-σ-continuous function and I is the σ-ideal generated by sets on which f is continuous.

Then I is a porosity ideal. To see this choose countable bases $\mathcal{O}$ for X and $\mathcal{P}$ for Y, let $U = \{u(O, P) : O \in \mathcal{O}, P \in \mathcal{P}\}$ where $u(O, P) = \{x \in O : f(x) \notin P\}$ and $\mathrm{por}(a) = \{x \in X : \forall O \in \mathcal{O}\ x \in O \to \exists P \in \mathcal{P}\ f(x) \in P \wedge u(O, P) \in a\}$. A review of definitions will show that the function f is continuous on some set X if and only if A is porous with respect to this porosity.

Thus the forcing P_I is proper and preserves nonmeagerness by Theorem 4.2.2. The continuous reading of names fails as witnessed by the function f and clearly the forcing has been designed in a minimal way for such a purpose. The Fact shows that the forcing is densely naturally isomorphic to P_J where J is the σ-continuity ideal obtained from the Pawlikowski function, and if initially f was the Pawlikowski function then the ideal I is homogeneous. Thus without loss of generality it is possible to consider only the case of the σ-ideal I generated by the sets of continuity of the Pawlikowski function.

The forcing P_I is proper and it preserves Baire category. It is not bounding since it does not have the continuous reading of names by its definition. I will show that it preserves outer Lebesgue measure. The argument uses a claim of independent interest. Let $f : X \to Y$ be a Borel function, let λ be the outer Lebesgue measure on 2^ω, and let P be the Solovay forcing of λ-positive Borel subsets of 2^ω ordered by inclusion.

Claim 4.2.12. *If $p \in P$ is a condition and $\dot{A}$ is a P-name such that $p \Vdash \dot{f} \restriction \dot{A} \cap V$ is continuous, then there are sets $B_n : n \in \omega$ in the ground model such that $p \Vdash \dot{A} \cap V \subset \bigcup_n \check{B}_n$ and $f \restriction B_n$ is continuous for every $n \in \omega$.*

Proof. For every clopen set $O \subset 2^\omega$ let $B_O = \{x \in X : \exists q_{x,O} \subset p \cap O\ \lambda(q_{x,O}) > \frac{3}{4}\lambda(O) \wedge q_{x,O} \Vdash \check{x} \in \dot{A}\}$. It will be enough to show that the sets $B_O : O$ a clopen set, have the properties stated in the claim. The Lebesgue density theorem shows that $p \Vdash \dot{A} \cap V \subset \bigcup_O \check{B}_O$. To show that $f \restriction B_O$ is continuous, assume for contradiction it is not and find a sequence $x_n : n \in \omega$ of points in B_O together with its limit $x \in B_O$ such that the set of values $\{f(x_n) : n \in \omega\}$ does not have $f(x)$ as an accumulation point. Clearly, the condition $\bigcap_m \bigcup_{n>m} q_{x_n,O} \cap q_{x,O}$ has λ-mass $> \frac{1}{4}\lambda(O)$ and it forces that infinitely many of the x_n's as well as x belong to the set $\dot{A}$, contradicting the assumption that $\dot{f} \restriction \dot{A} \cap V$ is continuous. $\square$

Now look at the forcing P_I. If it does not preserve outer Lebesgue measure, then by the results of Section 3.2.2 $I \perp \lambda$ holds and so there are Borel sets $B \subset X$ and $D \subset X \times 2^\omega$ such that $B \notin I$ the vertical sections of D are Lebesgue null while the horizontal sections of its complement are I-small. This means that there are P-names $\dot{A}_n : n \in \omega$ for sets such that $\dot{f} \restriction \dot{A}_n$ is continuous for every $n \in \omega$ and $B \cap V \subset \bigcup_n \dot{A}_n$. Together with the previous claim this contradicts the I-positivity of the set B.

4.2.4 Uniform convergence

Let X be a Polish space with a countable topology basis $\mathcal{O}$ and $f_n : X \to Y$, $n \in \omega$ be a sequence of Borel functions into a metric space Y with a fixed metric d, converging pointwise. Given a point $x \in X$ and a set $A \subset X$, consider the *oscillation* $\mathrm{osc}(x, A)$, the supremum of all real numbers $q \geq 0$ such that for every open neighborhood $O \subset X$ containing the point x, it is the case that $\inf_p \sup\{d(f_m(y), f_n(y)) : m > n > p, y \in O \cap A\} \geq q$. This notion comes from the work of Zalcwasser, and it is connected with the notion of uniform convergence [40], p. 279. It is not difficult to show that whenever $K \subset X$ is a compact set then $\forall x \in K \, \mathrm{osc}(x, K) = 0 \leftrightarrow$ the sequence $\langle f_n : n \in \omega \rangle$ converges uniformly on K. Let I be the σ-ideal generated by sets $A \subset X$ such that $\mathrm{osc}(x, A) = 0$ for all $x \in A$. I will show that this is a porosity ideal, and at least in the case of zero-dimensional space X compact sets are dense in P_I.

First, for every rational number $q > 0$, every basic open neighborhood $O \subset X$ and every number $p \in \omega$ let $u(q, O, p) = \{x \in O : \exists m > n > p \; d(f_m(x), f_n(x)) \geq q\}$, let $U = \{u(q, O, p)\}$ and let the porosity $\mathrm{por} : \mathcal{P}(U) \to \mathcal{B}(X)$ be defined by $\mathrm{por}(a) = \{x \in X : \forall q > 0 \, \exists O \in \mathcal{O} \, \exists p \in \omega \; x \in O \wedge u(q, O, p) \in a\}$. Clearly this is an inclusion-preserving function. It is straightforward to check that for every set $A \subset X$, $\forall x \in A \; \mathrm{osc}(x, A) = 0 \leftrightarrow \exists a \subset U \; A \subset \mathrm{por}(a) \setminus \bigcup a$ and so the σ-ideal I coincides with the one derived from the porosity por.

To show that compact sets are dense in the poset P_I, for every set $A \subset X$ define a game $G_{\mathrm{osc}}(A)$ between Players I and II. Player I gradually constructs a point $x \in X$, player II builds subsets $a_n : n \in \omega$ of the set U and player I wins if $x \in A \setminus \bigcup_n (\mathrm{por}(a_n) \setminus \bigcup a_n)$. I must specify the schedule for both players. Fix an enumeration $\{O_j : j \in \omega\}$ of a basis for the Polish space X and $\{q_j : j \in \omega\}$ of nonnegative rationals. At round $j \in \omega$, Player I must specify whether $x \in O_j$ or not, and Player II must indicate which of the sets $u(q_k, O_l, p)$ belong to which sets a_n, for $k, l, p, n \in j$. This defines the game.

Claim 4.2.13. *Player II has a winning strategy in the game $G_{\mathrm{por}}(A)$ if and only if $A \in I$.*

The right-to-left implication is immediate–if $A \in I$ then Player II can win ignoring opponent's moves entirely. On the other hand, if Player II has a winning strategy σ then consider the sets $A_n = \{x \in A : \text{if Adam produces the point } x \text{ then the strategy } \sigma \text{ will beat him by arranging } x \in \mathrm{por}(a_n) \setminus \bigcup a_n\}$. It is clear that $A = \bigcup_n A_n$ since the strategy σ was winning, so it will be enough to prove that $\forall x \in A_n \; \mathrm{osc}(x, A_n) = 0$. Fix a point $x \in A_n$ and consider the play against the strategy σ in which Player I produces the point x. For every rational number $q > 0$ there must be an initial segment τ of the play of length j and numbers $l, p \in j$ such that $x \in O_l$ and the strategy σ put the set $u(q, O_l, p)$ in the set a_n. Look at the open set that Player I committed to at this round, this is the open set $P = \bigcap\{O_k : k \in j \wedge x \in O_k\} \cap \bigcap\{X \setminus O_k : k \in j \wedge x \notin O_k\}$ containing the point x. I claim that for every point $y \in A_n \cap O_l \cap P$ it is the case that $\forall m > n > p \ d(f_m(y), f_n(y)) < q$, showing that $\operatorname{osc}(x, A) \leq q$. And indeed, every such a point $y \in A_n$ can be produced only with a play whose initial segment is τ and by the definition of the set A_n it must fall out of the set $u(q, O_l, p)$. The definition of the set $u(q, O_l, p)$ says then exactly that $\forall m > n > p \ d(f_m(y), f_n(y)) < q$.

The rest of the proof is standard. Suppose that $B \notin I$ is a Borel set, and use the claim and Borel determinacy, 1.4.2, to find a winning strategy σ for Player I in the game $G_{\operatorname{osc}}(B)$. Consider the set $C \subset X$ of all points that the strategy σ can output against some counterplay. Then

- $C \subset B$ since the strategy σ was winning;
- $C \notin I$ since the strategy σ remains winning in the game $G_{\operatorname{osc}}(C)$ for obvious reasons;
- C is compact since $C = g''Z$ where Z is the compact space of all counterplays of Player I and $g : Z \to X$ is the continuous function which assigns to each play the point the strategy σ produces against that play. Note that Z is compact because Player I has only finitely many possible moves at any given round, and the function g is continuous, because the preimage of any open set O_j is the clopen set of all plays against which the strategy σ was forced to admit at round j that the resulting point will be in the set O_j.

It follows that every Borel I-positive set contains a compact I-positive subset. This extends to analytic sets by Theorem 4.2.3 and in the context of AD to all sets.

4.2.5 Differentiability

Let $f : [0, 1] \to [0, 1]$ be a function. If $A \subset [0, 1]$ is a set and $x \in A$ is a point I will say that the function $f \upharpoonright A$ is differentiable at a point x if either x is isolated in A or else the numbers $\frac{f(y)-f(x)}{y-x} : y \in A$ converge as y tends to the point x. I will say that the function f is differentiable at A if $f \upharpoonright A$ is differentiable at all points of the set A.

Let I be the σ-ideal generated by sets on which the function f is differentiable. I will show that this is a porosity ideal. This implies that the associated forcing is proper. I am really in dark as to which further forcing properties it may have and how it depends on the function f.

Let $U = \{u(d, e) : d, e \subset [0, 1] \text{ are rational intervals}\}$ where $u(d, e) = \{x \in d : f(x) \in e\}$. The porosity function is defined by the following formula. If $a \subset U$ is a set then $x \in \operatorname{por}(a)$ if for every rational number q there is a cone centered at the point $(x, f(x))$ of rational slope and thickness $\leq q$ and a basic open interval c containing x such that for every pair of rational intervals d, e, if $d \subset c$ and $d \times e$ is disjoint from the cone then $u(d, e) \in a$}. It is clear that this is an inclusion-preserving function and its values are Borel sets. A review of the definitions shows that a set $A \subset [0, 1]$ is porous if and only if the function f is differentiable on it.

4.3 Capacities

If ϕ is a subadditive capacity on a Polish space X one can consider the σ-ideal $I_\phi = \{A \subset X : \phi(A) = 0\}$ and the associated forcing P_{I_ϕ}. It turns out that properness and other forcing properties of the poset depend heavily on the measure theoretic properties of the capacity ϕ.

4.3.1 Some measure-theoretic definitions

I will start this section with a brief restatement of several measure-theoretic definitions and facts.

Definition 4.3.1. *[40], 30.1. A* capacity *on a Polish space X is a function ϕ : $\mathcal{P}(X) \to \mathbb{R}^+$ satisfying the following demands:*

1. *(monotonicity) $\phi(0) = 0$, $A \subset B \to \phi(A) \leq \phi(B)$.*
2. *(continuity in increasing unions) If $A_n : n \in \omega$ is an inclusion increasing sequence of subsets of the space X then $\phi(\bigcup_n A_n) = \sup_n \phi(A_n)$.*
3. *(outer regularity on compact sets) if $K \subset X$ is a compact set then $\phi(K) = \inf\{\phi(O) : K \subset O, O$ open$\}$. This implies that ϕ is continuous in decreasing intersections of compact sets.*
4. *$\phi(K) < \infty$ for every compact set $K \subset X$.*

Note that the way it is stated above, a capacity is defined on all subsets of the space X. All capacities I will encounter are *outer* capacities in the sense that they are determined by their values on Borel sets through the property $\phi(A) = \inf\{\phi(B) : A \subset B, B$ Borel$\}$. It is not difficult to see that if ϕ satisfies the above properties on Borel sets then the outer capacity derived from it satisfies these properties as well.

In fact, every outer capacity is determined by its values on basic open sets from any basis closed under finite unions. We set $\phi(O) = \sup\{\phi(D) : D \subset O, D \in \mathcal{B}\}$ for open sets O, $\phi(K) = \inf\{\phi(O) : K \subset O, O$ open$\}$ for compact sets K, $\phi(B) = \sup\{\phi(K) : K \subset B, K$ compact $\}$ for Borel sets B, and $\phi(A) = \inf\{\phi(B) : A \subset B, B$ Borel$\}$. The first step is justified by the continuity of the capacity in increasing unions and the second by the outer regularity on compact sets. The third step is justified by Choquet's theorem, the central fact about capacities:

Fact 4.3.2. *[40], 30.13. Suppose that ϕ is a capacity on a Polish space X and $A \subset X$ is an analytic set. Then $\phi(A) = \sup\{\phi(K) : K \subset A, K \text{ compact}\}$.*

Many capacities used in this book share some other properties:

Definition 4.3.3. *The capacity ϕ is* subadditive *if $\phi(A \cup B) \leq \phi(A) + \phi(B)$.*

For a subadditive capacity ϕ the collection $I_\phi = \{A \subset X : \phi(A) = 0\}$ forms a σ-ideal, and I will naturally study the forcing properties of the forcing P_{I_ϕ}. A frequently encountered stronger property is that of strong subadditivity.

Definition 4.3.4. *The capacity ϕ on a Polish space X is* strongly subadditive *if $\phi(A) + \phi(B) \geq \phi(A \cup B) + \phi(A \cap B)$ whenever $A, B \subset X$. I will frequently use a trivial restatement of this inequality: $\phi(A \cup B) - \phi(A \cap B) \leq (\phi(A) - \phi(A \cap B)) + (\phi(B) - \phi(A \cap B))$.*

Strongly subadditive capacities share a number of properties of interest. Among them:

Fact 4.3.5. *[6] Let X be a compact zero-dimensional Polish space, $\mathcal{O}$ the collection of its clopen sets, and $\phi : \mathcal{O} \to \mathbb{R}^+$ a strongly subadditive submeasure on $\mathcal{O}$. Then ϕ can be extended in a unique fashion to an outer capacity on the space X. The extension capacity is given by $\psi(O) = \sup\{\phi(P) : P \subset O, P \in \mathcal{O}\}$ for open sets $O \subset X$, and $\psi(A) = \inf\{\psi(O) : A \subset O, O \subset X \text{ open}\}$ for all other sets $A \subset X$.*

Theorem 4.3.6. *Suppose that ϕ is an outer strongly subadditive capacity on a compact Polish space X.*

1. *[6] ϕ is outer regular on all sets, that is $\phi(A) = \inf\{\phi(O) : A \subset O, O \text{ open}\}$ for all sets A.*
2. *[6] ϕ is an envelope of measures, that is $\phi(B) = \sup\{\psi(B) : \psi \text{ is a measure less or equal to } \phi \text{ on all Borel sets}\}$.*
3. *In the Solovay model, every set has a Borel subset of the same capacity and ϕ is continuous in increasing wellordered unions.*
4. *(ZF+DC+AD+) Every set has a Borel subset of the same capacity and ϕ is continuous in increasing wellordered unions.*
5. *ϕ is continuous in increasing wellordered unions of length $< \text{add(null)}$. In particular, if $\text{add(null)} > \aleph_1$ then every coanalytic set has a Borel subset of the same capacity.*

6. *Whenever P is a forcing with the Sacks property and $\varepsilon > 0$ is a real number, then every open set of capacity $< \varepsilon$ in the extension is covered by such an open set in the ground model.*

I will tackle the problem of capacitability of coanalytic sets in ZFC in Theorem 4.3.21.

Proof. Fix a countable topology basis $\mathcal{O}$ for the Polish space X closed under finite unions and intersections.

For (4), suppose that $B \subset X \times 2^\omega$ is an arbitrary set. I must find an analytic subset $A \subset B$ such that $\phi(\text{proj}(A)) = \phi(\text{proj}(B))$. An application of Proposition 3.9.20 will then complete the proof of (4). Fix a countable basis $\mathcal{O}$ of the Polish space X closed under finite unions, fix a Borel bijection $\pi : 2^\omega \to X \times \omega^\omega$, and let $\varepsilon > 0$ be a real number. Consider the game $G_\varepsilon(B)$ between Players I and II in which Player I indicates basic open sets $P_n \in \mathcal{O}$ such that $P_0 \subset P_1 \subset \ldots, \phi(P_{n+1}) - \phi(P_n) < 2^{-3n-2}$ and $\phi(P_n) \leq \varepsilon$, and Player II gradually builds a binary sequence $x \in 2^\omega$; I will write $\text{proj}(\pi(x))$ for the projection of the point $\pi(x) \in X \times \omega^\omega$ into the first coordinate. Player II is allowed to postpone the decision-making process by an arbitrary number of rounds. Player II wins if $\text{proj}(\pi(x)) \notin \bigcup_n P_n$ and $\pi(x) \in B$. The following claim is key.

Claim 4.3.7. *If $\varepsilon > \phi(\text{proj}(B))$ then Player I has a winning strategy. If $\varepsilon < \phi(\text{proj}(B))$ then Player I does not have a winning strategy.*

Proof. If $\varepsilon > \phi(\text{proj}(B))$ then Player I can win simply by constructing an open set $P = \bigcup_n P_n$ such that $\phi(P) < \varepsilon$ and $\text{proj}(B) \subset P$. On the other hand, if $\varepsilon < \phi(\text{proj}(B))$ and σ is a strategy for Player I then write $\delta = (\phi(\text{proj}(B)) - \varepsilon)/2$ and consider the collection U of all basic open sets that σ can produce if Player II postpones his first decision to a number m such that $2^{-m} < \delta$. It will be enough to show that $\phi(\bigcup U) < \phi(\text{proj}(B))$. Then any point $\langle x, y \rangle \in B$ with $x \notin \bigcup U$ will provide a counterplay against the strategy σ in which Player II wins.

To show that $\phi(\bigcup U) < \phi(\text{proj}(B))$, for any number $n \geq m$ write U_n for the collection of all open sets the strategy σ can play at round n in some play in which Player II postponed his first decision to round m. Since $\bigcup U$ is the increasing union $\bigcup_n \bigcup U_n$ and the capacity ϕ is continuous in increasing unions, it will be enough to show that $\phi(\bigcup U_n) < \varepsilon + (1 - 2^{-n})\delta$. The proof of this statement proceeds by induction on n; it is clearly true for $n \leq m$. Now suppose that it has been verified for some $n \geq m$. To proceed to $n+1$ note that the set U_{n+1} consists of $< 2^{n+1}$ sets, each of which has ϕ-mass $< \phi(P) + 2^{-n} \cdot \delta$ for some set $P \in U_n$. A simple exercise in strong subadditivity then shows that $\phi(\bigcup U_{n+1}) < \phi(\bigcup U_n) + 2^n \cdot 2^{-3n-2} \leq \phi(\bigcup U_n) + 2^n \cdot 2^{-2n-2} \cdot \delta < \varepsilon + (1 - 2^{-n-1})\delta$ as required. $\square$

Now let $\varepsilon < \phi(B)$ be a real number. To complete the proof of (4), it is enough to find an analytic set $A \subset B$ such that $\phi(\text{proj}(A)) > \varepsilon$. Use a determinacy argument and the previous claim to find a winning strategy σ for Player II in the game $G_\varepsilon(B)$. Let A be the collection of all points the strategy σ can generate against a counterplay of Player I. The winning condition for Player II shows that $A \subset B$, moreover the set A is clearly analytic. Since the strategy σ remains winning for Player II in the game $G_\varepsilon(A)$, another reference to the Claim shows that $\phi(A) \geq \varepsilon$. (4) follows.

(5) is an immediate consequence of the following claim.

Claim 4.3.8. *Suppose that κ is an uncountable regular cardinal, ε, δ are positive real numbers, and $B_\alpha : \alpha \in \kappa$ are sets of ϕ-mass $< \varepsilon$. Then there is an infinite set $b \subset \kappa$ such that $\phi(\bigcup_{\alpha \in b} B_\alpha) < \varepsilon + \delta$. If moreover $\kappa < \text{add(null)}$ then the set b can be found of size κ.*

Proof. Let me first argue for the last sentence. For every ordinal $\alpha \in \kappa$ choose an open set $O_\alpha \subset X$ covering the set B_α, with $\phi(O_\alpha) < \varepsilon + \delta/2$. Let $f_\alpha : \omega \to \mathcal{O}$ be some function such that the sets $f_\alpha(n) : n \in \omega$ form an increasing sequence with union O_α and for every number $n \in \omega$, $\phi(f_\alpha(n+1)) < \phi(f_\alpha(n)) + 2^{-2n-2} \cdot \delta$. Since $\kappa < \text{add(null)}$, [2] shows that there is a tunnel $g : \omega \to \mathcal{P}(\mathcal{O})$ such that for every number $n \in \omega$, $|g(n)| < 2^n$, and for every ordinal $\alpha \in \kappa$ there is a number m_α such that $\forall n > m_\alpha\ f_\alpha(n) \in g(n)$. Use a counting argument to find a fixed number $m \in \omega$ and a fixed sequence $t \in \mathcal{O}^m$ such that the set $S = \{\alpha \in \kappa : f_\alpha \restriction m = t \wedge \forall n \geq m\ f_\alpha(n) \in g(n)\}$ is of size κ. It will be enough to show that the open set $\bigcup_{\alpha \in S} O_\alpha$ has ϕ-mass $< \varepsilon + \delta$. By the continuity of the capacity in increasing unions, it will be enough to show that the open sets $P_n = \bigcup_{\alpha \in S} f_\alpha(n)$ have capacity $< \varepsilon + \delta$ for every number $n \in \omega$. The proof of this statement proceeds by induction on n. It is clearly true for $n \leq m$. For the induction step, suppose it holds at some number $n \geq m$ and note that the collection $\{f_\alpha(n+1) : \alpha \in S\}$ has size $\leq 2^{n+1}$ by the choice of g and m. Since $\phi(f_\alpha(n+1)) < \phi(f_\alpha(n)) + 2^{-2n-2} \cdot \delta$, a simple exercise in strong subadditivity shows that $\phi(P_{n+1}) < \phi(P_n) + 2^{n+1} \cdot 2^{-2n-2} \cdot \delta \leq \phi(P_n) + 2^{-n-1} \cdot \delta \leq \varepsilon + (1 - 2^{-n-1})\delta$ as desired.

The second sentence now abstractly follows from the previous paragraph. Find a countable elementary submodel M of a large structure and force over M to increase add(null) past κ with a c.c.c. forcing. The generic extension $M[g]$ satifies the formula that there is an open set O of capacity $< \varepsilon + \delta$ and a set $b \subset \kappa$ of size κ such that $O_\alpha \subset O$ for every ordinal $\alpha \in b$. Now in V, the set b is still infinite, and an absoluteness argument shows that $\forall \alpha \in b\ B_\alpha \subset O$ as required. $\square$

Finally, (3) abstractly follows from (5). Suppose that κ is an inaccessible cardinal, $G \subset \text{Coll}(\omega, < \kappa)$ is a generic filter and $V(\mathbb{R} \cap V[G])$ is the derived choiceless Solovay model. Suppose that $B_\alpha : \alpha \in \kappa$ is some sequence of sets in the model $V(\mathbb{R} \cap V[G])$, all of ϕ-mass $\leq \varepsilon$. I must argue that $\bigcup_\alpha B_\alpha$ has ϕ-mass $\leq \varepsilon$.

Let h be a $V[G]$-generic filter for the amoeba forcing for measure. The model $V(\mathbb{R} \cap V[G][h])$ is again a Solovay model, there is a unique elementary embedding $j : V(\mathbb{R} \cap V[G]) \to V(\mathbb{R} \cap V[G][h])$ fixing the ground model pointwise, and the proof of (5) shows that in the model $V(\mathbb{R} \cap V[G][h])$ the set $\bigcup_\alpha j(B_\alpha)$ has ϕ-mass equal to ε. An elementarity argument then yields $V(\mathbb{R} \cap V[G]) \models \phi(\bigcup_\alpha B_\alpha) = \varepsilon$ as required.

The last item is left to the reader. $\qquad\square$

The investigation of the finer forcing properties of the forcing P_{I_ϕ} lead to the following measure theoretic definition, which is to guarantee that the forcing does not add splitting reals.

Definition 4.3.9. *A capacity ϕ on a Polish space X is* Ramsey *if for all real numbers $\varepsilon, \delta > 0$ and every infinite collection $A_n : n \in \omega$ of subsets of the space X of capacity $< \varepsilon$ there are distinct indices $n \neq m$ such that $\phi(A_n \cup A_m) < \varepsilon + \delta$.*

In the way of examples note that the outer Lebesgue measure is not Ramsey as any infinite collection of stochastically independent sets of measure $1/2$ will show. For most capacities I do not know how to decide whether they are Ramsey or not, see Question 7.3.5. Many Ramsey capacities are produced in Sections 4.3.5 and 4.3.6. The treatment of Ramsey capacities is in many ways parallel to the strongly subadditive capacities.

Theorem 4.3.10. *Let X be a compact zero-dimensional Polish space, $\mathcal{O}$ the collection of its clopen sets, and $\phi : \mathcal{O} \to \mathbb{R}^+$ a Ramsey submeasure on $\mathcal{O}$. Then ϕ can be extended in a unique fashion to an outer capacity ψ on the space X. The extension capacity is given by $\psi(O) = \sup\{\phi(P) : P \subset O, P \in \mathcal{O}\}$ for open sets $O \subset X$, and $\psi(A) = \inf\{\psi(O) : A \subset O, O \subset X \text{ open}\}$ for all other sets $A \subset X$. The capacity ψ is Ramsey.*

Proof. Fix an enumeration of the countable basis $\mathcal{O}$ consisting of clopen sets. The following two claims, providing equivalent restatements of Ramseyness, are the main ingredients of the proof.

Claim 4.3.11. *For all positive real numbers $\varepsilon, \delta > 0$ and every sequence $U_n : n \in \omega$ of basic open sets of ϕ-mass smaller than ε there is an infinite set $b \subset \omega$ such that ϕ-masses of finite unions of sets among $U_n : n \in b$ are smaller than $\varepsilon + \delta$.*

Proof. This is just a repeated use of Ramsey theorem. By induction on $m \in \omega$ build infinite sets $b_m \subset \omega$ and numbers $n_m \in \omega$ such that

- $\omega = b_0 \supset b_1 \supset \dots$
- $n_m = \min(b_{m+1})$;
- the numbers $\phi(\bigcup_{l \in m} U_{n_l} \cup O_k) : k \in b_m$ are bounded below δ.

Once this is done, the infinite set $b = \{n_m : m \in \omega\}$ has the desired properties. The induction hypotheses trivially hold at $m = 0$. To perform the inductive step at a given natural number m, consider the sets $V_k = \bigcup_{l \in m} U_{n_l} \cup U_k$ for all numbers $k \in b_m$. These form an infinite collection of sets of ϕ-mass less than $\varepsilon + \delta'$ for some $\delta' < \delta$. Coose a real number $\delta' < \delta'' < \delta$ and consider the partition $p : [b_m]^2 \to 2$ defined by $p(h, k) = 0$ iff $\phi(V_k \cup V_l) \geq \varepsilon + \delta_{m+2}$. The Ramsey property of the capacity implies that there are no infinite 1-homogeneous sets, and the Ramsey theorem implies that there must be an infinite 0-homogeneous set $b_{m+1} \subset b_m$. The induction step is completed by letting $n_m = \min(b_{m+1})$.   □

Claim 4.3.12. *For all positive real numbers* $\varepsilon, \delta > 0$ *there is a number* $n \in \omega$ *such that for every basic open set* $U \subset X$ *with* $\phi(U) < \varepsilon$ *there is a basic open set* $V \subset X$ *among the first n many basic open sets such that* $\phi(U \cup V) < \varepsilon + \delta/2$ *and for every basic open* $W \subset X$, *if* $\phi(W \cup V) < \varepsilon + \delta/2$ *then* $\phi(W \cup V \cup U) < \varepsilon + \delta$.

Proof. Suppose this fails for some $\varepsilon, \delta > 0$, and for every $n \in \omega$ find an open set $U_n \subset X$ which forms a counterexample; in particular, $\phi(U_n) < \varepsilon$. Use the previous claim to find an infinite subset $b \subset \omega$ such that the ϕ-masses of finite unions of sets among $\{U_n : n \in b\}$ are bounded below $\varepsilon + \delta/2$. Let $V_m : m \in \omega$ be an inclusion-increasing collection of basic open sets exhausting the open set $\bigcup_{n \in b} U_n$, and let $n_m : m \in \omega$ be numbers in the set b such that V_m is among the first n_m many basic open sets. Since V_m does not work for U_{n_m}, there are basic open sets W_m such that $\phi(V_m \cup W_m) < \varepsilon + \delta/2$ but $\phi(V_m \cup W_m \cup U_{n_m}) \geq \varepsilon + \delta$. Use the Ramsey property and the Ramsey theorem to find an infinite set $c \subset \omega$ so that the ϕ-masses of unions of pairs of sets among $\{V_m \cup W_m : m \in C\}$ are bounded below $\varepsilon + \delta$. Let $m \in c$, and find $k \in c$ such that $U_{n_m} \subset V_k$. Now $V_m \cup W_m \cup U_{n_m} \subset V_m \cup W_m \cup V_k \cup W_k$, and so $\phi(V_m \cup W_m \cup U_{n_m}) < \varepsilon + \delta$. Contradiction!   □

Now consider the function ψ defined in the statement of the theorem. To show that this is a capacity, it is only necessary to prove that it is continuous in increasing unions. I must first prove the continuity of ψ in increasing unions of open sets. If $O_n : n \in \omega$ are open sets forming an increasing union and $U \in \mathcal{O}$ is a basic clopen set, a subset of the union $\bigcup_n O_n$, then, since U is compact, U must be a subset of one of the sets O_n. This immediately implies that $\psi(\bigcup_n O_n) = \sup_n \psi(O_n)$. Suppose now that $A_n : n \in \omega$ is an increasing union of arbitrary sets of ψ-mass $< \varepsilon$, and suppose that $\delta > 0$ is a real number. To show that $\psi(\bigcup_n A_n) \leq \varepsilon + \delta$, choose open sets $O_n \subset X : n \in \omega$ covering the respective sets, with ϕ-masses bounded below δ. I will find an infinite set $b \subset \omega$ such that $\psi(\bigcup_{n \in b} O_n) \leq \varepsilon + \delta$; this will complete the proof of the continuity. The argument closely mimics the proof of Claim 4.3.11.

By induction on $m \in \omega$ build numbers $n_m \in \omega$ and sets $b_m \subset \omega$ so that

- $\omega = b_0 \supset b_1 \supset b_2 \supset \ldots$ are infinite sets;
- $n_m = \min(b_{m+1})$ for all numbers $m \in \omega$;
- for every number m, the numbers $\psi(\bigcup_{l \in m} O_{n_l} \cup O_k) : k \in b_m$ are bounded below $\varepsilon + \delta$.

In the end the set $b = \{n_m : m \in \omega\}$ is as required by the continuity of ψ in increasing unions of open sets. To perform the induction, suppose the set b_m as well as the numbers $n_l : l \in m$ have been found. Find real numbers $\varepsilon', \delta' > 0$ such that all sets $\bigcup_{l \in m} O_{n_l} \cup O_k : k \in b_m$ have mass $< \varepsilon'$ and $\varepsilon' + \delta' < \varepsilon + \delta$. Find a number $n' \in \omega$ that works as in the above claim. A pigeonhole argument shows that there must be a basic open set V among the first n' many sets in the basis, and an infinite set $b_{m+1} \subset b_m$ such that for every number $k \in b_{m+1}$, the set V works as in Claim 4.3.12 for inclusion-cofinally many clopen subsets of the set $\bigcup_{l \in m} O_{n_l} \cup O_k$. It follows immediately that for any two numbers $k, k' \in b_{m+1}$ it is the case that $\psi(\bigcup_{l \in m} O_{n_l} \cup O_k \cup O_{k'}) < \varepsilon' + \delta'$. Let $n_m = \min(b_{m+1})$. The induction hypotheses continue to hold.

Now that I have verified that ψ is a capacity, its uniqueness follows. Any other outer capacity extending ϕ must agree with ψ on compact sets by the definitions, consequently it must agree with ψ on all analytic sets by Choquet's theorem, and finally it must agree with ϕ on all other sets since it is an outer capacity. The proof of Ramseyness of ψ is left to the reader. □

Note that subadditivity of ϕ was not needed anywhere in the proof; in this book I have no use for capacities that are not subadditive though. The claim in the proof shows that the collection of all Ramsey capacities is Borel in a suitable sense.

Theorem 4.3.13. *Suppose that ϕ is an outer Ramsey capacity on a compact Polish space X.*

1. *ϕ is outer regular on all sets, that is $\phi(A) = \inf\{\phi(O) : A \subset O, O \text{ open}\}$ for all sets $A \subset X$.*
2. *In the Solovay model, every set has a Borel subset of the same capacity and ϕ is continuous in increasing wellordered unions.*
3. *(ZF+DC+AD+) Every set has a Borel subset of the same capacity and ϕ is continuous in increasing wellordered unions.*
4. *Whenever κ is a regular cardinal and MA_κ holds then ϕ is continuous in increasing wellordered unions of length κ. In particular, under MA_κ, every coanalytic set has a Borel subset of the same capacity.*

5. *Whenever Q is a forcing with the Laver property, $\varepsilon, \delta > 0$ are real numbers and $O \subset X$ is an open set oin the extension of ϕ-mass $< \varepsilon$, then there is a ground model open set $P \subset X$ such that $O \subset P$ and $\phi(P) < \varepsilon + \delta$.*

Proof. The first item is very similar to the previous theorem. Consider the function $\psi : \mathcal{P}(X) \to \mathbb{R}^+$ given by $\psi(A) = \inf\{\phi(O) : A \subset O$ and $O \subset X$ is an open set$\}$. It will be enough to show that ψ is a capacity: as ψ and ϕ agree on compact sets by the outer regularity of the capacity ϕ on such sets, Choquet's theorem 4.3.2 applied to both of them shows that they must agree on analytic sets, and as ϕ is an outer capacity, this implies that $\phi = \psi$ and so ϕ is outer regular.

To prove that ψ is a capacity, it is only necessary to verify its continuity in increasing unions. Suppose that $A_n : n \in \omega$ is an inclusion-increasing sequence of sets of ψ-mass $< \varepsilon$ for some fixed real number ε and $\delta > 0$ is a real number. To show that $\psi(\bigcup_n A_n) < \varepsilon + \delta$, find open sets $O_n : n \in \omega$ such that $A_n \subset O_n$ and $\phi(O_n) < \varepsilon$, and use the proof of Claim 4.3.11 to find an infinite set $b \subset \omega$ such that the finite unions of sets among $O_n : n \in b$ have ϕ-mass bounded below $\varepsilon + \delta$. By the continuity of ϕ in increasing unions of open sets it follows that $\bigcup_{n \in b} O_n$ is an open set covering $\bigcup_n A_n$ of ϕ-mass $< \varepsilon + \delta$, witnessing that $\bigcup_n A_n < \varepsilon + \delta$ as required.

The next three items use the following key property of the capacity ϕ only. It is a property shared by every outer regular capacity used in mathematical practice, and in the case of Ramsey capacities it is verified by an argument entirely parallel to Claim 4.3.12. Fix a countable topology basis $\mathcal{O}$ for the space X.

(*) for all positive real numbers $\varepsilon, \delta > 0$ and every open set $O \subset X$ with $\phi(O) < \varepsilon$ there is a basic open set $U \subset O$ such that for every open set $P \subset X$ with $\phi(U \cup P) < \varepsilon$ it is the case that $\phi(O \cup P) < \varepsilon + \delta$.

Items (2) and (4) use the following family of forcings. For every positive real $\varepsilon > 0$ let Q_ε be the partial order of open sets of ϕ-mass $< \varepsilon$ ordered by reverse inclusion. It is immediate that the union of sets in the generic filter is an open set of mass $\leq \varepsilon$. The key observation is that the partial order Q_ε is σ-linked. For every open set $O \in Q_\varepsilon$ find rational numbers $q(O) < \varepsilon, r(O) > 0$ such that $\phi(O) < q(O)$ and $q(O) + r(O) < \varepsilon$, and find a basic open set $U(O) \subset O$ that works as in (*) for these two rationals and the set O. Clearly, if $O, P \in Q_\varepsilon$ and $q(P) = q(O), r(P) = r(O)$ and $U(P) = U(O)$, then $\phi(O \cup P) < \varepsilon$ and the two conditions are compatible.

For (4), assume that MA_κ holds and suppose that $A_\alpha : \alpha \in \kappa$ is an inclusion increasing collection of sets of ϕ-mass $\leq \varepsilon$; I must show that $\phi(\bigcup_\alpha A_\alpha) < \varepsilon + \delta$ for any given real number $\delta > 0$. For every ordinal $\alpha \in \kappa$ fix an open set $O_\alpha \supset A_\alpha$ of ϕ-mass $< \varepsilon + \delta/3$. Consider the forcing $Q = Q_{\varepsilon + 2\delta/3}$. I claim that there is a condition in it forcing the generic open set to be a superset of κ many of the sets

$O_\alpha : \alpha \in \kappa$. If this was not the case then the c.c.c. of the forcing would imply that there is a fixed ordinal $\alpha \in \kappa$ such that $Q \Vdash$ for no ordinal $\beta \in \kappa$ greater than α is the open set O_β a subset of the generic filter. This is impossible though, because $O_{\alpha+1} \in Q$ is a condition forcing the opposite. Now use MA_κ to find a filter $g \subset Q$ such that for a cofinal set $b \subset \kappa$, $O_\alpha \bigcup g$. Clearly, $\phi(\bigcup_{\alpha \in b} O_\alpha) \leq \phi(\bigcup g) \leq \varepsilon + 2\delta/3$, and since $\bigcup_\alpha A_\alpha \subset \bigcup_{\alpha \in b} O_\alpha$, it must be the case that $\phi(\bigcup_\alpha A_\alpha) < \varepsilon + \delta$ as desired.

For (2), suppose that κ is an inaccessible cardinal, $G \subset \mathrm{Coll}(\omega, < \kappa)$ is a generic filter and $V(\mathbb{R} \cap V[G])$ is the derived choiceless Solovay model. Suppose that $B_\alpha : \alpha \in \kappa$ is some sequence of sets in the model $V(\mathbb{R} \cap V[G])$, all of ϕ-mass $\leq \varepsilon$. I must argue that $\bigcup_\alpha B_\alpha$ has ϕ-mass $\leq \varepsilon$. Let $\delta > 0$ be an arbitrary positive real number. Let h be a $V[G]$-generic filter for the forcing $Q_{\varepsilon+\delta}$ such that $\bigcup h$ covers cofinally many sets B_α. The model $V(\mathbb{R} \cap V[G][h])$ is again a Solovay model, there is a unique elementary embedding $j : V(\mathbb{R} \cap V[G]) \to V(\mathbb{R} \cap V[G][h])$ fixing the ground model pointwise, and in the model $V(\mathbb{R} \cap V[G][h])$ the set $\bigcup_\alpha j(B_\alpha)$ is covered by the set $\bigcup h$ which has ϕ-mass equal to $\varepsilon + \delta$. An elementary argument then yields $V(\mathbb{R} \cap V[G]) \models \phi(\bigcup_\alpha B_\alpha) = \varepsilon$ as required.

The integer game argument for (3) is somewhat more challenging. For simplicity assume that $X = 2^\omega$. Work under ZF+DC+AD+. Suppose that $\varepsilon, \delta > 0$ are positive real numbers and $B \subset 2^\omega \times \omega^\omega$ is a set such that $\phi(\mathrm{proj}(B)) > \varepsilon$. I must find an analytic subset $A \subset B$ such that $\phi(\mathrm{proj}(A)) \geq \varepsilon$. An application of Proposition 3.9.20 will then complete the proof. Enumerate all pairs of positive rationals and basic open sets by $\langle q_n, U_n \rangle : n \in \omega$. Consider the integer game between Players I and II in which Player I generates an open set $O \subset X$ of ϕ-mass $\leq \varepsilon + \delta$ and Player II produces a point $\langle x, y \rangle \in 2^\omega \times \omega^\omega$. Player II wins if $x \notin O$ and $\langle x, y \rangle \in B$. To finish the description of the game, I must set up a schedule for both players. At round n, Player I determines whether $U_n \subset O$ or not and whether $\phi(O \cup U_n) < q_n$ or not. If in the end his answers are inconsistent with the set O obtained, he produced an *inconsistent run*, and he loses. Player II can wait for an arbitrary finite number of moves (*place trivial moves*) before placing the next pair in $2 \times \omega$ on his sequence. This completes the description of the game.

I will show that Player II has a winning strategy σ. Once this is established, the set $B \subset 2^\omega \times \omega^\omega$ consisting of the points the strategy σ produces against all possible consistent runs by Player I is analytic by its definition and it is contained in A by the winning condition of the game. Moreover, $\phi(\mathrm{proj}(B)) \geq \varepsilon$, since Player I can produce any open set $O \subset 2^\omega$ of capacity $< \varepsilon$ as his resulting set, forcing the strategy σ to come up with a pair $\langle x, y \rangle \in A$ such that $x \notin O$. Note that then, $\langle x, y \rangle \in B$ by the definition of the set B, and so $\mathrm{proj}(B) \not\subset O$ and $\phi(\mathrm{proj}(B)) \geq \varepsilon$.

By a determinacy argument, it is enough to prove that Player I does not have a winning strategy. Suppose that σ is a strategy for Player I which produces only consistent runs (otherwise it is easy to defeat it by inducing an inconsistent run). Choose a real number $\delta > 0$. I will produce an open set $P \subset 2^\omega$ of ϕ-mass $< \varepsilon + \delta$ such that any point $\langle x, y \rangle \in 2^\omega \times \omega^\omega$ with $x \notin P$ can be produced by a counterplay against σ in such a manner that x does not belong to the resulting open set. This will certainly show that the strategy σ is not winning: just choose $\delta > 0$ so that $\phi(\text{proj}(A)) > \varepsilon + \delta$ and a point $\langle x, y \rangle \in A$ such that $x \notin P$ and construct the corresponding counterplay. In it, Player II won.

To construct the open set $P \subset X$, I will first introduce the following notation. For a finite play τ following the strategy σ let $O_\tau \subset X$ be the open set the strategy σ produces in the infinite extension of the play τ in which Player II posts no nontrivial moves past τ. Enumerate the set $(2 \times \omega)^{<\omega}$ as $t_n : n \in \omega$ in such a way that all initial segments of a given sequence are enumerated earlier than the sequence itself. Also, fix positive real numbers $\delta_n : n \in \omega$ whose sum is less than δ. By induction on $n \in \omega$, build finite plays τ_n against the strategy σ so that:

- the nontrivial moves of Player I in the play τ_n together form the sequence t_n;
- if $t_m \subset t_n$ then $\tau_m \subset \tau_n$;
- $\phi(\bigcup_{m \in n} O_{\tau_m}) < \varepsilon + \Sigma_{m \in n} \delta_m$.

In the end, clearly the set $P = \bigcup_n O_{\tau_n}$ is as required, since for any point $\langle x, y \rangle \in 2^\omega \times \omega^\omega$ with $x \notin P$, the play $\bigcup \{\tau_n : t_n \subset \langle x, y \rangle\}$ is a counterplay against the strategy in which Player II produces the point $\langle x, y \rangle$ and x does not belong to the resulting open set. Note also that $\phi(P) \leq \varepsilon + \Sigma_n \delta_n < \varepsilon + \delta$ by the last item of the induction hypothesis.

To perform the induction, start with $\tau_0 = 0$. Suppose that the plays $\tau_m : m \in n$ have been constructed. Use (*) to find a basic open set $U \subset \bigcup_{m \in n} O_{\tau_m}$ such that for every open set $O \subset 2^\omega$, if $\phi(O \cup U) < \varepsilon + \Sigma_{m \in n} \delta$ then $\phi(O \cup \bigcup_{m \in n} O_{\tau_m}) < \varepsilon + \Sigma_{m \in n+1} \delta_m$. To obtain the play τ_n, find a number $k \in n$ such that t_n is an immediate successor of τ_k. Consider the infinite play τ extending τ_k in which Player II places no nontrivial moves past τ_k and Player I follows the strategy σ, and its resulting open set $O_{\tau_k} \subset 2^\omega$. Since $\phi(O_{\tau_k} \cup U) < \varepsilon + \Sigma_{m \in n} \delta_m$, there must be a round in the play τ in which the strategy σ promised that for the resulting set $O \subset X$, $\phi(O \cup U) < q$ for some rational number $q < \varepsilon + \Sigma_{m \in n} \delta_m$. Now let τ_n be the extension of τ_k obtained by Player II placing trivial moves until that round, and then placing the pair $\langle b, l \rangle$ such that $t_k^\frown \langle b, l \rangle = t_n$. The choice of the basic open set U guarantees that the induction hypothesis (3) continues to hold.

The last item needs more careful work, as it is a property that is quite specific to Ramsey capacities. I will need the following strengthening of (*). Fix an enumeration of the countable topology basis $\mathcal{O} = \{U_i : i \in \omega\}$ for the space X closed under finite unions. The following holds and is verified by an argument entirely similar to Claim 4.3.12:

(**) For every $\varepsilon, \delta > 0$ there is a number $m = m(\varepsilon, \delta) \in \omega$ such that for every open set $O \subset X$ with $\phi(O) < \varepsilon$ there is a number $i \in m$ such that $\phi(O \cup U_i) < \varepsilon + \delta/2$ and moreover for every open set $P \subset X$ such that $\phi(P \cup U_i) < \varepsilon + \delta/2$ it is the case that $\phi(O \cup U_i \cup P) < \varepsilon + \delta$.

Now suppose Q is a forcing with Laver property, $\varepsilon, \delta > 0$ are real numbers and some condition $q_0 \in Q$ forces that $\dot{O} \subset X$ is an open set of ϕ-mass $< \varepsilon$. Fix an increasing sequence $\delta_i : i \in \omega$ of real numbers bounded below δ and for a number $n \in \omega$ define the following:

- $k_n = \Sigma_{i \in n} i!$;
- $m_n = \max\{m(\varepsilon + \delta_i, \varepsilon + \delta_{i+1}) : k_n \leq i < k_{n+1}\}$;
- $\dot{f}(n)$ is the Q-name for the pair $\langle \dot{f}_0(n), \dot{f}_1(n) \rangle$ such that $\dot{f}_0(n) = 1 \leftrightarrow U_n \subset \dot{O}$ and $\dot{f}_1(n) = \{\langle j, i \rangle : j \in m_{n+1}, k_{n+1} \leq i < k_{n+2} \text{ and } \phi(\dot{O} \cup U_j) < \varepsilon + \frac{\delta_i + \delta_{i+1}}{2}\}$.

Note that for every $n \in \omega$ there are only finitely many possibilities for the value of $\dot{f}(n)$ in the extension. Let $\dot{f}$ be the Q-name for a function with domain ω obtained from the names $\dot{f}(n) : n \in \omega$. By the Laver property of the forcing Q there must be a tree T and a condition $q_1 \leq q_0$ such that every node of T at the n-th level splits into at most $n+1$ many immediate successors and $q_1 \Vdash \dot{f} \in [\check{T}]$. Thinning out the tree T if necessary I may assume that for every node $t \in T$ there is a condition $q_t \leq q_1$ forcing $t \subset \dot{f}$. Consider the open set $P \subset X$, $P = \bigcup\{U_n : \exists t \in T \, \exists n \in \text{dom}(t) \, t_0(n) = 1\}$. It will be enough to show that $\phi(P) < \varepsilon + \delta$, since clearly $q_1 \Vdash \dot{O} \subset \check{P}$.

Let M be a countable elementary submodel of a large enough structure, and for every node $t \in T$ choose an M-generic filter $g_t \subset M \cap Q$ containing the condition q_1 such that $t \subset \dot{f}/g_t$. Let $O_t = \dot{O}/g_t$. Clearly, $P \subset \bigcup_t O_t$ and it will be enough to show that the latter set has ϕ-mass $< \varepsilon + \delta$. Enumerate the tree T with possible repetitions as $t_i : i \in \omega$ in such a way that $T_n = \{t_i : k_n \leq i < k_{n+1}\}$–note that $|T_n| \leq n!$. Now by induction on $l \in \omega$ prove that the open set $P_j = \bigcup_{i \in j} O_{t_i}$ has ϕ-mass $< \varepsilon + \delta_l$. This is obvious for $l = 1$. Suppose that a number $l \in \omega$ is given and $\phi(P_l) < \varepsilon + \delta_l$. Write $\delta'_l = \frac{\delta_l + \delta_{l+1}}{2}$, and find the number $n \in \omega$ such that $t_l \in T_n$, and find a number $j \in m_n$ such that $\phi(U_j \cup P_l) < \varepsilon + \delta'_l$ and for every open set $R \subset X$ such that $\phi(U_j \cup R) < \varepsilon + \delta'_l$ it is the case that $\phi(P_l \cup U_j \cup R) < \varepsilon + \delta_{j+1}$. Find the number $i \in j$ such that t_j is an immediate successor of the node t_i. Then $O_{t_i} \subset P_k$ and consequently $\phi(U_j \cup O_{t_i}) < \varepsilon + \delta'_l$. The definition of the name $\dot{f}$ and the fact that $\dot{f}(n-1)/g_{t_i} = \dot{f}(n-1)/g_{t_l}$ then imply that $\phi(U_j \cup O_{t_l}) < \varepsilon + \delta'_l$ and the properties of the basic open set U_j imply that $\phi(P_l \cup U_j \cup O_{t_l}) < \varepsilon + \delta_{l+1}$, in particular, $\phi(P_{l+1}) < \varepsilon + \delta_{l+1}$. $\qquad \square$

It turns out that the notion of Ramsey capacity has several interesting features which reach beyond the forcing applications. The Ramsey capacities behave well in the operation of *meet* of submeasures.

Definition 4.3.14. *Let ϕ, ψ be submeasures on a Polish space X. The meet $\phi \wedge \psi$ of these two submeasures is a set function defined by $(\phi \wedge \psi)(A) = \inf\{\phi(B) + \psi(C) : B \cup C = A\}$.*

It is not difficult to see that $\phi \wedge \psi$ is a submeasure, and it indeed is the meet of the two submeasures in the lattice of submeasures ordered by setwise domination. It is in general not true that a meet of two capacities is a capacity, however the following is true:

Proposition 4.3.15. *Suppose that ϕ, ψ are outer regular subadditive capacities on a Polish space X.*

1. If one of them is a Ramsey capacity then $\phi \wedge \psi$ is a capacity.
2. If both are Ramsey capacities then $\phi \wedge \psi$ is a Ramsey capacity.

Proof. It is clear that $\phi \wedge \psi$ is an outer regular countably subadditive submeasure. For the first item, it is necessary to verify the continuity under increasing unions. Suppose that the capacity ϕ is Ramsey. Suppose that $A_n : n \in \omega$ is an increasing collection of sets, $\varepsilon \geq 0$ is a real number, and $(\phi \wedge \psi)(A_n) \leq \varepsilon$ for every number $n \in \omega$. Let $\delta > 0$ be a real number. To show that $(\phi \wedge \psi)(\bigcup_n A_n) < \varepsilon + \delta$ choose sets B_n, C_n such that $A_n = B_n \cup C_n$ and $\phi(B_n) + \psi(C_n) < \varepsilon + \delta/4$. Find an infinite set $a \subset \omega$ such that $\sup_{n \in a} \phi(B_n) + \sup_{n \in a} \psi(C_n) < \varepsilon + \delta/3$ and use the Ramseyness of the capacity ϕ and Claim 4.3.11 to find an infinite set $b \subset a$ such that $\phi(\bigcup_{n \in b} B_n) < \sup_{n \in a} \phi(B_n) + \delta/2$. Now the set $A = \bigcup_n A_n$ can be covered by the set $\bigcup_{n \in b} B_n$ of small ϕ-mass and the set $\bigcup_{n \in a} D_n$ where $D_n = A_n \setminus \bigcup_{m \in a} B_m$. The sets $D_n : n \in a$ have small ψ mass since they are covered by the sets $C_n : n \in a$, and most importantly, they form an increasing sequence, so even their union D has a small ψ mass, $\psi(D) < \sup_{n \in a} \psi(C_n) + \delta/2$. It follows that $(\phi \wedge \psi)(\bigcup_n A_n) < \varepsilon + \delta$ as desired.

The proof of the second item is similar. Suppose that ϕ, ψ are Ramsey capacities and $A_n : n \in \omega$ are sets of $\phi \wedge \psi$ mass $< \varepsilon$, and $\delta > 0$ is a real number. To find an infinite set $c \subset \omega$ such that $(\phi \wedge \psi)(\bigcup_{n \in c} A_n) < \varepsilon + \delta$ first find sets B_n, C_n such that $A_n = B_n \cup C_n$ and $\phi(B_n) + \psi(C_n) < \varepsilon$ and then use the Ramseyness to find an infinite set $b \subset \omega$ such that $\phi(\bigcup_{n \in b} B_n) < \sup_{n \in b} \phi(B_n) + \delta/2$ and an infinite set $c \subset b$ such that $\phi(\bigcup_{n \in c} C_n) < \sup_{n \in b} \phi(C_n) + \delta/2$. It is immediate that the infinite set c works as desired. $\square$

Every strongly subadditive Ramsey capacity also has a natural compact metric space associated to it. Let $A, B \subset X$ be arbitrary sets and define $d(A, B) = \phi(A \cup B) - \min\{\phi(A), \phi(B)\}$. The strong subadditivity of the capacity ϕ can be used to show that this is a premetric, that is, $d(A, A) = 0$ for every set A and d satisfies the triangle inequality. Let $A \equiv B \leftrightarrow d(A, B) = 0$; then $\equiv$ is an equivalence relation, and consider the metric space Y of $\equiv$-equivalence classes. The outer regularity of the capacity ϕ shows that every set is equivalent to a G_δ set and the equivalence classes of basic open sets are dense, i.e. Y is separable. More importantly, if the capacity is Ramsey then the space Y is precompact, that is, every sequence $[A_n] : n \in \omega$ of points in the space Y contains a Cauchy subsequence: thinning out the sequence if necessary I may assume that the capacities of the sets involved converge to some real number $\varepsilon \geq 0$, and then Claim 4.3.11 together with an obvious diagonalization argument yields a subsequence such that the capacities of the unions of its tails converge to ε as well. A review of the definitions will show that such a subsequence must be Cauchy. This means that the completion $\bar{Y}$ of the space Y is a compact metric space. The metric d is a natural one in cases such as the Newtonian capacity and it would be interesting to find out what is the relationship of the compact space $\bar{Y}$ to other objects studied in potential theory.

4.3.2 General theorems

The key concern in this section is the properness of the forcing P_{I_ϕ}, where ϕ is a subadditive capacity on a Polish space X. The following definition will be critical.

Definition 4.3.16. *The capacity ϕ is stable if for every Borel set $A \subset X$ there is a Borel set $\tilde{A} \supset A$ of the same capacity such that for every Borel set $B \subset X$, $\phi(B) > 0$, $B \cap \tilde{A} = 0$ it is the case that $\phi(A \cup B) > \phi(A)$.*

As an example consider the Newtonian capacity in the Euclidean space $\mathbb{R}^3$ [40], 30.A. The unit sphere A has the same capacity as the open unit ball or the closed unit ball. It turns out that the stability property is witnessed by $\tilde{A} = $ the closed unit ball; if a positive capacity set is added to it, it will increase the capacity of the union. The sections below include a long list of stable capacities.

Theorem 4.3.17. *[84] Suppose that ϕ is a subadditive stable capacity. The forcing P_{I_ϕ} is proper.*

I will give two quite different proofs. One of them is combinatorial in nature and yields a natural fusion argument. The other proof is game theoretical, and it works only for outer regular capacities; on the other hand it requires the existence of the tilde set $\tilde{A} \supset A$ only for open sets A. The game theoretic argument yields dichotomy type information.

Proof. The combinatorial proof depends on a small claim.

Claim 4.3.18. *Suppose $B \in P_{I_\phi}$ is a condition and $D \subset P_{I_\phi}$ is an open dense set. Then there is a countable set $D' \subset D$ consisiting of subsets of B such that $\phi(\bigcup D') = \phi(B)$.*

Proof. Fix the condition B and the open dense set $D \subset P_{I_\phi}$. By induction on $\alpha \in \omega_1$ choose sets $C_\alpha \subset B$ in the open dense set D so that $\phi(\bigcup_{\beta \in \alpha} C_\beta) < \phi(\bigcup_{\beta \in \alpha+1} C_\beta)$ if possible. By the separability of the real line this process must end after countably many steps, and I will show that the only reason for it to end is that $\phi(\bigcup_{\beta \in \alpha} C_\beta) = \phi(B)$. Then $D' = \{C_\beta : \beta \in \alpha\} \subset D$ is the countable set required in the claim.

Well, suppose that writing $A = \bigcup_{\beta \in \alpha} C_\beta$ it is the case that $\phi(A) < \phi(B)$. Then $B \setminus \tilde{A}$ is a ϕ-positive Borel set and as such contains a subset $C_\alpha \in D$. By the definitory property of the tilde set $\tilde{A}$ it must be the case that $\phi(A \cup C_\alpha) > \phi(A)$ and I extended the induction process by one more step as required. □

The properness of the forcing P_{I_ϕ} immediately follows. If M is a countable elementary submodel of a large enough structure and $B \in P_I \cap M$ is a set of some positive capacity $\varepsilon > 0$, choose a decreasing sequence of rational numbers $q_n : n \in \omega$ which are less than ε and do not converge to zero, enumerate open dense sets in the model M by $D_n : n \in \omega$ and by induction on $n \in \omega$ build compact sets $B \supset C_0 \supset C_1 \supset \ldots$ in the model M such that $\phi(C_n) > q_n$ and the set C_{n+1} is covered by finitely many sets in the collection $D_n \cap M$. If this construction succeeds then $C = \bigcap_n C_n$ is a compact set of capacity $\geq \inf_n q_n$ by the continuity of the capacity in decreasing intersections of compact sets, and C is the desired master condition for the model M.

The set C_0 is obtained through Choquet's capacitability theorem 4.3.2. Suppose that C_n has been obtained. By the claim, in the model M there is a countable set $D' \subset D_n$ consisting of subsets of C_n such that $\phi(\bigcup D') = \phi(C_n)$. As the capacity is continuous in increasing unions, there is a finite set $D'' \subset D'$ in the model M such that $\phi(\bigcup D'') > q_{n+1}$. By the elementarity of the model again and the capacitability theorem, there is a compact set $C_{n+1} \subset \bigcup D''$ of ϕ-mass $> q_{n+1}$. This concludes the inductive construction and the combinatorial proof.

For the game theoretic proof, let ϕ be an outer regular subadditive stable capacity on the Polish space X. Fix a countable basis $\mathcal{O}$ of the space closed under finite unions. Let P be a forcing adding a point $\dot{x} \in \dot{X}$.

Consider an infinite game $G = G(P, x)$ between Players I and II. In the beginning, Player I indicates an initial condition $p_{ini} \in P$ and Player II answers with a positive real number $\varepsilon > 0$. After that, Player I produces a sequence $D_k : k \in \omega$ of open dense subsets of the forcing P as well as a set $A \subset X$ with $\phi(A) \leq \varepsilon$. Player II produces a sequence $p_{ini} \geq p_0 \geq p_1 \geq \dots$ such that $p_k \in D_k$ and p_k decides the membership of the point x in the k-th basic open subset of the space X in some fixed enumeration. Player II wins if, writing $g \subset P$ for the filter his conditions generate, the point $\dot{x}/g$ falls out of the set A.

In order to complete the description of the game, I must describe the exact schedule for both players. At round $k \in \omega$, Player I indicates the open dense set $D_k \subset P_{I_b}$ and a set $A_k \in \mathcal{O}$ so that $\phi(A_k) \leq \varepsilon$ and $k_0 \in k_1$ implies $A_{k_0} \subset A_{k_1}$ and $\phi(A_{k_1}) - \phi(A_{k_0}) \leq 2^{-k_0}$. In the end, obtain the set A as $A = \bigcup_k A_k$. The continuity of the capacity in increasing unions shows that $\phi(A) \leq \varepsilon$. Note that apart from the open dense sets, Player I has only countably many moves at his disposal. Still, he can produce a superset of any given set of ϕ-mass $< \varepsilon$ as his set A. Player II is allowed to tread water, that is, to wait for an arbitrary finite number of rounds (place *trivial moves*) before placing the next condition p_k on his sequence.

Lemma 4.3.19. *Player II has a winning strategy in the game G if and only if $P \Vdash \dot{x}$ falls out of all ground model coded ϕ-null sets.*

Proof. The key point is that the payoff set of the game G is Borel in the (large) tree of all legal plays, and therefore the game is determined by [49]. A careful computation will show that the winning condition for Player I is in fact an F_σ set.

For the left-to-right direction, if there is some condition $p \in P$ and a ϕ-null Borel set $B \subset X$ such that $p \Vdash \dot{x} \in \dot{B}$, then Player I can win by indicating $p_{ini} = p$, and after Player II chooses his number $\varepsilon > 0$, Player I produces an open set $A \subset X$ of mass $< \varepsilon$ such that $B \subset A$, on the side creating an increasing sequence $M_i : i \in \omega$ of countable elementary submodels of some large structure and playing in such a way that the sets $D_k : k \in \omega$ enumerate all open dense subsets of the poset P in the model $M = \bigcup_i M_i$, and $\{p_k : k \in \omega\} \subset M$. In the end, this must bring success: this way, Player II's filter $g \subset P$ is M-generic containing the condition p, by the forcing theorem applied in the model M, $M[g] \models \dot{x}/g \in B$, and by Borel absoluteness $\dot{x}/g \in A$ as desired.

The right-to-left direction is harder. Suppose that $P \Vdash \dot{x}$ falls out of all ground model coded ϕ-null sets, and σ is a strategy for Player I. By the determinacy of the game G, it will be enough to find a counterplay against the strategy σ winning for Player II. Let $p_{ini} \in P$ be the initial condition indicated by the strategy σ. There must be a number $\varepsilon > 0$ such that for every Borel set $A \subset X$ with $\phi(A) \leq \varepsilon$ there is a condition $q \leq p$ forcing $\dot{x} \notin \dot{A}$. If this failed, for every real number $\varepsilon > 0$ there would be a Borel set $A_\varepsilon \subset X$ of ϕ mass $< \varepsilon$ such that $p \Vdash \dot{x} \in \dot{A}_\varepsilon$; then $p \Vdash \dot{x} \in \bigcap_\varepsilon A_\varepsilon$, and since the latter set is ϕ-null, this would contradict the assumptions on the name $\dot{x}$. The real number ε will be Player II's initial response.

The rest of the play is obtained in the following way. Consider the tree T of all partial plays τ of the game G respecting the strategy σ such that they end at some round k with Player II placing a condition $p \in P$ as his last move such that

(*) for every Borel set $B \supset A_k$, if $\phi(B) - \phi(A_k) \leq 2^{-k}$ and $\phi(B) \leq \varepsilon$ then there is a condition $q \leq p$ such that $q \Vdash \dot{x} \notin B$.

Note that every infinite play whose initial segments form an infinite branch through the tree T Player II won in the end, because for no number k and no condition $p \in g$ it could be the case that $p \Vdash \dot{x} \in \dot{A}_k$ by the condition (*) and therefore $\dot{x}/g \notin A$. Now the play consisting of the initial moves described in the previous paragraph is in the tree T by the choice of the real number $\varepsilon > 0$, and so it will be enough to show that every node of the tree can be extended to a longer one.

Suppose $\tau \in T$ is a finite play of length $\bar{k}$, ending with a nontrivial move $p \in P$ of Player II and a move $A_{\bar{k}}$ of Player I, satisfying the property (*). Consider the infinite play extending τ in which Player I follows the strategy σ and Player II places only trivial moves past τ. Write A_k for the k-th basic open set the strategy produced during this play and $A = \bigcup_k A_k$. Clearly, $\phi(\tilde{A}) = \phi(A) \leq \phi(A_{\bar{k}}) + 2^{-\bar{k}}$, and by the property (*) there is a condition $q \leq p$ forcing $\dot{x} \notin \tilde{A}$. Let $r \leq q$ be a condition in the appropriate open dense set indicated by Player I, deciding whether the point $\dot{x}$ belongs to the appropriate basic open subset of X or not. This will be the next nontrivial move of Player II past τ in the required play in the tree T extending τ, I just have to decide at which round to place that move in order to make the condition (*) hold.

Assume for contradiction that for no round $k > \bar{k}$ the condition (*) will be satisfied after Player II places the move r at the round k. Then for every number $k > \bar{k}$ there is a Borel set $B_k \supset A_k$ such that $\phi(B_k) \leq \phi(A_k) + 2^{-k}$ such that $r \Vdash \dot{x} \in \dot{B}_k$.

Claim 4.3.20. $\phi(\bigcap_k B_k \cup A) = \phi(A)$.

Proof. Note $A = \bigcup_k A_k$ is an increasing union. If the claim failed, by the continuity of the capacity in increasing unions there would have to be a number $i > \bar{k}$ such that $\phi(\bigcap_k B_k \cup A_i) > \phi(A) + 2^{-i}$. However, $B_i \supset \bigcap_k B_k \cup A_i$ and $\phi(B_i) \leq \phi(A_i) + 2^{-i} \leq \phi(A) + 2^{-i}$, contradiction. $\square$

By the properties of the tilde operation, it must be the case that $\phi(\bigcap_k B_k \setminus \tilde{A}) = 0$. At the same time, $r \Vdash \dot{x} \in \bigcap_k B_k \setminus \tilde{A}$. This contradicts the assumption that $P \Vdash \dot{x}$ falls out of all ground model coded ϕ-null sets! $\square$

Now suppose that M is a countable elementary submodel of a large structure and $B \in P_{I_\phi} \cap M$ is a condition. To prove that the set $C = \{x \in B : x \text{ is } M\text{-generic}\}$ is positive, for any ϕ-null set $E \subset X$ I must produce a point $x \in C \setminus E$. Let σ be a winning strategy in the game $G(P_{I_\phi}, \dot{x}_{gen})$ for Player II and let $\varepsilon \in M$ be the real number it produces after Player I chooses $p_{ini} = B$. Choose an open set $A \subset X$ covering the set

A such that $\phi(A) < \varepsilon$ and simulate a play of the game in which Player II follows his winning strategy and Player I starts out with the condition B, enumerates all open dense subsets of the poset P_{I_ϕ} in the model M, and in the end produces the set $A \subset X$. Note that all initial segments of this play are in the model M and therefore the resulting point $x \in X$ is an M-generic element of $B \setminus E$ as required. $\qquad\square$

Theorem 4.3.21. *Suppose that ϕ is an outer regular subadditive stable capacity on a Polish space X. Then:*

1. *I_ϕ satisfies the third dichotomy, in fact every analytic set contains a Borel subset of the same capacity.*
2. *In the choiceless Solovay model, I_ϕ satisfies the first dichotomy and ϕ is continuous in increasing wellordered unions. In fact every set contains a Borel subset of the same capacity.*
3. *(ZF+DC+AD+) I_ϕ satisfies the first dichotomy and ϕ is continuous in increasing wellordered unions. Every set contains a Borel subset of the same capacity.*
4. *In the constructible universe, if the forcing P_{I_ϕ} is nowhere c.c.c. then there is a coanalytic ϕ-positive set without a ϕ-positive Borel subset.*

Proof. The first item is an immediate consequence of Choquet's theorem 4.3.2. For the second item, use Proposition 3.9.19. Work in the choiceless Solovay model and suppose that $A \subset X \times 2^\omega$ is a set such that $\phi(\mathrm{proj}(A)) \geq \varepsilon$. I must find an analytic subset $B \subset A$ such that $\phi(\mathrm{proj}(B)) \geq \varepsilon$. By the standard homogeneity arguments I may assume that the set A is in fact definable from ground model parameters. Move to the ground model and for every open set $O \subset X$ of capacity $< \varepsilon$ find a forcing P_O with a name $\langle \dot{x}_O, \dot{y}_O \rangle$ such that $P_O \Vdash \dot{x}_O \notin \dot{O}$ and $\mathrm{Coll}(\omega, <\kappa) \Vdash \langle \dot{x}_O, \dot{y}_O \rangle \in \dot{A}$. Let P be the side-by-side sum of these forcings and let $\dot{x}, \dot{y}$ be the side-by-side sum of the names $\dot{x}_O, \dot{y}_O$. Consider the set $B = \{\langle x, y \rangle : \exists g \subset P \ g \text{ is } V\text{-generic and } \langle x, y \rangle = \langle \dot{x}_O/g, \dot{y}_O/g \rangle\}$. The set B is analytic by its definition, and it is a subset of the set A by the usual homogeneity argument. It will be enough to show that $\phi(\mathrm{proj}(B)) \geq \varepsilon$.

Back in the ground model, look at the forcing P and the name $\dot{x}$. Clearly $\varepsilon(1_P) \geq \varepsilon$ and the review of the game theoretic proof above reveals that Player II has a winning strategy σ in the Borel game $G(P, \dot{x})$ which answers the condition $p_{ini} = 1_P$ with the number $\varepsilon_{ini} = \varepsilon$. The strategy σ remains winning in the Solovay model, and if Player I challenges it with an open set $O \subset X$ of capacity $< \varepsilon$ and a list of all open dense subsets of P in the ground model, it produces a point $\langle x, y \rangle \in B$ such that $x \notin O$. Thus $\phi(\mathrm{proj}(B)) \geq \varepsilon$ as desired.

The proof of the third item uses an integer game and Proposition 3.9.20. Work in the context of ZF+DC+AD+. Let $A \subset X \times 2^\omega$ be an arbitrary set such that $\phi(\mathrm{proj}(A)) > \varepsilon$. I must produce an analytic subset $B \subset A$ such that $\phi(\mathrm{proj}(B)) \geq \varepsilon$.

Let $\pi : 2^\omega \to X \times 2^\omega$ be a Borel bijection and for any set $C \subset X \times 2^\omega$ consider the integer game $G(C, \varepsilon)$ played as follows. Player I gradually produces an open set $O \subset X$ of ϕ-mass $\leq \varepsilon$ and Player II produces a binary sequence $r \in 2^\omega$; Player II wins if, writing $\langle x, y \rangle = \pi(r)$ it is the case that $\pi(r) \in C$ and $x \notin O$.

In order to complete the description of the game I must specify the schedule for both players. At round n Player I must play a basic open set O_n such that $O_0 \subset O_1 \subset O_2 \subset \ldots$, the ϕ-masses of these sets are $\leq \varepsilon$, and in the end, $O = \bigcup_n O_n$ and $\phi(O) < \phi(O_n) + 2^{-n}$. Player II in turn can wait for an arbitrary finite number of rounds before placing the next bit on his sequence r. The following claim is key.

Claim 4.3.22. *If $\phi(\mathrm{proj}(C)) < \varepsilon$ then Player I has a winning strategy. If Player I has a winning strategy then $\phi(\mathrm{proj}(C)) \leq \varepsilon$.*

Proof. The first sentence records the obvious fact that if $\phi(\mathrm{proj}(C)) < \varepsilon$ then Player I can win by producing some open set O covering the projection $\mathrm{proj}(C)$, ignoring Player II's moves altogether. On the other hand, suppose $\phi(\mathrm{proj}(C)) > \varepsilon$ and σ is a strategy for Player I. I must produce a counterplay winning for Player II. For every finite play t respecting the strategy σ consider the set O_t which the strategy σ outputs if the play t is completed without further nontrivial move by Player II. Moreover, for every bit $b \in 2$ consider the plays tnb resulting from t by Player II placing one more nontrivial move b at round n past t, and consider the set $R_{tb} = \bigcap_n \tilde{O}_{tnb} \setminus \tilde{O}_t$. I claim that $\phi(R_{tb}) = 0$.

To see this, by the properties of the tilde operation it is enough to show that $\phi(O_t \cup \bigcap_n \tilde{O}_{tnb}) = \phi(O_t)$. The proof of this equality follows almost literally the argument in Claim 4.3.20.

Now, since $\phi(\mathrm{proj}(C)) > \varepsilon$, the projection $\mathrm{proj}(C)$ is not covered by the sets $\tilde{O}_0 \cup \bigcup_{t,b} R_{tb}$, which has mass $\leq \varepsilon$. There must then be a point $\langle x, y \rangle \in A$ such that $x \notin \tilde{O}_0 \cup \bigcup_{t,b} R_{tb}$ and a binary sequence $r \in 2^\omega$ such that $\pi(r) = \langle x, y \rangle$. I will show that Player II has a winning counterplay against the strategy σ in which he produces the binary sequence r. Note that I just need to decide at which rounds Player I places the bits of the sequence r. By induction on $m \in \omega$ build finite plays $0 = t_0 \subset t_1 \subset t_2 \subset \ldots$ against the strategy σ such that:

- t_{m+1} obtains from t_m by Player I waiting for some time and then adding one more bit of the sequence r as the last move of t_{m+1};
- $x \notin \tilde{O}_{t_m}$.

Once this has been done then the play $\bigcup_m t_m$ is winning for Player II, since by the second item of the induction hypothesis the point $x \in X$ does not belong to any of the open sets the strategy σ produced in the course of the play. The induction itself is trivial to perform. Once the finite play t_m has been found, let $b \in 2$ be the next bit on the sequence $r \in 2^\omega$ and note that $x \notin R_{t_m,b}$. Therefore there must be a number n such that $x \notin \tilde{O}_{t_m nb}$. The play $t_{m+1} = t_m nb$ then satisfies the induction hypothesis, and this completes the proof of the claim. □

Now back to the set $A \subset X \times 2^\omega$. Since $\phi(\text{proj}(A)) > \varepsilon$, the previous claim together with a determinacy argument show that Player II has a winning strategy σ in the game $G(C, \varepsilon)$. Let $B = \pi''\{r \in 2^\omega$: the strategy σ produces the sequence r against some counterplay}. The set B is analytic by virtue of its definition, it is a subset of the set A since the strategy σ was winning, and moreover, the strategy σ remains a winning strategy in the play $G(B, \varepsilon)$ for obvious reasons. One more application of the claim shows that $\phi(\text{proj}(B)) \geq \varepsilon$ as required.

The proof of (4) closely follows the proof of Theorem 4.5.6(4). I omit it. □

The preservation properties of the forcings P_{I_ϕ} are closely connected to fundamental measure theory facts.

Theorem 4.3.23. *Suppose that ϕ is a subadditive capacity on a Polish space X such that the forcing P_{I_ϕ} is proper. Then the forcing P_{I_ϕ} is bounding.*

Proof. This is just a conjunction of Choquet's capacitability theorem 4.3.2 and Theorem 3.3.2. It is enough to show that compact sets are dense in every Polish topology τ extending the original one. But ϕ remains a capacity with respect to that topology τ, and therefore every analytic ϕ-positive set has a τ-compact subset of arbitrarily close ϕ-mass! □

Theorem 4.3.24. *Suppose that ϕ is a strongly subadditive capacity such that the forcing P_{I_ϕ} is proper. Then the forcing P_{I_ϕ} preserves outer Lebesgue measure.*

Proof. This is again just a conjunction of Choquet's envelope of measures theorem 4.3.6(2) and Theorem 3.6.2. The strongly subadditive capacity ϕ is an envelope of measures and therefore the ideal $I_\phi = \{A \subset X :$ for all measures $\mu \leq \phi \ \mu(A) = 0\}$ is polar. □

Theorem 4.3.25. *Suppose that ϕ is a subadditive Ramsey capacity such that the forcing P_{I_ϕ} is proper. Then P_{I_ϕ} does not add a splitting real.*

Proof. Suppose $B \in P_{I_\phi}$ is a condition and $\dot{a}$ is a name for a subset of ω. Without loss of generality assume that the set B is compact and there is a Borel function $f : B \to \mathcal{P}(\omega)$ such that $B \Vdash \dot{a} = f(\dot{x}_{gen})$. Assume that there is no infinite set $b \subset \omega$ and no condition $C \subset B$ such that $C \Vdash \check{b} \subset \dot{a}$. I must produce an infinite set $d \subset \omega$ and a condition $D \subset B$ such that $D \Vdash \dot{a} \cap \check{d} = 0$.

In order to do this, choose a strictly decreasing sequence $\varepsilon_n : n \in \omega$ of positive real numbers smaller that $\phi(B)$ which converges to a nonzero number, and by induction on $n \in \omega$ construct a decreasing sequence $C_n : n \in \omega$ of compact subsets of B of capacity $\phi(C_n) > \varepsilon_n$ and distinct natural numbers $m_n : n \in \omega$ such that for all points $x \in C_n$, the set $f(x) \subset \omega$ does not contain the number m_n. In the end, the compact set $\bigcap_n C_n$ has nonzero ϕ-mass and it forces $\dot{a} \cap \{m_n : n \in \omega\} = 0$ as desired.

To perform the induction step, consider the infinite collection of sets $A_m : m \in \omega$ defined by $A_m = \{x \in C_n : m \notin f(x)\}$. If infinitely many of them had capacity $\leq \varepsilon_{n+1}$ then there would be an infinite subcollection $A_m : m \in b$ for some set $b \subset \omega$ such that $\bigcup_{m \in b} A_m < \varepsilon_n$, and then the nonzero capacity Borel set $C_n \setminus \bigcup_{m \in b} A_m$ forces $b \subset \dot{x}$, contradiction. Thus there is a number m_n larger than all $m_k : k \in n$ such that $\phi(A_{m_n}) > \varepsilon_{n+1}$, and then the compact set $C_{n+1} \subset A_{m_n}$ can be extracted with a reference to Choquet's capacitability theorem 4.3.2. $\qquad\square$

Theorem 4.3.26. *Suppose that ϕ is a strongly subadditive stable capacity. The forcing P_{I_ϕ} can be embedded into a σ-closed*c.c.c. iteration.*

In fact the assumption of strong subadditivity can be weakened to $A \subset B, A \subset C$, $\phi(A) = \phi(B) = \phi(C)$ implies $\phi(A) = \phi(B \cup C)$, a property which clearly follows from strong subadditivity, but it is also satisfied for some other capacities such as the Steprāns capacities of Section 4.3.4.

Proof. Fix a countable topology basis $\mathcal{O}$ of the underlying Polish space X closed under finite unions. Consider the partial order P of all pairs $p = \langle a_p, b_p \rangle$ satisfying the following demands:

- a_p is a countable set of ϕ-positive compact sets, closed under ϕ-positive finite intersections and ϕ-positive subtraction of a basic open set;
- b_p is a countable collection of sets $z \subset a_p$ such that $\forall C \in a_p \forall O \in \mathcal{O} \; \phi(O \cup (C \cap \bigcup z)) = \phi(O \cup C)$.

The ordering is that of coordinatewise extension. The partial order P is clearly σ-closed. If $G \subset P$ is a generic filter then $Q \in V[G]$ is the partial order $\bigcup_{p \in G} a_p$ ordered by inclusion. I will show that Q is a c.c.c. forcing and if $H \subset Q$ is a $V[G]$-generic filter then the upwards closure K of H in P_{I_ϕ} is a V-generic filter. This will complete the proof of the theorem. The following claim is critical:

Claim 4.3.27. *Suppose that $p = \langle a_p, b_p \rangle$ is a condition in the forcing P, and consider the set $B = \bigcap_{z \in b_p} \bigcup z \subset X$. Then:*

1. *$\phi(B) > 0$ and for every set $C \in a_p$, and basic open set $O \in \mathcal{O}$, $\phi(O \cup C) = \phi(O \cup (C \cap B))$;*
2. *for every closed set $C \subset B$ of positive capacity there is a condition $q \leq p$ such that $C \in a_q$.*

The claim implies that K is forced to be a V-generic filter: if $\langle p, C \rangle$ is a condition in the iteration $P * Q$ and $D \subset P_{l_\phi}$ is an open dense set, strengthening the condition p if necessary I may assume that $C \in a_p$, and then for every ϕ-positive closed set $C' \subset C \cap \bigcap_{z \in b_p} \bigcup z$ in the open dense set D there is a condition $q \le p$ in the forcing P such that $C' \in a_q$ by the Claim. Clearly the condition $\langle q, C' \rangle \le \langle a_p, C \rangle$ in the iteration $P * \dot{Q}$ forces the filter $K \subset P_{l_\phi}$ to meet the open dense sets $D \subset P_{l_\phi}$.

Towards the c.c.c. of the forcing $\dot{Q}$, suppose for contradiction that $p \in P$ is a condition forcing $\dot{A} \subset \dot{Q}$ to be an uncountable maximal antichain. In the P-extension, for every condition $C \in Q$ and basic open set $O \in \mathcal{O}$ let $\alpha(O, C) = \sup\{\phi(O \cup (\bigcup y \cap C)) : y \subset A \text{ countable}\}$. Note that the supremum is in fact attained, and I will first argue that $\alpha(C) = \phi(O \cup C)$.

If not, there must be a condition $q \le p$ and a set $y \subset a_q$ such that $C \in a_q$ and q forces $\check{y} \subset \dot{A}$ and $\alpha(O, C) = \phi(O \cup (\bigcup y \cap C)) < \phi(O \cup C)$. Now let $B = C \cap \bigcap_{z \in b_q} \bigcup z$ be the set from the claim. Note $\phi(O \cup C) = \phi(O \cup (B \cap C))$. Find a ϕ-positive compact set $C' \subset O \cup (B \cap C) \setminus$ the tilde of $O \cup (\bigcup y \cap C)$ and a condition $r \le q$ such that $C' \in a_r$. Strengthening the condition r if necessary I may assume that there is a set $C'' \in a_r$ such that $r \Vdash \check{C}'' \in \dot{A}$ and C'' is compatible with C'. By the definition of the tilde operation then, $\phi(O \cup \bigcup(y \cup \{C''\}) \cap C) > \phi(\bigcup y \cap C)$, contradicting the choice of the set y.

Now suppose that M is a countable elementary submodel containing $p, \dot{A}$, let $g \subset P \cap M$ be an M-generic filter containing the condition p, and let $q \le p$ be the natural lower bound of this filter. The previous paragraph shows that the pair $r = \langle a_q, b_q \cup \{\dot{A}/g\} \rangle$ is a condition in the forcing P and it clearly forces that $\dot{A}/g \subset P$ is a maximal antichain. Therefore $\dot{A} = \dot{A}/g$, contradicting the presumed uncountability of the antichain $\dot{A}$.

Thus all that remains is to prove the claim. For the first item, suppose that $C \in a_p$ is a set, $O \in \mathcal{O}$ is a basic open set, and let $O_n : n \in \omega$ be a collection of basic open sets such that $\phi(O \cup \bigcup_n O_n) < \phi(O \cup C)$. It will be enough to show that the set $D = \bigcap_{z \in b_p} \bigcup z \cap (C \setminus \bigcup_n O_n) \setminus O$ is nonempty. In order to do this, enumerate the set b_p by $z_m : m \in \omega$ and by induction on $m \in \omega$ find a descending chain $C_0 \supset C_1 \supset \ldots$ of compact sets in the collection a_p such that $C_0 = C \setminus O$, C_{m+1} is a set disjoint from O_m and it is a subset of some element of z_m, and $\phi(O \cup \bigcup_n O_n \cup C_m) > \phi(O \cup \bigcup_n O_n)$. In the end clearly $\bigcap_m C_m$ is a nonempty subset of D as required.

To perform the inductive construction assume that the compact set C_m has been constructed. Since the capacity ϕ is continuous in increasing unions, there must be a number $\bar{n} > m$ such that $\phi(O \cup \bigcup_{n \in \bar{n}} O_n \cup C_m) > \phi(O \cup \bigcup_n O_n)$. From the second item of the definition of the poset P it follows that $\phi(O \cup \bigcup_{n \in \bar{n}} O_n \cup (C_m \cap \bigcup z_m)) > \phi(O \cup \bigcup_n O_n)$. The strong subadditivity of the capacity ϕ now implies that there must be an element $E \in z_m$ such that $\phi(O \cup \bigcup_n O_n) < \phi(O \cup \bigcup_n O_n \cup (C_m \cap E \setminus O_n))$. This completes the induction step and the proof of the first item.

The second item of the claim is now immediate. If $C \subset B$ is a ϕ-positive compact set then let a_q be the closure of the set $a_p \cup \{C\}$ under positive finite intersections and positive subtractions of a basic open set. The sets in a_q are then either already in a_p or subsets of the set C. Let $q = \langle a_q, b_p \rangle$. The verification of the second condition in the definition of the poset P is trivial. $\qquad \square$

Theorem 4.3.28. *Suppose that ϕ is a strongly subadditive stable capacity. Then*

1. *every function $f : \omega_1 \to 2$ in the P_{I_ϕ} extension has an infinite ground model subfunction;*
2. *every intermediate extension of the P_{I_ϕ}-extension is c.c.c.;*
3. *($\mathrm{add}(\mathrm{null}) > \aleph_1$) every intermediate extension is given by a single Solovay real.*

Proof. The proof of (1) closely follows the argument for Theorem 4.1.6. It is necessary to argue that for every collection $\{B_\alpha : \alpha \in \omega_1\}$ of Borel sets of ϕ-mass $< \varepsilon$ and for every real number $\delta > 0$ there is an infinite subcollection $\{B_\alpha : \alpha \in b\}$ whose union has capacity $< \varepsilon + \delta$, which is part of the content of Claim 4.3.8.

The second item follows from Theorem 4.3.21 and Proposition 3.9.2 in the presence of large cardinals, however it is possible to perform a manual construction to avoid the large cardinal assumptions. The strong subadditivity is not needed here.

In the proof of (3), I must first show that every c.c.c. intermediate extension is countably generated. This follows the lines of the proof of Theorem 4.1.7. I must use the assumption $\mathrm{add}(\mathrm{null}) > \aleph_1$ to show that for every collection $\{B_\alpha : \alpha \in \omega_1\}$ of Borel sets of ϕ-mass $< \varepsilon$ and for every real number $\delta > 0$ there is an uncountable subcollection $\{B_\alpha : \alpha \in b\}$ whose union has capacity $< \varepsilon + \delta$. This is again proved in Claim 4.3.8. Finally, I must show that the intermediate c.c.c. countably generated extension given by a measure algebra. In order to do that, note that the c.c.c. forcing is associated with $\mathbf{\Pi}_1^1$ on $\mathbf{\Sigma}_1^1$ ideal, it is bounding and c.c.c., and so by Corollary 3.3.7 it is a Maharam algebra. Now the forcing P_{I_ϕ} preserves outer Lebesgue measure by Theorem 4.3.24 (another use of strong subadditivity), the intermediate c.c.c. forcing must preserve outer Lebesgue measure too, and as [17] shows, the Solovay forcing is the only Maharam algebra forcing which preserves outer Lebesgue measure! $\quad \square$

4.3.3 Potential theory

Theorem 4.3.29. *(Murali Rao) All capacities in use in potential theory are stable.*

The proof of this theorem requires a bit of space and explanations. I will use a general approach to potential spaces exposed in [1], Section 2.3.

Definition 4.3.30. *Let M be a space with a positive measure ν, and let $n \in \omega$. A kernel on $\mathbb{R}^n \times M$ is an transform function $g : \mathbb{R}^n \times M \to \mathbb{R}$ such that $g(\cdot, y)$ is* *lower semicontinuous for every point* $y \in M$ *and* $g(x, \cdot)$ *is* ν-*measurable for each point* $x \in \mathbb{R}^n$.

Definition 4.3.31. *For every* ν-*measurable function* $f : M \to \mathbb{R}$ *let* $\mathcal{G}f : \mathbb{R}^n \to \mathbb{R}$ *be the function defined by* $\mathcal{G}f(x) = \int_M g(x, y)f(y)d\nu(y)$.

Now let $p \geq 1$ be a real number. Associated with it is the uniformly convex Banach space $L^p(\nu)$ and its subset $L^p_+(\nu)$ consisting of non-negative functions. We are ready to define the capacity $c = c_{g,p}$ on $\mathbb{R}^n$:

Definition 4.3.32. *For every set* $E \subset \mathbb{R}^n$ *let* $\Omega_E = \{f \in L^p_+(\nu) : \forall x \in E \; \mathcal{G}f(x) \geq 1\}$ *and let* $c_{g,p}(E) = \inf\{\int_M f^p d\nu : f \in \Omega_E\}$.

It turns out that the function $c_{g,p}$ is an outer regular subadditive capacity, see [1], Propositions 2.3.4–6 and 2.3.12. It is not immediately clear if it has to be strongly subadditive, even though in many cases including the Newtonian capacity it is [44]. Most capacities in potential theory are obtained in this way; I just mention the most notorious examples.

Example 4.3.33. The Newtonian capacity results from a Newton kernel and $p = 2$. The Newton kernel is a special case of Riesz kernels with $\alpha = 2$, see below. This is perhaps not the simplest way of viewing this classical capacity. A simpler definition can be found in [40], 30.B.

Example 4.3.34. The Riesz capacities result from Riesz kernels. If $0 < \alpha < n$ is a real number, the Riesz kernel $I_\alpha : \mathbb{R}^n \to \mathbb{R}$ is given by

$$I_\alpha(x) = a_\alpha \int_0^\infty t^{\frac{\alpha-n}{2}} e^{-\frac{\pi|x|^2}{t}} \frac{dt}{t} = \frac{\gamma_\alpha}{|x|^{n-\alpha}}$$

for certain constants a_α, γ_α; the above setup will yield the α-th Riesz capacity by letting $M = \mathbb{R}^n$, $\nu =$ the Lebesgue measure, and $g(x, y) = I_\alpha(x - y)$.

Example 4.3.35. The Bessel capacities result from Bessel kernels. If $\alpha > 0$ then the Bessel kernel $G_\alpha : \mathbb{R}^n \to \mathbb{R}$ is given by

$$G_\alpha(x) = a_\alpha \int_0^\infty t^{\frac{\alpha-n}{2}} e^{-\frac{\pi|x|^2}{t} - \frac{t}{4\pi}} \frac{dt}{t}.$$

Then proceed similarly as in the case of Riesz capacities.

I will now show that the capacities obtained in this way are stable. The key tool is the following description of the closure $\bar{\Omega}_E$ of the set Ω_E in the space $L^p(\nu)$:

Fact 4.3.36. *[1], Proposition 2.3.9. Let* $E \subset \mathbb{R}^n$ *be a set. Then* $\bar{\Omega}_E = \{f \in L^p_+(\nu) :$ *for all but* ϕ-*null set of* $x \in E, \; \mathcal{G}f(x) \geq 1\}$.

Since the set $\bar{\Omega}_E \subset L^p(\nu)$ is closed and convex, the uniform convexity of the Banach space $L^p(\nu)$ implies that there is a *unique* up to L^p equivalence function $f \in \bar{\Omega}_E$ with the smallest norm. The function f is called the *capacitary function* of the set E, and clearly $\int_M f^p d\nu = \phi(E)$. We will write $f = f_E$. Note that the transform $\mathcal{G}f_E$, called the *capacitary potential* of the set E does not depend on the choice of f_E within its equivalence class.

For every set $A \subset \mathbb{R}^n$ let $\tilde{A} = A \cup \{x \in \mathbb{R}^n : \mathcal{G}f_E(x) \geq 1\}$. The following claim immediately implies that this set works as demanded by Definition 4.3.16.

Claim 4.3.37. *Let $A \subset B \subset \mathbb{R}^n$ be arbitrary sets. Then $\phi(A) < \phi(B)$ if and only if $\phi(B \setminus \tilde{A}) > 0$.*

Proof. On one hand, if $\phi(B \setminus \tilde{A}) = 0$ then $f_A \in \bar{\Omega}_B$ and therefore $\phi(B) \leq \int_M f_A^p d\nu = \phi(A)$. On the other hand, suppose $\phi(A) = \phi(B)$. Since $A \subset B$, it is the case that $f_B \in \bar{\Omega}_A$. Since $\phi(A) = \phi(B)$, it is the case that the norms of the functions f_B and f_A coincide. By the uniqueness of the function of minimal norm in the set $\bar{\Omega}_A$ it must be the case that $f_A = f_B$ up to the L^p equivalence. Thus $B \setminus \tilde{A} \subset \{x \in B : \mathcal{G}f_A(x) < 1\} = \{x \in B : \mathcal{G}f_B(x) < 1\}$ and the capacity of the latter set is zero by the definition of the set $\bar{\Omega}_B$. The claim follows. $\square$

Slight variations of the above definitions are in use in potential theory. A typical small change is the replacement of the Banach space $L^p(\nu)$ with $l^q L^p(\nu)$ in the case of Besov capacities and Lizorkin–Triebel capacities [1], Chapter 4. The results and proofs mentioned above apply again in these cases. The above definitions can be further generalized to yield capacities on spaces other than $\mathbb{R}^n$.

4.3.4 Steprāns capacities

In [74] Steprāns implicitly constructed a wide family of stable capacities and studied forcings associated with them. In this section I will expose the scheme for generating these capacities and prove the necessary theorems.

Let T be a finitely branching tree. For every node $t \in T$ let a_t denote the set of all its immediate successors in the tree T. For every node $t \in T$ choose a *norm* ϕ_t on $\mathbb{R}^{a_t}$ satisfying the following properties.

1. (respects absolute value) $|f| \leq |g| \rightarrow \phi_t(f) \leq \phi_t(g)$.
2. (subadditivity) $\phi_t(|f| + |g|) \leq \phi_t(f) + \phi_t(g)$.
3. (multiplicativity) $\phi_t(\varepsilon f) = \varepsilon \phi_t(f)$ for every real number $\varepsilon \geq 0$.
4. (normalized) $\phi_t(\text{the unit function}) = 1$.

For every natural number $n \in \omega$ the sequence $\vec{\phi} = \langle \phi_t : t \in T \rangle$ generates a norm ψ_n on $\mathbb{R}^{T_n}$ defined by induction on $n \in \omega$: $\psi_1(f) = \phi_0(f)$ and $\psi_{n+1}(f) = \psi_n(g)$ where $g \in \mathbb{R}^{T_n}$ is defined by $g(t) = \phi_t(f \upharpoonright a_t)$. Note that the norms ψ_n inherit the above four properties. The associated outer regular capacity η on the Polish space $[T]$ is defined in the natural way. For a clopen subset $O \subset [T]$ find a number $n \in \omega$ with a function $f : T_n \to 2$ such that for every $x \in [T]$, $x \in O \leftrightarrow f(x \upharpoonright n) = 1$ and let $\eta(O) = \psi_n(f)$. Since the norms on the sequence $\vec{\phi}$ are normalized, this does not depend on the choice of the number n. For an open set $B \subset [T]$ let $\eta(B) = \sup\{\eta(O) : O \subset B : O$ clopen$\}$, and for all other sets $A \subset [T]$ let $\eta(A) = \inf\{\eta(B) : B$ open and $A \subset B\}$. It is not difficult to see that this is a countably subadditive outer regular submeasure. I must show that this definition indeed yields a stable capacity.

Theorem 4.3.38. *The function η is a stable capacity.*

Proof. The key point of the argument is the definition of the submeasures $\eta_t : t \in T$. These are defined in the same way as η except the definition starts at the node t instead of $0 \in T$. The multiplicativity of the norms on the sequence $\vec{\phi}$ implies that for every set $B \subset O_t$ its is the case that $\eta_t(B) \cdot \eta(O_t) = \eta(B)$, and the definitions imply that for every set $B \subset [T]$ and every number $n \in \omega$, $\eta(B) = \psi_n(g)$ where $g \in \mathbb{R}^{T_n}$ is the function defined by $g(t) = \eta_t(B)$.

The following important claim is reminiscent of the Lebesgue density theorem.

Claim 4.3.39. *Whenever $B \subset [T]$ is an η-positive set and $\varepsilon > 0$ is a real number then there is a node $t \in T$ such that $\eta_t(B) > 1 - \varepsilon$.*

Proof. Suppose this fails for some set $B \subset X$, $\eta(B) > 0$, and a real number $\varepsilon > 0$. By induction on $n \in \omega$ build sets $K_n : n \in \omega$ and $O_n : n \in \omega$ so that:

- $K_n \subset T$ is a set of mutually incomparable nodes, and $O_n = \bigcup\{O_t : t \in K_n\}$. $K_0 = \{0\}$ and $O_0 = X$.
- Every node in K_n has some extension in K_{n+1}, and every node in K_{n+1} has some initial segment in K_n, so $O_{n+1} \subset O_n$.
- For every node $t \in K_n$ the set $O_{n+1} \cap O_t$ covers the set $B \cap O_t$, and $\eta(O_{n+1} \cap O_t) < (1 - \varepsilon)\eta(O_t)$. Such a set certainly exists by the assumption.

The multiplicativity of the norms can then be used to show that $\eta(O_n) < (1 - \varepsilon)^n$. Thus $\eta(\bigcap_n O_n) = 0$. On the other hand, the last item implies that $B \subset \bigcap_n O_n$, a contradiction. $\square$

To show that η is a capacity it is only necessary to verify that it is continuous in increasing unions. Let $B = \bigcup_n B_n$ be an increasing union, and assume for contradiction that for a real number $\varepsilon > 0$, $(1 - \varepsilon)\eta(B) > \sup_n \eta(B_n)$. Consider the set

$K = \{t \in T : \exists n \ \eta_t(B_n) > 1 - \varepsilon\}$ and its subset $L \subset K$ consisting of those nodes of K which have no proper initial segment in K.

Observe that $\eta(\{B \setminus \bigcup_{t \in L} O_t\}) = 0$. If this failed then by the countable sub-additivity of the submeasure η there must be a number $n \in \omega$ such that $\eta(\{B_n \setminus \bigcup_{t \in L} O_t\}) > 0$ and by the previous claim there must be a node $s \in T$ such that $\eta_s(\{B_n \setminus \bigcup_{t \in L} O_t\}) > 1 - \varepsilon$. This node cannot extend any element of the set L, nevertheless it must be an element of the set K, a contradiction.

It follows that the open set $\bigcup_{t \in L} O_t$ has η-mass at least that of B. The definition of the submeasure η provides for an existence of a finite set $M \subset L$ such that $(1 - \varepsilon)\eta(\bigcup_{t \in M} O_t) > \sup_n \eta(B_n)$. For every node $t \in M$ find a number $m_t \in \omega$ such that $\eta_t(B_{m_t}) > 1 - \varepsilon$ and let $m = \max_{t \in M} m_t$. Since the sets $B_n : n \in \omega$ form an inclusion-increasing sequence, this means that $\eta_t(B_m) > 1 - \varepsilon$ for every node $t \in M$, and the multiplicativity of the norms implies that $\eta(B_m) > (1 - \varepsilon)\eta(\bigcup_{t \in M} O_t) > \sup_n \eta_n(B_n)$, a contradiction. Thus the submeasure η indeed is a capacity.

To see that the capacity η is stable, let $A \subset [T]$ be a set. For every number $\varepsilon > 0$ let $A_\varepsilon = \{x \in [T] : \exists n \ \eta_{x \restriction n}(A) > 1 - \varepsilon\}$ and let $\tilde{A} = A \cup \bigcap_\varepsilon A_\varepsilon$. I claim that this set works as demanded by Definition 4.3.16. First note that $\eta(\tilde{A}) = \eta(A)$: every set A_ε is open and by the multiplicativity it has capacity $\leq (1 - \varepsilon)\eta(A)$, and therefore $\eta(\bigcap_\varepsilon A_\varepsilon) \leq \eta(A)$. Moreover by Claim 4.3.39, $\eta(A \setminus \bigcap_\varepsilon A_\varepsilon) = 0$ and therefore $\eta(A \cup \bigcap_\varepsilon A_\varepsilon) = \eta(\bigcap_\varepsilon A_\varepsilon) \leq \eta(A)$ as required. Now suppose that $B \subset X \setminus \tilde{A}$ is a positive capacity set. By the countable additivity of the capacity η, there is a real number $\varepsilon > 0$ such that the set $B \setminus A_\varepsilon$ has positive capacity. Use Claim 4.3.39 to find a sequence $t \in T$ such that $\eta_t(B) > 1 - \varepsilon$. It is now clear that $\eta_t(B) > 1 - \varepsilon \geq \eta_t(A)$ and it follows that $\eta(A \cup B) > \eta(A)$ as required. □

The properties of the forcings P_{t_η} are unclear. This family of forcings includes the Solovay forcing: the capacity η is a measure if each of the norms ϕ_t is just a convex combination of the outputs of the normed function. On the other hand many of the forcings are nowhere c.c.c. The capacities obtained are in general not strongly sub-additive even though Fremlin [21] showed that they satisfy certain weakenings of the strong subadditivity. I will show that the Steprāns capacities are envelopes of measures, which implies that the resulting forcings preserve the outer Lebesgue measure.

Proposition 4.3.40. *Let η be a Steprāns capacity. Then η is an envelope of measures.*

Proof. Let T be the finitely branching tree supporting the capacity η and let $\phi_t : t \in T$ be the norms from which the capacity η is defined. Suppose that $K \subset [T]$ is a compact set; I must find a measure μ such that $\mu \leq \eta$ and $\mu(K) = \eta(K)$. By tree induction on $t \in T$ assign real numbers w_t to nodes in the tree T so that $w_0 = \eta(K)$ and $w_s = w_t \cdot \eta_s(K)/\Sigma_{u \in a_t} \eta_u(K)$ whenever s is an immediate successor of the node t. It is clear that $w_t = \Sigma_{s \in a_t} w_s$ holds for every node $t \in T$. Let μ be the unique measure on $[T]$ such that $\mu(O_t) = w_t$ for all nodes $t \in T$.

Since the set K is compact, it is clear that $\mu(K) = w_0 = \eta(K)$. To show that $\mu \leq \eta$ it is enough to prove $\mu(O) \leq \eta(O)$ for every clopen set $O \subset [T]$. However, the multiplicativity of the norms can be used to show that in fact $\mu(O) \leq \eta(O \cap K)$.

$\square$

4.3.5 A construction of Ramsey capacities

In this section I will produce a great number of Ramsey capacities ϕ such that the forcing P_{I_ϕ} is proper. The properness will be guaranteed by the fact that they are pavement measures, with a reference to Section 4.5. As an artifact of the construction the forcings P_{I_ϕ} collapse Lebesgue measure. The following definition is motivated by the study of pathological submeasures [7], [15]; it is related to the Maharam problem.

Definition 4.3.41. *Let a be a finite set and ψ a submeasure on it. The* pathologicity index path(ψ) *be the smallest real number ε such that there is a collection $\{b_i : i \in j\}$ of subsets of a and nonnegative real numbers $c_i : i \in j$ with $\Sigma_{i \in j} c_i = 1$, $\psi(b_i) < \varepsilon$ and $\Sigma_i c_i \chi(b_i) \geq 1/2$.*

It is easy to see that for every measure μ on the set a there is a set of μ-mass at least $1/2$ and ψ-mass at most path(ψ), and this is how I will use the pathologicity. The choice of the constant $1/2$ is rather arbitrary as long as it is larger than zero. The key point is that there are submeasures such that path(ψ) is much smaller than the ψ-mass of the whole space.

Example 4.3.42. A direct attack at the definition. Let $u, v \in \omega$ be natural numbers such that $u/2 \leq v < u$ and consider the pavement submeasure ψ on the set $a = [u]^v$ defined by the pavers $b_i = \{y \in a : i \in y\}$ for $i \in u$ and the weight function $w(b_i) = \frac{1}{u-v+1}$. In other words, $\psi(b) = \frac{1}{u-v+1}$ times the smallest number of sets of the form $b_i : i \in u$ necessary to cover the set $b \subset a$. It is immediate that $\psi(a) = 1$ and path $(\psi) = \frac{1}{u-v+1}$ since $\Sigma_{i \in u} \frac{1}{u} \chi(b_i) = \frac{v}{u} \geq \frac{1}{2}$. Manipulating the numbers u, v one can clearly bring path (ψ) arbitrarily close to 0.

Example 4.3.43. Davies and Rogers used a considerably more sophisticated submeasure in their construction of a Hausdorff measure which attains only the values 0 and ∞ [11]. I will restate their example in probabilistic terms. Let n be a large natural number and $\varepsilon < 1/2\sqrt{n}$ be a positive real number. Let X be the n-dimensional sphere in $\mathbb{R}^{n+1}$, and for every $x \in X$ write $C(x) \subset X$ for the cap $\{y \in X : x \cdot y \geq \varepsilon\}$. Writing μ for the usual probability measure on X, a computation shows that $\mu(C(x)) > 1/4$. Throw a large number of mutually independent darts on the sphere X, obtaining a finite set $a \subset X$. As the number of darts grows, the probability of the following event tends to 1:

- a cannot be covered by n many caps of the form $C(x) : x \in X$. This uses Lyusternik–Shnirelman–Borsuk theorem, which says that whenever $D_i : i \in n$ is a collection of closed sets covering the sphere X then one of them contains a pair of antipodal points.
- For every point $x \in a$ the set $a \cap C(x)$ has size at least $|a|/4$. This uses some Monte Carlo type considerations.

Now let a be a set of darts satisfying these two items, and let ϕ be the submeasure on the set a defined by $\phi(b) = \max\{1, m(b)/|a|\}$, where $m(b)$ is the smallest number of caps of the form $C(x) : x \in a$ such that their union covers the set b. The first item shows that $\phi(a) = 1$, and the second item shows that the average of the characteristic functions of the caps $C(x) : x \in a$ is greater than $1/4$ at every point in a. Note that $\phi(C(x)) = 1/|a|$ is very small.

Towards the construction of the Ramsey capacities, let T be a finitely branching tree and $\psi_t : t \in T$ be submeasures on the set a_t of all immediate successors of the node t in the tree T such that $\psi_t(a_t) = 1$ and $\mathrm{path}(\psi_t) \cdot |T_{|t|}| \to 0$. Let $X = [T]$ and define a submeasure $\phi : \mathcal{P}(X) \to \mathbb{R}^+$ by $\phi(A) = \inf\{\Sigma_{t \in T}\psi_t(b_t) : b_t \subset a_t$ and for every branch $x \in A$ there is $t \in T$ with $x \restriction |t| + 1 \in b_t\}$. It is clear that ϕ is a pavement submeasure and therefore the forcing P_{I_ϕ} is proper by Theorem 4.5.2. The following theorem is the central result of this section.

Theorem 4.3.44. ϕ *is a Ramsey capacity.*

Proof. I will first introduce helpful notation. An *assignment* is a function F with $\mathrm{dom}(F) = T$ and $\forall t \in T$ $F(t) \subset a_t$. I will write $O_F = \{x \in X : \exists t \, x \restriction (|t| + 1) \in F(t)\}$. The proof of the theorem depends on a small claim.

Claim 4.3.45. *Suppose that $\varepsilon, \delta > 0$ are real numbers. There is a natural number $n = n(\varepsilon, \delta)$ such that for every assignment F with $\Sigma_{t \in T}\phi_t(F(t)) < \varepsilon$ there is an assignment G such that $\Sigma_{t \in T_{\leq n}}\phi_t(G(t)) < \varepsilon + \delta/2$, $\Sigma_{t \in T_{>n}}\phi_t(G(t)) < \delta/2$, and $O_F \subset O_G$.*

Proof. Just choose a number $k \in \omega$ such that $(1/2)^k \varepsilon < \delta/3$ and a number $n = n(\varepsilon, \delta) \in \omega$ such that $k \cdot |T_n| \cdot \max\{\mathrm{path}(\psi_t) : t \in T_n\} < \delta/3$. I claim that the number n works. Let F be an assignment with $\Sigma_{t \in T}\phi_t(F(t)) < \varepsilon$. For every node $s \in T_{n+1}$ write $g(s) = \Sigma_{t \supseteq s}\phi_t(F(t))$.

Fix a node $u \in T_n$ and consider the function $g \restriction a_u$. I will argue that there is a set $c_u \subset a_u$ such that $\phi_u(c_u) < \delta/3|T_n|$ and $\Sigma_{s \in a_u \setminus c_u}g(s) < (1/2)^k\Sigma_{s \in a_u}g(s)$.

The pathologicity of the submeasure ϕ_u kicks in exactly at this point. Let $c_i, b_i :$ $i \in j$ witness the fact that $\mathrm{path}(\psi_u) < \delta/3k|T_n|$. Thus $b_i \subset a_u : i \in j$ are sets of ψ_u-submeasure $< \delta/3k|T_n|$ and $c_i \in [0, 1] : i \in j$ are real numbers with unit sum such that $\Sigma_i c_i \chi(b_i) \geq 1/2$. Since the corresponding convex combination of the numbers $\Sigma_{s \in b_i} g(s) : i \in j$ gives a sum at least $\Sigma_{s \in a_u} g(s)/2$, there must be an index $i_0 \in j$ such that $\Sigma_{s \in b_{i_0}} g(s) \geq \Sigma_{s \in a_u} g(s)/2$. Repeat this process k times, each time reaping at least one half of the sum of the remaining numbers $g(s)$ by some set $b_{i_m} \subset a_u$. In the end, let $c_u = \bigcup_{m \in k} b_{i_m}$.

The conclusion of the proof is now at hand. The desired assignment G is obtained from F by letting $G(u) = F(u)$ for all $u \in T_{<n}$, $G(u) = F(u) \cup c_u$ for all $u \in T_n$, $G(t) = 0$ for every node $t \in T_{>n}$ such that the restriction of the node t to the $(n+1)$-st level falls into one of the sets $c_u : u \in T_n$, and $G(t) = F(t)$ for all the other nodes $t \in T_{>n}$. $\qquad\square$

The proof that ϕ is Ramsey is now straightforward. If $\varepsilon, \delta > 0$ are real numbers and $A_k : k \in \omega$ are sets of ϕ-mass $< \varepsilon$, first choose assignments $F_k : k \in \omega$ witnessing this, and then find the number $n = n(\varepsilon, \delta)$ and assignments $G_k = G(F_k, \varepsilon, \delta)$ as in the claim. There are only finitely many possibilities for $G_k \upharpoonright T_{\leq n}$ and so for two distinct numbers $k \neq l$ it will be the case that $G_k \upharpoonright T_{\leq n} = G_l \upharpoonright T_{\leq n}$. It is clear that $\phi(A_k \cup A_l) < \varepsilon + \delta$ as witnessed by the assignment $G : t \mapsto G_k(t) \cup G_l(t)$.

To show that ϕ is a capacity, it is enough to argue that it is continuous in increasing unions. Let $A_k : k \in \omega$ be an $\subset$-increasing sequence of sets of ϕ-mass $\leq \varepsilon$; I must show that $\phi(\bigcup_k A_k) \leq \varepsilon$. Fix a real number $\delta > 0$. To show that $\phi(\bigcup_k A_k) < \varepsilon + \delta$ choose assignments $F_k : k \in \omega$ witnessing that $\phi(A_k) < \varepsilon + \delta/2$ and by induction on $m \in \omega$ find

- infinite sets $u_m \subset \omega$ such that $\omega = u_0 \supset u_1 \supset u_2 \supset \dots$
- an increasing sequence of numbers $n_m \in \omega$;
- assignments $F_k^m : k \in u_m$ such that $F_k^m \upharpoonright T_{\leq n_m}$ is the same for all $k \in u_m$, this common part denoted by G_m, $G_0 \supset G_1 \supset \dots$, $\Sigma_{|t| \leq n_m} G_m(t) < \varepsilon + \Sigma_{l \in m} \delta \cdot 2^{-l}$ and $\Sigma_{|t| > n_m} F_k^m(t) < \delta \cdot 2^{-m}$ for all $k \in u_m$.

This is easy to do using the claim repeatedly. In the end, let $k_m = \min(u_m)$ and consider the assignment $G : t \mapsto \bigcup_m F_{k_m}(t)$; this is slightly bigger than $\bigcup_m G_m$. The induction hypotheses imply that $\Sigma_t \psi_t(G(t)) < \varepsilon + \delta$ and the assignment G witnesses $\phi(A_{k_m}) < \varepsilon + \delta$ for all numbers m and so even $\phi(A) < \varepsilon + \delta$ as desired. $\qquad\square$

Example 4.3.46. A capacity connected to the (f, h)-bounding property of forcings introduced by Shelah [2], 7.2.13 is obtained through this scheme in the following way. Let $f, h \in \omega^\omega$ be functions such that $f(n)/2 \leq h(n) \leq f(n)$ and $\prod_{m \in n} f(m)^{h(m)} \cdot$

$\frac{1}{f(n)-h(n)+1} \to 0$. Let T be the tree of all finite sequences t such that for every number $n \in \mathrm{dom}(t)$, $t(n) \subset f(n)$ is a set of size $h(n)$. For every node $t \in T$ let m_t be the submeasure obtained in Example 4.3.42 with $u = f(|t|)$ and $v = h(|t|)$, and let $c_{f,h}$ be the resulting capacity. It is not difficult to show that if a forcing is not f, h-bounding then it makes the ground model reals into a $c_{f,h}$-mass zero set. The rather restrictive condition $f(n)/2 \le h(n)$ can be replaced by a number of much weaker requirements.

The construction of capacities in this section carries with it the following forcing effect.

Proposition 4.3.47. *Suppose that ϕ is a capacity constructed using the above scheme. Then $I_\phi \perp \mathrm{null}$.*

Proof. Let T be the tree used in the construction with submeasures $\psi_t : t \in T$. For simplicity suppose that the submeasures ψ_t used to generate the capacity depend only on the length of the sequence t and write ψ_k for the submeasure used at sequences of length k. For each number $k \in \omega$ let $\mathcal{A}_k = \{A_i^k : i \in j_k\}$ and $c_i^k : i \in j_k$ witness the pathologicity of the submeasure ψ_k. Let U be the tree of all finite sequences s such that for every number $k \in \mathrm{dom}(s)$, $s(k) \in \mathcal{A}_k$. Equip U with the probability measure μ obtained as a product of the probability measures n_k on $\mathcal{A}_k$ which assign each singleton $\{A_i^k\}$ mass c_i^k.

Now consider the Borel set $D \subset [T] \times U$ defined by $\langle v, u \rangle \in D$ if and only if $\exists^\infty k\ v(k) \notin u(k)$. Note that $n_k(\{A_i^k : v(k) \notin A_i^k\}) \le 1/2$ and therefore the vertical sections of the set D have zero μ-mass. On the other hand, $\psi_k(u(k)) \le \varepsilon(\psi_k)$ and so $\phi(\{v : v(k) \in u(k)\}) \le \varepsilon(\psi_k) \cdot |T_k|$ which tends to 0 as k diverges to ∞. Consequently the horizontal sections of the complement of the set D have ϕ-mass 0. $\qquad\square$

4.3.6 Another Ramsey capacity

In this section I will provide a family of examples of nontrivial stable strongly subadditive Ramsey capacities. The treatment is parallel to the Steprāns capacities.

Fix a function $f : \mathbb{R}^+ \to \mathbb{R}^+$ which is differentiable, increasing, concave, and $f(0) = 0$, $f(1) = 1$. Typical examples include $f(x) = x^r$ for some real number $0 < r < 1$ or $f(x) = \log_2(1+x)$. Let T be a finitely branching tree. I will derive a strongly subadditive capacity ϕ_f on $[T]$ in a way that closely mimics the construction of Steprāns capacities.

By induction on $n \in \omega$ define a norm $|g|_n$ for all functions $g \in (\mathbb{R}^+)^{T_n}$. Let $|g|_0 = g(0)$ and $|g|_{n+1} = |h|_n$ where the function $h \in (\mathbb{R}^+)^{T_n}$ is defined by $h(t) = f(\text{arithmetical average of the values } g(s) \text{ where } s \in T_{n+1} \text{ is an immediate successor of } t)$.

Claim 4.3.48. *Let $n \in \omega$ be a number and g, h, k, l functions from T_n to $\mathbb{R}^+$.*

1. *If $g \leq h$ everywhere then $|g|_n \leq |h|_n$.*
2. *$|g+h|_n \leq |g|_n + |h|_n$.*
3. *If $g \leq h, k$ and $h, k \leq l$ everywhere and $g+l \leq h+k$ everywhere then $|g|_n + |l|_n \leq |h|_n + |k|_n$.*

Proof. This is proved by an easy induction on $n \in \omega$. For the third item note that if f is differentiable, increasing and concave, $u \leq v, x$ and $v, x \leq y$ and $u+y \leq v+x$, then $f(u) + f(y) \leq f(V) + f(x)$. To see this, fix u, y and minimize $f(V) + f(x)$ subject to the condition $u \leq v, x \leq y$ and $u+y \leq v+x$. Since the function f is increasing I can assume that $x \leq v$ and x is the smallest possible given v, which is $x = u+y-v$. Now differentiating the function $f(V) + f(u+y-V)$ with respect to v gives a nonpositive number since f' is a decreasing function and $u+y-v \leq v$. Thus the maximum of $f(V) + f(u+y-V)$ is attained at the boundary point $v = y$ as desired. $\square$

The capacity ϕ_f is now defined as follows. For a clopen set $O \subset [T]$ find a number $n \in \omega$ and a function $g : T_n \to 2$ such that $x \in O \leftrightarrow g(x \restriction n) = 1$ and let $\phi_f(O) = |g|_n$. This definition does not depend on the choice of the number n since $f(1) = 1$. For an open set $A \subset [T]$ let $\phi_f(A) = \sup\{\phi_f(O) : O \subset A$ is clopen$\}$ and for all other sets $B \subset [T]$ let $\phi_f(B) = \inf\{\phi_f(A) : B \subset A$ and A is open$\}$. The second item of the claim shows that the function ϕ_f is strongly subadditive on clopen sets. Fact 4.3.5 then shows that the natural extension of ϕ_f to all subsets of $[T]$ gives a strongly subadditive capacity. It is not difficult to see that if the function f is continuous at 0 then in every tree T which branches sufficiently fast the capacity ϕ_f assigns singletons zero mass.

I will investigate the special case of the capacities $\phi_r = \phi_f$ where $f(x) = x^r$ for a real number $0 < r < 1$. Let T be a finitely branching tree such that each node on its n-th level has many immediate successors.

Theorem 4.3.49. *Let $0 < r < 1$ be a real number. Then ϕ_r is a stable Ramsey capacity.*

Proof. First, some notation. Fix the real number r and write $\phi_r = \phi$. I will need the capacities ϕ_t and norms $|.|_{t,n}$ for every node $t \in T$ defined just as ϕ itself except starting at the node $t \in T$ instead of 0. The above work can be performed in the same manner for these capacities to show that they are strongly subadditive. It is clear from the definitions that for every set $A \subset [T]$ and every number $n \in \omega$, $\phi(A) = |g|_n$ where the function g assigns a node $t \in T_n$ the value $\phi_t(A)$.

Arguments for both the stability and Ramseyness use a certain multiplicativity property of the capacity ϕ. This is recorded in the following claim; the proof is an elementary induction on $n \in \omega$.

Claim 4.3.50. *For every number $n \in \omega$, every function $g : T_n \to \mathbb{R}^+$ and every positive real number a, $a^{(r^n)} |g|_n = |ag|_n$.*

For the stability, it is enough to define the tilde operation on open sets. Then, for an arbitrary Borel set B, the set $\tilde{B} = \bigcup_n \tilde{O}_n$ will work, where $O_n : n \in \omega$ are open supersets of B whose capacity tends to that of B.

Suppose that $O \subset [T]$ is an open set and let $\tilde{O} = O \cup \{x \in [T] : \forall \varepsilon > 0 \; \exists n \in \omega \; \phi_{x \restriction n}(O) > (1 - \varepsilon)^{(r^{-n})}\}$. I will show that this set works as in Definition 4.3.16.

First, I must argue that $\phi(\tilde{O}) = \phi(O)$. Choose a real number $\varepsilon > 0$ and work to prove that $\phi(O) \geq (1 - \varepsilon)\phi(O \cup P_\varepsilon)$, where $P_\varepsilon = \bigcup \{O_u : \phi_u(O) > (1 - \varepsilon)^{(r^{-|u|})}\}$. This will be enough since the set $O \cup P_\varepsilon$ covers $\tilde{O}$ by the definitions. For every node $t \in T$ and every natural number $n > |t|$ let $f_{t,n} : T_n \restriction t \to \mathbb{R}^+$ be the function defined by $f_{t,n}(s) = \phi_s(O)$, and $g_{t,n} : T_n \to \mathbb{R}^+$ be the function defined by $g_{t,n}(s) = f_{t,n}(s)$ except when there is some node u, $t \subseteq u \subseteq s$ such that $\phi_u(O) > (1 - \varepsilon)^{(r^{-|u|})}$, in which case assign $g_{t,n}(s) = 1$. Using the Claim 4.3.50 repeatedly, by induction on $n - |t|$ argue that $|g_{t,n}|_{t,n} < (1 - \varepsilon)^{(r^{|t|})} |f_{t,n}|_{t,n}$. A review of definitions shows that $\phi(O \cup P_\varepsilon) = \sup_n |g_{0,n}|_{t_n}$, and therefore $\phi(O) \geq (1 - \varepsilon)\phi(O \cup P_\varepsilon)$ as desired.

Now I must show that for every ϕ-positive set $C \subset [T]$ disjoint from $\tilde{O}$ it is the case that $\phi(O \cup C) > \phi(O)$. This uses the following claim similar to Claim 4.3.39.

Claim 4.3.51. *For every ϕ-positive Borel set $C \subset [T]$ and every real number $\varepsilon > 0$ there is a node $u \in T$ such that $\phi_u(C) > (1 - \varepsilon)^{(r^{-|u|})}$.*

Proof. Since it has been already established that ϕ is a capacity, it is harmless to assume that the set C is compact. Fix a real number $\varepsilon > 0$ and suppose the claim fails for C and ε. Find a natural number n such that $(1 - \varepsilon)\phi(P) < \phi(C)$ where $P \subset [T]$ is the clopen set of all sequences sharing their initial segment of length n with some element of the set C. Such a number n exists by the outer regularity of the capacity ϕ and a compactness argument. By downward tree induction on the node t, starting at nodes at level n, argue that $(1 - \varepsilon)^{(r^{-|t|})} \phi_t(P) > \phi_t(C)$. This yields a contradiction at $t = 0$. $\qquad\square$

Use the claim repeatedly to find a point $x \in C$ such that $\forall \varepsilon > 0 \; \exists n \in \omega \; \phi_{x \restriction n}(C) > (1 - \varepsilon)^{(r^{-n})}$. Since $x \notin \tilde{O}$, there must then be a number $n \in \omega$ such that $\phi_{x \restriction n}(C) > \phi_{x \restriction n}(O)$. A review of definitions shows that this implies $\phi(O \cup C) > \phi(O)$.

Now for the proof of the Ramsey property, let $\varepsilon, \delta > 0$ be real numbers. Find a number $n \in \omega$ such that $\varepsilon \cdot 2^{(r^n)} < \varepsilon + \delta$ and find a number k such that the function

$g : T_n \to \mathbb{R}^+$ given by $g(t) = 3/k$ has such a small norm that $\varepsilon \cdot 2^{(r^n)} + |g|_n < \varepsilon + \delta$. Now let $m = |T_n|^k + 1$. I claim that among any collection of sets $A_i : i \in m$ of capacity $< \varepsilon$ there are two whose union has capacity $< \varepsilon + \delta$. By a counting argument there must be numbers $i \neq j \in m$ and a function $h : T_n \to k$ such that for all nodes $t \in T_n$, both $h(t)/k \leq \phi_t(A_i) \leq (h(t) + 1)/k$ and $h(t)/k \leq \phi_t(A_j) \leq (h(t) + 1)/k$. Consider the functions $g_i, g_j, g : T_n \to \mathbb{R}^+$ given by $g_i(t) = c_t(A_i), g_j(t) = c_t(A_j)$ and $g(t) = \phi_t(A_i \cup A_j)$. The strong subadditivity of the capacities c_t shows that $g(t) \leq g_i(t) + g_j(t) + 2/k$ and consequently $g(t) \leq 2g_i(t) + 3/k$. A computation using the previous two claims and the choice of the numbers n, k shows that $\phi(A_i \cup A_j) = |g(t)|_n \leq |2g_i(t)|_n + |3/k|_n = 2^{(r^n)}|g_i(t)|_n + |3/k|_n = 2^{(r^n)}\phi(A_i) + |3/k|_n \leq 2^{(r^n)}\varepsilon + |3/k|_n < \varepsilon + \delta$ as desired. $\square$

4.3.7 An improper capacity forcing

For a long time I could not find an example of a capacity ϕ such that the forcing P_{I_ϕ} is not proper. The present section provides a host of such examples, showing that the subject of study is quite varied and difficult. I will begin with a simple measure theoretic proposition.

Proposition 4.3.52. *Suppose that $\phi_n : n \in \omega$ is a collection of capacities on a Polish space X such that $\phi_n(X) = 1$. Let $\psi : \mathcal{P}(X) \to \mathbb{R}^+$ be defined by $\psi(A) = \Sigma_n 2^{-n-1} c_n(A)$. Then*

1. *ψ is a capacity;*
2. *if all capacities ϕ_n are outer regular then so is ψ;*
3. *if all capacities ϕ_n are strongly subadditive then so is ψ;*
4. *if all capacities ϕ_n are Ramsey then so is ψ.*

Proof. The argument begins with the investigation of the sum of two capacities.

Claim 4.3.53. *Suppose that ϕ_0, ϕ_1 are two (outer regular, strongly subadditive, Ramsey) capacities. Then $\phi_0 + \phi_1$ is a (outer regular, strongly subadditive, Ramsey) capacity.*

This is a triviality; I will argue for the outer regularity and Ramseyness of $\phi_0 + \phi_1$. Let $\psi = \phi_0 + \phi_1$. Suppose first that ϕ_0, ϕ_1 are both outer regular capacities. If $A \subset X$ is an arbitrary set and $\eta(A) < \varepsilon$ then use the outer regularity of ϕ_0 and ϕ_1 to find open sets $O_0, O_1 \subset X$, both covering the set A, such that $\phi_0(O_0) + \phi_1(O_1) < \varepsilon$. Then clearly $\psi(O_0 \cap O_1) < \varepsilon$, proving the outer regularity of ψ. Now suppose that ϕ_0, ϕ_1 are both Ramsey capacities, $\varepsilon, \delta > 0$ and $A_n : n \in \omega$ are subsets of the space X such that $\psi(A_n) < \varepsilon$ for all $n \in \omega$. Use a counting argument to find an infinite set $a \subset \omega$ and real numbers $\varepsilon_0, \varepsilon_1$ such that $\varepsilon_0 + \varepsilon_1 = \varepsilon$ and

$\phi_0(A_n) < \varepsilon_0 + \delta/4$, $\phi_1(A_n) < \varepsilon_1 + \delta/4$ for all numbers $n \in a$. Use the Ramseyness of the capacity c_0 and the Ramsey theorem to find an infinite set $b \subset a$ such that $\phi_0(A_n \cup A_m) < \varepsilon_0 + \delta/2$ for all numbers $n \in m$. Use the Ramseyness of the capacity c_1 to find numbers $n \neq m \in b$ such that $\phi_1(A_n \cup A_m) < \varepsilon_1 + \delta/2$. Then $\psi(A_n \cup A_m) < \varepsilon + \delta$ as required.

The infinite sum operation in the proposition is treated similarly. I will argue for the outer regularity and Ramseyness, the other cases being similar. Suppose that all capacities $\phi_n : n \in \omega$ are outer regular, $A \subset X$ is an arbitrary set and $\psi(A) < \varepsilon$. Find a number $n \in \omega$ such that $\psi(A) < \varepsilon - 2^{-n}$, and consider the capacity $\psi_n = \Sigma_{m \in n} 2^{-m-1} \phi_m$. By the previous paragraph this is an outer regular capacity and so there is an open set $O \supset X$ such that $\psi_n(O) < \varepsilon - 2^{-n}$. However, $\psi \leq d_n + 2^{-n}$ and so $\psi(O) < \varepsilon$ as required. Now suppose that $\phi_n : n \in \omega$ are all Ramsey capacities, let $\varepsilon, \delta > 0$ be real numbers and let $A_m : m \in \omega$ be sets such that $\psi(A_m) < \varepsilon$. Find a number $n \in \omega$ such that $2^{-n} < \delta$. As before, consider the capacity ψ_n. It is Ramsey, smaller than ψ, and therefore there must be numbers $m \neq k$ such that $\psi_n(A_m \cup A_k) < \varepsilon + \delta - 2^{-n}$. Since $\psi \leq \psi_n + 2^{-n}$, it must be the case that $\psi(A_m \cup A_k) < \varepsilon + \delta$ as required. $\qquad \square$

In order to produce a subadditive (outer regular, strongly subadditive, Ramsey) capacity ψ such that the forcing P_{I_ψ} is not proper, it will now be enough to find a nondecreasing sequence $\phi_n : n \in \omega$ of unit mass subadditive (outer regular, strongly subadditive, Ramsey) capacities such that the respective σ-ideals I_{ϕ_n} do not stabilize in the sense of Proposition 2.2.6. Consider the subadditive (outer regular, strongly subadditive, Ramsey) capacity $\psi = \Sigma_n 2^{-n-1} \phi_n$. The σ-ideal $I_\psi = \{A \subset X : d(A) = 0\}$ is clearly the intersection of the nonstabilizing decreasing sequence of σ-ideals $I_{\phi_n} : n \in \omega$, and therefore the forcing P_{I_ψ} is not proper.

To build the capacities $\phi_n : n \in \omega$ as in the previous paragraph I will use the construction from Section 4.3.6. Let $r_n : n \in \omega$ be a decreasing sequence of reals smaller than one, bounded away from zero. Let $\phi_n = \phi_f$ where $f(x) = x^{r_n}$, as constructed in Section 3.11. It is immediate that $\phi_n : n \in \omega$ is an increasing sequence of strongly subadditive Ramsey capacities. To prove that the sequence of their respective null ideals does not stabilize it will be enough to prove the following.

Proposition 4.3.54. *Every ϕ_{n+1}-positive compact set has a compact subset which is ϕ_{n+1}-positive and ϕ_n-null.*

Proof. It will be enough to prove the following claim.

Claim 4.3.55. *For every compact ϕ_{n+1}-positive compact set K and every real number $\varepsilon > 0$ there is a compact set $L \subset K$ such that $\phi_{n+1}(L) > \phi_{n+1}(K)$ and $\phi_n(L) < \varepsilon$.*

Once the claim is proved, it is easy to build a descending sequence of compact sets below any given ϕ_{n+1}-positive set whose intersection has positive ϕ_{n+1}-mass but zero ϕ_n-mass. Note that capacities are continuous in decreasing intersections of compact sets.

To prove the claim, use the fact that the sequence $r_{n+1}^m : m \in \omega$ goes to zero faster than $r_n^m : m \in \omega$ to find a natural number m and real numbers α, β such that $(1 - \varepsilon)^{(r_{n+1}^{-m})} < \alpha < \beta < \varepsilon^{(r_n^{-m})}$. From the proof of Proposition 4.3.49 recall the capacities $(\phi_{n+1})_t$ associated with a node $t \in T_m$. For each such node $t \in T_m$ find a compact set $K_t \subset K \cap [T \restriction t]$ such that $\alpha \cdot (\phi_{n+1})_t(K) < (\phi_{n+1})_t(K_t) < \beta \cdot (\phi_{n+1})_t(K)$ and let $L = \bigcup_t K_t$. I claim that the set L works as desired. And really, Claim 4.3.50 shows that $\phi_{n+1}(L) > \alpha^{(r_{n+1}^m)} \cdot \phi_{n+1}(K) > (1 - \varepsilon) \cdot \phi_{n+1}(K)$, and since for every $t \in T_m$ $(\phi_n)_t(K_t) \leq (\phi_{n+1})_t(K_t) \leq \beta$ holds, $\phi_n(L) \leq \beta^{(r_n^m)} < \varepsilon$ as required. $\qquad\square$

In fact, the above way of constructing a capacity ϕ such that the forcing P_{I_ϕ} is not proper is in a precise sense canonical, as the following theorem shows.

Theorem 4.3.56. *Suppose that ϕ is a strongly subadditive capacity. The following are equivalent:*

1. *P_{I_ϕ} is not proper;*
2. *there are a positive set C and an increasing sequence of strongly subadditive capacities $\phi_n : n \in \omega$ below ϕ such that the respective ideals I_{ϕ_n} strictly decrease below C and $I_\phi = \bigcap_n I_{\phi_n}$.*

There are two attractive corollaries. The first shows that in the special case of σ-ideals derived from strongly subadditive capacities the properness of the quotient forcing is a projective statement. This partially answers certain concerns from Chapter 2, in particular Question 7.1.1. I do not know if the complexity bound is optimal.

Corollary 4.3.57. *The set $\{\phi : \phi$ is a strongly subadditive capacity on 2^ω such that the forcing P_{I_ϕ} is proper$\}$ is Π_3^1.*

Proof. In order to make sense of this note that strongly subadditive capacities on 2^ω form a Polish space under suitable coding. Let $\mathcal{O}$ denote the set of all clopen subsets of 2^ω. If ϕ is a strongly subadditive submeasure on $\mathcal{O}$ then the set function $\tilde{\phi} : \mathcal{P}(2^\omega) \to \mathbb{R}^+$ defined by $\tilde{\phi}(O) = \sup\{\phi(P) : P \in \mathcal{O}, P \subset O\}$ for open sets $O \subset 2^\omega$ and $\tilde{\phi}(A) = \inf\{\tilde{\phi}(O) : A \subset O, O \text{ open}\}$ for all other sets $A \subset 2^\omega$, is a strongly subadditive capacity and it is in fact the unique strongly subadditive capacity extending ϕ. Thus the strongly subadditive capacities may be identified with the closed subset of $\mathbb{R}^{\mathcal{O}}$ consisting of strongly subadditive submeasures on $\mathcal{O}$.

The corollary is then a simple complexity calculation using the third item of the theorem. □

The second corollary shows that improperness of the quotient forcing is in fact equivalent to a stronger statement. I do not know if in ZFC it is equivalent to collapsing c to $\aleph_0$.

Corollary 4.3.58. *If the forcing P_{I_ϕ} is not proper, then under some condition, the forcing P_{I_ϕ} adds a countable set of ordinals not covered by a ground model countable set of ordinals.*

Proof. The proof of the equivalence of (1) and (2) goes through a third statement equivalent to both. It is perhaps not as quotable as either (1) or (2), but it is still interesting in that it shows that the definition of stability of capacities was in some sense natural. Here it is:

(*) there is a ϕ-positive compact set $K \subset 2^\omega$ and an open set $O \subset 2^\omega$ such that for every compact ϕ-positive set $K' \subset K$ it is the case that $\phi(K' \cup O) > \phi(O)$, but for every real number $\varepsilon > 0$ there is an open set $P_\varepsilon \supset O$ such that $\phi(P_\varepsilon) < \phi(O) + \varepsilon$ and every compact ϕ-positive subset $K' \subset K$ has a compact ϕ-positive subset $K'' \subset K'$ such that $\phi(K'' \cup P_\varepsilon) = \phi(P_\varepsilon)$.

Let me first show that (*) is equivalent to the improperness of the forcing P_{I_ϕ}. Suppose that (*) holds as witnessed by some K and O. For every rational number $\varepsilon > 0$ choose a witnessing set $P_\varepsilon \supset O$. Let $D_\varepsilon = \{K'' \subset K : \phi(K'' \cup P_\varepsilon) = \phi(P_\varepsilon)\}$; this set is dense below the condition K by (*). If the forcing P_{I_ϕ} was proper, there would have to be a compact ϕ-positive set $L \subset K$ which is covered by a countable (indeed, finite) subset $D'_\varepsilon \subset D_\varepsilon$, this for every positive rational number ε. By the strong subadditivity, $\phi(\bigcup D'_\varepsilon \cup P_\varepsilon) = \phi(P_\varepsilon)$, and so $\phi(\bigcap_\varepsilon (\bigcup D'_\varepsilon \cup P_\varepsilon)) = \phi(O)$. Another application of strong subadditivity shows that $\phi(L \cup O) - \phi(O) \le \phi((\bigcap_\varepsilon \bigcup D'_\varepsilon) \cup (\bigcap_\varepsilon P_\varepsilon)) - \phi(O)$, which is equal to zero by the previoius sentence and so $\phi(L \cup O) = \phi(O)$, contradicting the assumption on the sets K and O.

On the other hand, if (*) fails then the forcing P_{I_ϕ} is proper. Let me just give the briefest of arguments here, because the proof is essentially identical to the game theoretic one for Theorem 4.3.17. I will just indicate the necessary changes. The game G will be played for the forcing P_{I_ϕ} and its canonical generic point only. In the construction of the counterplay against a putative winning strategy σ of Player I, Player II will place a move K_n at a round $m \in \omega$ so that, writing $O_m \subset 2^\omega$ for the basic open set the strategy σ played at the round m, for every open set $P \supset O_m$ such that $\phi(P) \le \phi(O_m) + 2^{-m}$ there is a positive compact set K'_n such that every positive compact set $K''_n \subset K'_n$ satisfies $\phi(K''_n \cup P) > \phi(P)$. The existence of such a compact set K_n and a number m follows from the failure of (*).

To show that (*) is equivalent to (2) above, first assume that (*) holds as witnessed by some sets K and $O \subset 2^\omega$. For every natural number $n \in \omega$ find a set $P_{2^{-n}} \supset O$ as in (*) and let $\psi_n(A) = \phi(P_{2^{-n}} \cup A) - \phi(P_{2^{-n}})$ for every set $A \subset K$. It is essentially trivial to argue that the functions $\psi_n : n \in \omega$ are strongly subadditive capacities smaller than ϕ. The statement (2) will then be satisfied with $\phi_n = \Sigma_{m \in n} 2^{-m} \cdot \psi_m$. On the other hand, if (2) holds then the forcing P_{I_ϕ} is not proper by Proposition 2.2.6, and therefore (*) follows by the previous paragraph. $\qquad\square$

4.4 Hausdorff measures and variations

This section deals with forcings obtained from a generalization of the ideal σ-generated by sets of finite Hausdorff measure.

Definition 4.4.1. *Let X be a Polish space, U a countable collection of its Borel subsets, let* diam $: U \to \mathbb{R}^+$ *be a function referred to as a* diameter function, *and let* $w : \mathcal{P}(U) \to \mathbb{R}^+$ *be a* weight function *satisfying the following demands:*

1. *w is Borel;*
2. *if $a \subset U$ is a set and $w(a)$ is a finite number then the diameters of the sets in a converge to 0;*
3. *w is weakly subadditive: there is a function $f \in \omega^\omega$ such that for all sets $a, b \subset U$, if $w(a), w(b) < n$ then $w(a \cup b) < f(n)$.*

The Hausdorff submeasure *ϕ derived from U, w is then given by $\phi_\delta(A) =$ $\inf\{w(a) : a \subset U$ consists of sets of diameter $< \delta$ and $A \subset \bigcup a\}$ and $\phi(A) =$ $\sup_{\delta > 0} \phi_\delta(A)$.*

The definition subsumes Hausdorff measures on compact metric spaces and many other submeasures. In fact the terminology is a little misleading since the function ϕ does not have to be a submeasure if w is not subadditive. Nevertheless, in typical examples this situation does not arise. I will be interested in the forcing properties of the poset $P_{I_{\sigma\phi}}$ where $I_{\sigma\phi}$ is the σ-ideal generated by sets of finite ϕ-mass.

Note that any presentation of a Hausdorff submeasure forcing is itself a Hausdorff submeasure forcing. Increasing the Polish topology on the space X to make the sets in the collection U clopen and choosing a continuous bijection $\pi : C \to X$ for some closed set $C \subset \omega^\omega$ it becomes obvious that every Hausdorff submeasure forcing has a presentation on the Baire space in which the sets in the collection U are open.

4.4.1 General theorems

Theorem 4.4.2. *[18] Suppose that ϕ is a Hausdorff submeasure on a Polish space X. The forcing $P_{I_{\sigma\phi}}$ is proper.*

Proof. For simplicity assume that $X = \omega^\omega$ and the sets in the collection U are open. Suppose that ϕ is a Hausdorff submeasure, P is a partial order and $\dot{x}$ is a P-name for an element of the Baire space. The Hausdorff game $G(\sigma\phi, P, \dot{x})$ is a game of length ω between Players I and II played in the following fashion. In the beginning Player I indicates an initial condition p_{ini} and then he produces one-by-one open dense subsets $\{D_n : n \in \omega\}$ of the poset P, and dynamically on a fixed schedule a Borel set $A \subset X$ of σ-finite ϕ-mass. Player II plays one by one decreasing conditions $p_{ini} \geq p_0 \geq p_1 \geq \ldots$ so that $p_n \in D_n$ and p_n decides the n-th digit of the point $\dot{x}$. He is allowed to hesitate for any number of rounds before placing his next move. Player II wins if, writing g for the filter Player II obtained, it is the case that $\dot{x}/g \notin A$.

To make this precise, Player I produces subsets $\{a_k^l : k, l \in \omega\}$ of the set U so that all elements of a_k^l have diameter $\leq 2^{-l}$, and $w(a_k^l) \leq k$. The Borel set $A \subset X$ above is then extracted as $\bigcup_k \bigcap_l \bigcup a_k^l$. Note that this is indeed a set in the ideal I as each set $\bigcap_l \bigcup a_k^l$ has Hausdorff submeasure at most k. Enumerating the set U as $\{u_i : i \in \omega\}$, Player I must indicate at round n which among the sets $u_i : i \in n$ fall into which set $a_k^l : k, l \in n$. Note that in this way Player I's moves related to the set A can be coded as natural numbers. Note also that given any set $B \in I_{\sigma\phi}$, Player I can play so that his resulting set $A \in I_{\sigma\phi}$ is a superset of B.

Lemma 4.4.3. *The following are equivalent:*

1. *$P \Vdash \dot{x}$ is not contained in any ground model Borel $I_{\sigma\phi}$-small set;*
2. *Player II has a winning strategy in the game $G(\sigma\phi, P, \dot{x})$.*

Theorem 4.4.2 immediately follows. An application of the lemma to the poset $P_{I_{\sigma\phi}}$ and its generic real shows that Player II has a winning strategy σ in the game $G(\sigma\phi, P_{I_{\sigma\phi}}, \dot{x}_{gen})$. Now suppose M is a countable elementary submodel of a large enough structure containing the ideal I and the strategy σ, and let $B \in M \cap P_I$ be a condition. We must show that the set $\{x \in B : x \text{ is } M\text{-generic}\}$ is $I_{\sigma\phi}$-positive. Well, if A is a Borel $I_{\sigma\phi}$-small set then consider the play of the game in which Player II follows the strategy σ and Player I indicates $B = p_{ini}$, enumerates the open dense sets in the model M, and dynamically produces a superset of the set A. Clearly, all moves of the play will be in the ground model, therefore the filter $g \subset M \cap P_I$ Player II creates will be M-generic, and the resulting real $\dot{x}_{gen}/g \in B$ will be M-generic and outside the small set A as desired.

One direction of the lemma is easy. If some condition $p \in P$ forces the point $\dot{x}$ to belong to some ground model coded I-small Borel set B, then Player I has a simple winning strategy. He will indicate $p_{ini} = p$, dynamically produce a suitable superset $A \supset B$, $A \in I$, and on the side he will find an inclusion increasing sequence $\{M_n : n \in \omega\}$ of countable elementary submodels of some large enough structure such that the n-th Player II's move p_n belongs to the model M_n, and he will make sure to enumerate all open dense subsets of the poset P that occur in the model $N = \bigcup_n M_n$. This is certainly easily possible. In the end Player II's filter g will be N-generic, by the forcing theorem $N[g] \models \dot{x}/g \in A$, by Borel absoluteness $\dot{x}/g \in A$, and Player I won.

For the other direction of the lemma note that the payoff set of the game is Borel and therefore the game is determined – Fact 1.4.2. Thus it will be enough to obtain a contradiction from the assumption that $P \Vdash \dot{x}$ is not contained in any ground model coded Borel I-small set and yet Player I has a winning strategy σ. A small claim will be used repeatedly:

Claim 4.4.4. *For every condition $p \in P$ and every number $k \in \omega$ there is a number $l(p, k) \in \omega$ such that for every set $a \subset U$ of weight $\leq k$ consisting of sets of diameter $\leq 2^{-l(p,k)}$ there is a condition $q \leq p$ forcing $\dot{x} \notin \bigcup a$.*

Proof. Suppose this fails for some p, k, and for every natural number $l \in \omega$ find a set $a_l \subset U$ of weight $\leq k$ consisting of sets of diameter $\leq 2^{-l}$ such that $p \Vdash \dot{x} \in \bigcup a_l$. But then, $p \Vdash \dot{x} \in \bigcap_l \bigcup a_l$, and the latter set is certainly in the ideal $I_{\sigma\phi}$, being of ϕ-mass $\leq k$. Contradiction! $\qquad \square$

First, some notation. Fix an enumeration $U = \{u_i : i \in \omega\}$ from which Player I's schedule is derived. Fix a function $g \in \omega^\omega$ such that for every number n, for every collection of $\leq n$ many subsets of U of weight $\leq n$, their union has weight $\leq g(n)$. Such a function exists by the weak subadditivity of the weight function w.

Player II will obtain a winning counterplay against the strategy σ by induction. His moves will be denoted by p_n, played at rounds i_n, and on the side he will produce numbers l_n. The intention is that the resulting point $\dot{x}/g$ will fall out of all sets $\bigcup a_k^{l_k} : k \in \omega$. For the convenience of notation let τ_n be the initial segment of the counterplay ending after the round i_n. The induction hypothesis is the following:

- for every number $k \in n$ and every index $i \in i_n$, if the strategy σ placed the set u_i in $a_k^{l_k}$ then $p_n \Vdash \dot{x} \notin \dot{u}_i$;
- $l_n > l(p_n, g(n))$, and the diameters of all sets $u_i : i > i_n$ are less than 2^{-l_n}.

This will certainly conclude the proof. To argue that Player II won this run of the game $G(\sigma\phi, P, \dot{x})$, note that whenever $u = u_i \in a_k^{l_k}$ is a Borel set, then every condition Player I played after round i forces $\dot{x} \notin u$ by the first item of the induction hypothesis. Since the set $u \subset \omega^\omega$ is open, this means that $\dot{x}/g \notin u$. This in turn means that $\dot{x}/g \notin \bigcup_k a_k^{l_k} \supset A$, and Player II won.

To get p_0, l_0, i_0 just find a condition $p_0 \in D_0$ below p_{ini}, let i_0 be some number such that all sets $u_i : i > i_n$ have diameter less than $l(p_0, g(0))$, and let l_0 be a large enough number. The induction hypotheses are satisfied. Now suppose that the objects p_k, l_k, i_k have been constructed. For every number $k \in n$ let $b_k \subset \{u_i : i \geq i_n\}$ be the collection of all sets in U indexed by a number $i \geq i_n$ which the strategy σ places into the set $a_k^{l_k}$ if Player II makes no nontrivial move past p_n. Let $b_n \subset U$ be the collection of all sets that the strategy σ places into the set $a_k^{l_n}$ if Player II makes no nontrivial move past p_n. Let $b = \bigcup_{k \leq n+1} b_k \subset U$. Note that this collection consists of sets of diameter $\leq l_n$ and it has weight $\leq g(n)$ by the weak subadditivity of the weight function w. Therefore, there must be a condition $q \leq p_n$ forcing $\dot{x} \notin \bigcup b$. Let $p_{n+1} \leq p_n$ be a condition deciding the $(n+1)$-st digit of the point $\dot{x} \in \omega^\omega$. Let $l_{n+1} > l(p_{n+1}, g(n+1))$ and let i_{n+1} be a number large enough so that all sets $u_i : i \geq i_{n+1}$ have diameters $< 2^{-l_n}$. The induction hypothesis continues to hold. $\square$

Theorem 4.4.5. *[18] Suppose that ϕ is a Hausdorff submeasure on a Polish space X.*

1. *The σ-ideal $I_{\sigma\phi}$ satisfies the third dichotomy.*
2. *In the choiceless Solovay model, $I_{\sigma\phi}$ satisfies the first dichotomy and it is closed under wellordered unions.*
3. *(ZF+DC+AD+) $I_{\sigma\phi}$ satisfies the first dichotomy and it is closed under wellordered unions.*

Proof. For the first item note that the ideal $I_{\sigma\phi}$ is generated by Borel sets. Consider the partial order $Q_{I_{\sigma\phi}}$ of $I_{\sigma\phi}$-positive analytic subsets of the space X ordered by inclusion. By Proposition 2.1.11, this forcing adds a generic point $\dot{x}_{gen} \in X$ which belongs to all sets in the generic filter. Suppose $A \subset X$ is an analytic $I_{\sigma\phi}$-positive set. To find a Borel $I_{\sigma\phi}$-positive subset $B \subset A$, let M be a countable elementary submodel of a large enough structure and let $B = \{x \in A : x$ is M-generic for the forcing $P_{I_{\sigma\phi}}\}$. The proof of Theorem 4.4.2 shows that this is an $I_{\sigma\phi}$-positive set, and as it is a one-to-one continuous image of the G_δ set of M-generic filters on the poset $Q_{I_{\sigma\phi}} \cap M$, it is Borel. The first item follows.

For the second item, work in the Solovay model and suppose $A \subset X \times 2^\omega$ is an arbitrary set such that $\text{proj}(A) \notin I_{\sigma\phi}$; I must find an analytic set $B \subset A$ such that $\text{proj}(B) \notin I_{\sigma\phi}$. The proof will then be completed by a reference to Proposition 3.9.17. By the usual homogeneity arguments I can assume that the set A is definable from parameters in the ground model. In the ground model, find a forcing P of size $< \kappa$ and P-names $\dot{x}, \dot{y}$ for elements of the space X and 2^ω respectively such that $P \Vdash \dot{x} \notin \bigcup(I_{\sigma\phi} \cap V)$ and $\mathrm{Coll}(\omega, < \kappa) \Vdash \langle \dot{x}, \dot{y} \rangle \in \dot{A}$. Use Lemma 4.4.3 to find a winning strategy σ for Player II in the game $G(\sigma\phi, P, \dot{x})$ and run back to the Solovay model. There the strategy σ still remains a winning strategy in the ground model version of the game $G(\sigma\phi, P, \dot{x})$ by a wellfoundedness argument. This shows that the set $\{\dot{x}/g : g \subset P \text{ is a } V\text{-generic filter}\}$ is $I_{\sigma\phi}$-positive, and it is the projection of the analytic set $B = \{\langle \dot{x}/g, \dot{y}/g \rangle : g \subset P \text{ is } V\text{-generic}\}$ which by the usual homogeneity argument is a subset of the set A. The second item follows.

The last item uses an integer game. Fix a Borel bijection $\pi : 2^\omega \to X \times 2^\omega$ and let $A \subset X \times 2^\omega$ be an arbitrary set. I must show that if $\mathrm{proj}(A) \notin I_{\sigma\phi}$ then there is an analytic set $B \subset A$ such that $\mathrm{proj}(B) \notin I_{\sigma\phi}$. Then, a reference to Proposition 3.9.18 will conclude the argument for the last item. Define an integer variation $G(A)$ of the game introduced in Theorem 4.4.2: Player I produces a countable sequence of subsets $A_k : k \in \omega$ of the space X such that $\phi(A_k) \leq k$, and Player II produces an infinite binary sequence $y \in 2^\omega$. Player II wins if $\pi(y) \in A$ and the first coordinate of the point $\pi(y)$ falls out of the set $\bigcup_n A_n$. To complete the description of the game, I must specify the schedule for both players. Player I creates sets $a_k^l \subset U$ for all numbers $k, l \in \omega$ such that $w(a_k^l) \leq k$ and sets in a_k^l have diameter $< 2^{-l}$; the sets A_k are then obtained as $A_k = \bigcap_l \bigcup a_k^l$. At round n Player I must specify all the finitely many sets of diameter $\geq 2^{-n}$ that belong to $a_k^l : k, l \leq n$. On the other hand, Player II plays the sequence y bit by bit and he can wait for arbitrary finite number of rounds before placing the next bit on the sequence.

The following claim is key.

Claim 4.4.6. *Player I has a winning strategy in the game $G(A)$ if and only if* $\mathrm{proj}(A) \in I_{\sigma\phi}$.

With the claim in hand, suppose that $\mathrm{proj}(A) \notin I_{\sigma\phi}$. A determinacy argument gives a winning strategy σ for Player II in the game $G(A)$. Let $B = \{\pi(y) : y \in 2^\omega \text{ is} \text{ the result of the strategy } \sigma \text{ playing against some counterplay}\}$. Since the strategy σ is winning for Player II, clearly $B \subset A$. By the virtue of its definition, B is analytic. Finally, since the strategy σ remains winning in the game $G(B)$ for Player II, it must be the case that $\mathrm{proj}(B) \notin I_{\sigma\phi}$. The third item of the present theorem follows.

The argument for the claim is standard. On one hand, if $\mathrm{proj}(A) \in I_{\sigma\phi}$ the Player I can win ignoring adversary moves altogether. On the other hand, if σ is a strategy for Player I then I will produce a set $C \in I_{\sigma\phi}$ such that for every point $x \in X \setminus C$ and every point $z \in 2^\omega$ there is a play against the strategy in which Player II produces a sequence $y \in 2^\omega$ such that $\pi(y) = \langle x, z \rangle$ and wins. This will certainly prove the claim, because if $\mathrm{proj}(A) \notin I_{\sigma\phi}$ then Player II can win against the strategy σ by choosing a point $x \in \mathrm{proj}(A) \setminus C$, a point $z \in 2^\omega$ such that $\langle x, z \rangle \in A$, and finding a winning play.

To define the set $C \in I_{\sigma\phi}$, I need to introduce new notation. If t is a finite play against the strategy σ, $n \in \omega$ is a natural number and $b \in 2$ is a bit, then tnb denotes the play against the strategy σ which starts out with t and then proceeds to round n without Player II making any nontrivial move except at round n, where he places the bit b. For numbers $k, l \in \omega$ and a finite play t against the strategy let $a_k^l(t) \subset U$ be the collection of those sets in U which the strategy σ places into the set a_k^l past the play t in the infinite extension of the play t in which Player II makes only trivial moves after t. For a finite play t and a finite sequence $\vec{l}$ of natural numbers let $C(t, \vec{l}) = \bigcap_n \bigcup_{k \in \mathrm{dom}(\vec{l})} \bigcup (a_k^{\vec{l}(k)}(tn0) \cup a_k^{\vec{l}(k)}(tn1))$. It is immediate that this is a set of finite ϕ-mass. For a finite play t and a number $k \in \omega$ let $C(t, k) = \bigcap_{l>n} a_k^l(t)$; this is again a set of finite ϕ-mass. Finally, let $C = \bigcup_{t,k} C_{t,k} \cup \bigcup_{t,\vec{l}} C_{t,\vec{l}}$ and argue that the set C works as announced.

To show this, suppose that $y \in 2^\omega$ is a point such that $\pi(y) \in A$ and the first coordinate of the point $\pi(y)$ does not belong to the set C. By induction on $n \in \omega$ build plays t_n and numbers $\vec{l}(n)$ so that:

- $0 = t_0 \subset t_1 \subset \dots$, the plays t_n follow the strategy σ, $t_{n+1} = t_n mb$ for suitable number m and bit b, and Player II in the course of these plays builds the sequence $y \in 2^\omega$;
- $x \notin \bigcup a_k^{\vec{l}(k)}(t_n)$ for all numbers $k \leq n$.

In the end, Player II will win, because the second item implies that for every $k \in \omega$, $x \notin \bigcup a_k^{\vec{l}(k)}$. Suppose that the play t_n has been constructed, together with the numbers $\vec{l}(k) : k \in n$. First, find the number $\vec{l}(n)$: as $x \notin C(t_n, n)$, there must be a number $\vec{l}(n) > n$ such that $x \notin \bigcup a_n^{\vec{l}(n)}$. And second, find the number $m \in \omega$ and the bit $b \in 2$ such that $t_{n+1} = t_n mb$. The bit b is determined by the sequence y. Now since $x \notin C(t, \vec{l})$, there must be a number m such that $x \notin \bigcup_{k \leq n} \bigcup a_k^{\vec{l}(k)}(t_n mb)$, and then let $t_{n+1} = t_n mb$. The induction hypotheses continue to hold. $\qquad \square$

There is a large natural class of Hausdorff submeasures for which I can prove a stronger forcing preservation theorem.

Definition 4.4.7. *The weight function w is* lower semicontinuous *if the diameters of the sets in U converge to 0 and for every set $a \subset U$, $w(a) = \sup\{w(b) : b \subset a$ is finite$\}$.*

Theorem 4.4.8. *[18] If the Hausdorff submeasure ϕ is derived from a lower semicontinuous weight function then the forcing $P_{I_{\sigma\phi}}$ is bounding and does not add splitting reals.*

Proof. This will be a corollary to the following fact of independent interest. Let J be a universally Baire σ-ideal such that the forcing P_J is proper. The following are equivalent:

1. P_J has the Laver property;
2. for every Hausdorff submeasure ϕ derived from a lower semicontinuous weight function, $I_{\sigma\phi} \not\perp J$.

The proof of this equivalence requires large cardinal assumptions in its full generality. For the purposes of the theorem such assumptions are not necessary though, because I will apply it only in the case of the Mathias ideal J. Then P_J is in the forcing sense equivalent to the Mathias forcing and it has the Laver property. This shows in particular that the ideal $I_{\sigma\phi} \not\perp J$, which by the results of Section 3.4 is equivalent to $P_{I_{\sigma\phi}}$ not adding unbounded or splitting reals.

The $(2)\rightarrow(1)$ implication is easier. Suppose that J is a σ-ideal on a Polish space Y such that P_J is proper and fails to have the Laver property, as witnessed by some ground model functions $f, g \in \omega^\omega$ and a name $\dot{h}$ for a function in the extension dominated by f. Then $I \perp J$ for some σ-finite ideal I derived from a lower semicontinuous Hausdorff submeasure. Namely, let $X_n = [f(n)]^{g(n)}$ and let θ_n be the submeasure on X_n defined by $\theta_n(a) =$ the smallest possible size of a set $z \subset X_n$ such that a contains no superset of z, and use the method of Section 4.4.3 to obtain a σ-finite ideal I on the space $X = \Pi_n X_n$. It is not difficult to see that the forcing P_I adds a function $\dot{x}_{gen} \in \Pi_n X_n$ such that every ground model function h dominated by f satisfies $h(n) \in \dot{x}_{gen}(n)$ for all but finitely many n. To see that $I \perp J$, find a J-positive Borel set $C \subset Y$ and a Borel function $k : C \rightarrow \omega^\omega$ such that $C \Vdash \dot{h} = k(\dot{r}_{gen})$ and let $D \subset X \times C$ be the Borel set $D = \{\langle x, y \rangle : \text{for all but finitely many } n, k(y)(n) \in x(n)\}$. Now the vertical sections of the set D are J-small since for every given point $x \in X$ the condition C forces $\dot{h}$ to avoid the prediction by x infinitely many times. And the horizontal sections of the complement of the set D are I-small since for every given point $y \in Y$ the poset P_I forces the function $\dot{x}_{gen}$ to predict $k(y)$ at all but finitely many values.

Of course from the point of view of forcing preservation it is the implication $(1)\rightarrow(2)$ that is most interesting. Suppose $C \in P_J$ and $B \in P_I$ are positive Borel sets and $D \subset C \times B$ is a Borel set with I-small vertical sections. I must produce a J-positive horizontal section of the complement of the set D.

First, fix some instrumental objects. Use P_J uniformization 2.3.4 to see that thinning out the set C if necessary we may assume that there are Borel maps $a_k^l : C \to \mathcal{P}(U) : k, l \in \omega$ such that for every element $y \in C$, the set $a_k^l(y)$ has weight $\leq k$ and consists of sets of diameter $\leq 2^{-l}$, and the vertical section D_y of the set D above y is covered by the σ-finite set $\bigcup_k \bigcap_l \bigcup a_k^l(r)$. Fix also an enumeration $U = \{u_i : i \in \omega\}$, a function $g \in \omega^\omega$ such that unions of $\leq n$ many subsets of U of weight $\leq n$ have weight $\leq g(n)$, fix a winning strategy σ for Player II in the Laver game 3.10.11 associated with the poset P_J, and let M be a countable elementary submodel of a large enough structure containing the strategy σ as well as other relevant objects.

By induction on $n \in \omega$ build plays $\tau_0 \subset \tau_1 \subset \tau_2 \subset \ldots$ of the Laver game of the respective length $i_0, i_1, i_2, \ldots$ observing the strategy σ, conditions $B = B_0 \supset B_1 \supset B_2 \supset \ldots$ in the forcing P_J and numbers $l_0, l_1, l_2, \ldots$. The intention is that the resulting set $C_\tau \subset Y$ of the play $\tau = \bigcup_n \tau_n$ is J-positive, the intersection $\bigcap_n B_n$ is a singleton containing some unique $x \in X$ and the set $C_\tau \times \{x\}$ is a subset of the complement of D, secured by the fact that for every $y \in C_\tau$, $x \notin \bigcup_k \bigcup a_k^{l_k}(y)$. The induction hypotheses are:

- The finite plays as well as the sets B_n are in the model M, and B_{n+1} belongs to the n-th open dense subset of the forcing P_J in the model M in some fixed enumeration.
- $l_n > l(B_n, g(n))$, and for every number $i \leq i_n$ it is the case that $\operatorname{diam}(u_i) \geq 2^{-l_n}$ and for every number $i > i_n$ it is the case that $\operatorname{diam}(u_i) \leq l(B_n, g(n))$. Here $l(B_n, g(n))$ refers to the number identified in Claim 4.4.4.
- For every number j, $i_n \leq j < i_{n+1}$, Player I places the following move in τ_{n+1} at round j. Consider the equivalence relation E_n^j given by $r \ E_n^j \ s$ if and only if for every number i, $i_n \leq i \leq j$, and for every number $k \leq n$, $u_i \in a_k^{l_k}(r) \leftrightarrow u_i \in a_k^{l_k}(s)$. Player I plays the partition of the set $C \subset Y$ into the finitely many Borel E_n^j equivalence classes, asking the opponent to choose $n+1$ many of them. The strategy σ answers with a set C_j, a union of at most $n+1$ many equivalence classes. Let $b_n^j = \{u_i \in U : i_n \leq i \leq j, \exists k \leq n \ \exists y \in C_j \ u_i \in a_k^{l_k}(y)\}$. Note that the collection b_n^j consists of sets of diameter at most $l(B_n, g(n))$ and has weight at most $g(n)$ since it is a union of $n+1$ many sets of weight $\leq n$.
- Whenever $i \in i_n$ and $\exists k \leq n \ \exists y \in \bigcap_{j \in i_n} C_j \ u_i \in a_k^{l_k}(y)$ then $B_n \cap u_i = 0$.

This will certainly be enough. The first item implies that the intersection $\bigcap_n B_n$ will be a singleton by Proposition 2.1.2 applied in the model M. The resulting set $C_\tau \subset Y$ of the play τ is J-positive. The third item then implies that $x \in \bigcap_{n \in \omega} B_n$

and $r \in \bigcap_{j \in \omega} B_j$ then $\langle x, r \rangle \notin C$ as required, since $x \notin \bigcup_k \bigcup a_k^{l_k}$. The second item is present only to keep the induction going.

Now suppose the play τ_n, the set B_n and the numbers l_n, i_n have been found. Consider the infinite run of the Laver game extending τ_n according to the third inductive item, and the collection $b_n = \bigcup_{j \in \omega} b_n^j$. This collection consists of sets of diameter $\geq l(B_n, g(n))$ and by the lower semicontinuity of the Hausdorff submeasure in question, it has weight at most $g(n)$. Therefore the Borel set $B_{n.5} = B_n \setminus \bigcup b_n$ is I-positive. Note that this set already satisfies the last item of the induction hypothesis at $n + 1$ no matter what the number i_{n+1} will be. Now find an arbitrary set $B_{n+1} \subset B_{n.5}$ in the n-th open dense set in the model M and consider the number $l(B_{n+1}, g(n + 1))$. Let i_{n+1} be some number such that $\text{diam}(u_i) \leq 2^{-l(A_{n+1}, g(n+1))}$ for every $i > i_n$, and let $l_{n+1} > l$ be some number such that $\text{diam}(u_i) \geq l_{n+1}$ for every $i \leq i_n$. This completes the inductive step and the proof of the theorem. $\qquad\square$

4.4.2 Hausdorff measures

Suppose X is a compact metric space with metric d, U is the collection of all balls with rational radius with centers at some given countable dense set, diam is the usual metric diameter function, $h : \mathbb{R}^+ \to \mathbb{R}^+$ is a continuous nondecreasing *gauge function* with $h(0) = 0$ and $w(a) = \Sigma_{u \in a} h(\text{diam}(u))$. Then Definition 4.4.1 results in the usual h-dimensional Hausdorff measure.

It is interesting to investigate when Theorem 4.4.8 can be applied to the forcing $P_{l_{\sigma\phi}}$. Clearly the difficulty is that in the above setup for a Hausdorff measure there are infinitely many distinct balls of a given radius. Nevertheless, we have

Proposition 4.4.9. *Suppose that the function h satisfies the doubling condition, that is, there is a number $r \in \mathbb{R}$ such that $h(2\varepsilon) < rh(\varepsilon)$ for every $\varepsilon \in \mathbb{R}^+$. Let ϕ be the resulting Hausdorff measure. Then there is a Hausdorff submeasure ψ derived from a lower semicontinuous weight function such that $I_{\sigma\psi} = I_{\sigma\phi}$.*

Proof. Use the compactness of the space X to find a finite 2^{-n}-net $A_n \subset X$ for every natural number $n \in \omega$; that is, every point of the space X is 2^{-n}-close to some point in the set A_n. Let U be the set of all balls with centers in the set A_n and radius 2^{-n} for some number n, such a ball will be assigned diameter 2^{-n} and weight $h(2^{-n})$, and for every set $a \subset U$ let $w(a) = \Sigma_{u \in a} h(\text{diam}(u))$. This is clearly a lower semicontinuous weight function. Let ψ be the resulting Hausdorff submeasure. It is not difficult to use the doubling condition to show that $\phi(B) \leq \psi(B) \leq r\phi(B)$ and therefore the σ-finite ideals associated with the submeasures ϕ, ψ coincide. $\qquad\square$

Corollary 4.4.10. *If the gauge function satisfies the doubling condition then the forcing $P_{l_{\sigma\phi}}$ is bounding and does not add splitting reals.*

This should be compared with a classical result of Howroyd [26]:

Fact 4.4.11. *If the gauge function h satisfies the doubling condition then every analytic φ-positive set has a compact subset of φ-positive finite measure.*

This, together with Example 3.6.4, yields the following related corollary:

Corollary 4.4.12. *If the gauge function h satisfies the doubling condition then the forcing $P_{I_{\sigma\phi}}$ is bounding and preserves outer Lebesgue measure.*

Measure theorists have long studied the question whether in every Hausdorff measure every Borel set of positive measure must have a Borel subset of positive finite measure. This turns out to be false and Davies and Rogers [11] provided a classical counterexample: a compact metric space and a gauge function such that the resulting Hausdorff measure ψ achieves only 0 and ∞ as its values. In such a case, $I_{\sigma\psi} = I_{\psi} = I_{\eta}$ where η is the associated Hausdorff content capacity: $\eta(A) = \inf\{\Sigma_n h(\text{diam}(u_n)) : u_n \subset X, A \subset \bigcup_n u_n\}$.

Proposition 4.4.13. *The forcing $P_{I_{\sigma\psi}}$ from the Davies–Rogers example is proper, bounding, collapses outer Lebesgue measure, and adds no splitting reals.*

Proof. In all cases there are several possible proofs of the given property. The properness follows from the work in this section as well as from work in Section 4.5, since $I_{\sigma\psi} = I_{\eta}$ and η is a pavement submeasure. A review of the Davies and Rogers example shows that in the metric space in question there are only countably many possible distances forming a set of real numbers which converges to 0, and for each possible distance $\varepsilon > 0$ there is a finite set of clopen sets of diameter ε such that every other set of diameter ε is a subset of one of them. Thus the Hausdorff measure is in fact obtained from a lower semicontinuous weight function, and the work in the previous section shows that the forcing is bounding and adds no splitting reals. Another way of looking at this is that the Hausdorff content is a capacity [56], Theorem 47, and use 4.3.23 to show that the forcing is bounding. A careful review of the example shows that in fact the Hausdorff content in question is a special case of the scheme of Section 4.3.5 which produces Ramsey capacities, and therefore by Theorem 4.3.25 the forcing $P_{I_{\eta}} = P_{I_{\sigma\psi}}$ does not add splitting reals. To show that the forcing collapses outer Lebesgue measure it is possible to either review the example again and see that in fact Davies and Rogers prove that the ideal $I_{\sigma\psi} = I_{\eta}$ is not polar and use Proposition 3.6.10 with the Hausdorff content η. Another way is to use Proposition 4.3.47 which shows that the construction of capacities in Section 4.3.5 yields forcings which collapse outer Lebesgue measure. □

In fact, both P_{I_η} adding no splitting reals and collapsing outer Lebesgue measure abstractly implies that there are no sets of finite positive Hausdorff measure. To see this, first review the definitions and observe $I_\psi = I_\eta$. If there was a Borel set B of positive finite Hausdorff measure then the forcing $P_{I_\psi} \upharpoonright B$ is in the forcing sense equivalent to the Solovay forcing by the measure isomorphism theorem, and so it adds splitting reals and preserves outer Lebesgue measure.

4.4.3 Fat tree forcings

In [58], Shelah and Roslanowski isolated a class of forcings referred to as $\mathbb{Q}_1$. In this section I show that these forcings correspond to a certain class of Hausdorff submeasures. This opens a new way of looking at them, and as one application it shows that they do not add splitting reals by Theorem 4.4.8.

Suppose that T is a finitely branching tree. For each node $t \in T$ let a_t denote the set of the immediate successors of the tree T, and choose a collection $\vec{\phi} = \{\phi_t : t \in T\}$ where $\phi_t : \mathcal{P}(a_t) \to \mathbb{R}^+$ are functions which are monotone and uniformly weakly subadditive: there is a function $f \in \omega^\omega$ such that for every node $t \in T$, union of two subsets of a_t of weights $\leq n$ has weight $\leq f(n)$. These objects can be used to generate a Hausdorff submeasure on the Polish space $X = [T]$ in the following way. Let $U = \{u_{t,a} : t \in T, a \subset a_t\}$ where $u_{t,a} = \{x \in X : x \upharpoonright |t| + 1 \in a\}$, let $\operatorname{diam}(u_{t,a}) = 2^{-|t|}$, and for a set $b \subset U$ let $w(b) = \sup\{\phi_t(a) : u_{t,a} \in b\}$. Let ψ be the resulting Hausdorff submeasure on the space X. I will show that the forcing $P_{I_{\sigma\psi}}$ has a dense subset naturally isomorphic to the poset of $\vec{\phi}$-fat trees, where an infinite tree $S \subset T$ is $\vec{\phi}$-fat if for every number $k \in \omega$ there is $l \in \omega$ such that for every node $t \in S$ of length $> l$ the set $\phi_t(\{s \in S : s$ is an immediate successor of $t\}) > k$.

Proposition 4.4.14. *Suppose that $A \subset X$ is an analytic set. Then $A \notin I_{\sigma\psi}$ if and only if A contains all branches of some $\vec{\phi}$-fat tree. Under AD this extends to all subsets of the space X.*

Proof. For the ease of notation assume that the underlying tree T is just $\omega^{<\omega}$. Fix an analytic set $A \subset X$ together with a tree $Z \subset (\omega \times \omega)^{<\omega}$ projecting to it, and consider the game G between Players I and II. There is a counter c associated with the game, its initial value is zero. The initial move of Player I is a sequence $t \in T$. After that, at each round $i \in \omega$ Player I indicates an immediate successor $t_i \in T$ of the node played in the previous round, and time to time he increases the value of the counter c by one and plays a natural number n_j. Player II responds by playing a set $b_i \subset a_{t_i}$ such that $\phi_t(b_i) \leq c$. At round $i+1$, Player I must make sure that $t_{i+1} \notin b_i$. Player I wins if he increased the counter infinitely many times and the path through the tree T together with the sequence $n_j : j \in \omega$ he obtained form a branch through the tree Z. Note that before he increases the value of the counter the first time, Player I has a complete freedom of choice of moves. The following claim is key.

Claim 4.4.15. *Player II has a winning strategy if and only if $B \in I_{\sigma\psi}$. Player I has a winning strategy if and only if there is a fat tree $S \subset T$ such that $[S] \subset B$.*

Once the claim has been proved, the proposition follows by a determinacy argument using Fact 1.4.2.

The second equivalence in the claim is more or less obvious. Suppose that $S \subset T$ is a fat tree such that $[S] \subset A$. Then in the proper forcing $\mathbb{Q}_1$, $S \Vdash \dot{x}_{gen} \in \dot{A}$ by Shoenfield absoluteness, there is a name $\dot{y} \in \omega^\omega$ such that $S \Vdash \langle \dot{x}_{gen}, \dot{y} \rangle$ form a branch through the tree Z, and by the continuous reading of names of $\mathbb{Q}_1$ [58] there is a fat tree $U \subset S$ and a continuous function $f : U \to \omega^\omega$ such that $U \Vdash \dot{y} = \dot{f}(\dot{x}_{gen})$ and even such that $x, f(x)$ forms a branch through the tree Z for every path $x \in [U]$. Then Player I can win in the game by playing along the tree U, increasing the counter to $k+1$ and playing the number n_k every time he hits a node $t_i \in U$ such that all extensions $s \in U$ of t_i split into at least ϕ_s-mass $k+1$ many immediate successors and all paths $x \in [U]$ with $t_i \subset x$ have $f(x)(k) = n_k$. On the other hand, if Player I has a winning strategy σ then let $S \subset T$ be the tree of all nodes the strategy σ can arrive at; a simple compactness argument shows that every branch $x \in [S]$ is a result of the strategy σ acting against some counterplay, and therefore $[S] \subset B$. Another compactness argument shows that in fact S is an $\dot{\phi}$-fat tree.

Similarly, the right-to-left direction of the first equivalence in the claim is easy. Suppose that $A \in I_{\sigma\psi}$ is a set, and for every number n find a set $A_n \subset X$ of ψ-mass $\leq n$ such that $A \subset \bigcup_n A_n$. Find natural numbers $c_0 \in c_1 \in c_2 \in \dots$ such that for every number n and every node $t \in T$ the union of n-many subsets of a_t of ϕ_t-mass $\leq n$ has ϕ_t-mass $\leq c_n$. This is possible by the uniform weak subadditivity of the functions $\phi_t : t \in T$. The winning strategy of Player II is now described by the following rule: when Player I increases the value of the counter c to c_n after playing some node $t_i \in T$, Player II finds sets $b_{t,n} \subset a_t : t \in T, |t| > |t_i|$ such that $\phi_t(b_t) \leq n$ and $A_n \subset \bigcup_t u_{t,b_{t,n}}$, and continues playing in such a fashion that the set $\bigcup_{m \in n} b_{t_j,m}$ (which is of ϕ_{t_j}-mass $\leq c_n$) is a subset of his move b_{t_j} after the play reached the node t_j, this for all $j > i$. A review of the definitions reveals that this is a winning strategy for Player II.

This leaves us with the left-to-right direction of the first equivalence. Suppose that σ is a strategy for Player II. I will produce countably many sets of finite ψ-mass such that every point $x \in [T]$ belonging to neither of them and every sequence $\langle n_j : j \in \omega \rangle \in \omega^\omega$ can be produced by Player I in a play against the strategy σ. This will certainly suffice, since then if $A \notin I_{\sigma\psi}$, Player I can defeat the strategy by playing a suitable point in the set A and a witness sequence. For every partial play τ against the strategy σ and every natural number n let $A_{\tau,n} \subset X$ be the following set. For every number $l \in \omega$ and every node $t \in T$ of length $> l$ let $b_t^l(\tau, n) \subset a_t$ be the set the strategy σ indicates in the last move of a finite play which extends τ, finishes with the node t, and in which Player I raises the counter just once more at round l and plays the number n at that round. Note that there is at most one such play. Let $a^l(\tau) = \{u_{t, b_t^l(\tau, n)} : |t| > l\}$ and let $A_{\tau,n} = \bigcap_l \bigcup a^l(\tau)$. Note that $\psi(A_{\tau,n})$ is a finite number, bounded by the value of the counter at the end of the play τ plus two. I will show that the countably many sets $A_{\tau,n}$ have the required property.

Suppose that $x \in X$ is some point not in $\bigcup_{\tau,n} A_{\tau,n}$, and let $y \in \omega^\omega$. I will produce a winning counterplay against the strategy σ which ends in the point x. To construct the counterplay, I just have to show when Player I increases the value of the counter. By induction on number $k \in \omega$ build increasing sequence of natural numbers $i_k \in \omega$ such that the play τ_k lasting $i_k + 1$ rounds in which Player I follows the sequence x, raises the value of the counter at rounds $i_m : m \leq k$ and in which Player II follows the strategy σ is a legal play of the game, and $x \notin \bigcup a^{i_k}(\tau_{k-1}, y(k))$. The induction starts with letting $\tau_{-1} = 0$. To find the number i_0 note that $x \notin A_0$ and so there must be a number l such that $x \notin \bigcup a^l(0)$; let $i_0 = l$. Suppose τ_k has been constructed. To find the number i_{k+1} (and the play τ_{k+1}) first note that since $x \notin \bigcup a^{i_k}(\tau_{k-1})$, Player I can legally extend the play τ_k for arbitrary finite number of rounds following the sequence $x \in X$ as long as he does not increase the value of the counter. Now, since $x \notin A_{\tau_k, y(k)}$, there is a number $l \in \omega$ such that $x \notin \bigcup a^l(\tau_k)$. Just let $l = i_{k+1}$ and the induction hypotheses will be satisfied. $\qquad \square$

What are the finer forcing properties of partial orders of fat trees? Theorem 4.4.8 shows that they are all bounding and add no splitting reals. I will show that a number of them preserve various capacities. Let $f \in \omega^\omega$ be a function such that $f(n) > \prod_{m \in n} f(m)$ and let $g : \omega \times \omega \to \omega$ be a function such that for any given numbers $n, i \in \omega$ $g(n, i+1) > 2 \cdot g(n, i)$. Consider the space $X = \prod_j f(j)$, $X = \text{proj}[T]$ where T is the tree of all finite sequences such that for every $j \in \phi(t)$, $t(j) \in f(j)$. For every node $t \in T$ let $\phi_t(a) = $ the least number k such that $|a| > g(|t|, k)$, for $a \subset a_t$. The resulting $\vec{\phi}$-fat tree forcing is known as $PT_{f,g}$ [2]. I will show that depending on the choice of the functions f, g the poset $PT_{f,g}$ preserves outer Lebesgue measure and some Ramsey capacities.

Proposition 4.4.16. *Suppose that for every number $k \in \omega$, $\Sigma_i (\prod_{j \in i} f(j) \cdot \frac{g(k,i)}{g(k+1,i)}) < \infty$. Then $I_{\sigma\psi}$ is a polar σ-ideal.*

The scary arithmetical condition just says that the function $g(\cdot, i)$ grows very fast when compared with the numbers $f(j) : j \in i$. Together with the previous proposition and Theorem 3.6.2 this implies that the forcing $PT_{f,g}$ preserves outer Lebesgue measure. It is interesting to compare this argument with the original proof in [2].

Proof. This is essentially a triviality at this stage. Suppose that S is an $\bar{\phi}$-fat tree. I must produce a probability measure μ on T such that all subsets of finite ψ-mass have zero μ-mass. Just let μ be the unique probability measure such that for every node $t \in S$ the μ-masses of sets $O_s \cap [S]$, where $s \in S$ is an immediate successor of the node t, are all equal.

Suppose $A \subset [S]$ has finite ψ-mass, smaller than some natural number k. For an arbitrary real number $\varepsilon > 0$, I must show that $\mu(A) < \varepsilon$. Find a number $l \in \omega$ such that past level l, every node $s \in S$ has more than $g(|s|, k+1)$-many immediate successors, and $\Sigma_{i>l}(\Pi_{j \in i} f(j) \cdot \frac{g(i,k)}{g(i,k+1)}) < \varepsilon$. Since $\psi(A) < k$, there must be a choice of sets $b_t \subset a_t$ for every node $t \in S$ longer than l such that $|b_t| \leq g(|t|, k)$ and $A \subset \bigcup_t u_{t,b_t}$. The choice of the number l shows though that $\mu(\bigcup_t u_{t,b_t}) < \varepsilon$, completing the proof. $\qquad\square$

Theorem 4.4.17. *Suppose that η is a finite outer regular Ramsey capacity. Then there are functions f, g such that the forcing $PT_{f,g}$ preserves η. In fact, for real numbers $\varepsilon, \delta > 0$, every set of capacity $< \varepsilon$ in the extension has an open superset of capacity $< \varepsilon + \delta$ coded in the ground model.*

In fact it is clear from the argument that it is possible to preserve a countable collection of outer regular Ramsey capacities. With a suitable choice of the capacity η, such as that generated from submeasures of Example 4.3.42, the whole situation is closely related to the f, h-bounding property of the forcings [2], 7.2.13.

Proof. Fix a finite Ramsey outer regular capacity η on a Polish space Y. Observe that for every real number $\delta > 0$ and every natural number $p \in \omega$ there is a number $q = q(p, \delta)$ such that for every real number $\varepsilon > 0$ and every collection of q many sets of capacity $< \varepsilon$ there are p many of them whose union has capacity $< \varepsilon + \delta$. This is proved by an argument parallel to (*) in the proof of Theorem 4.3.13.

Now I am ready to state the arithmetical condition on the functions f, g which ensures the preservation of the capacity c in the forcing $PT_{f,g}$. Find an infinite sequence $\delta_i : i \in \omega$ of positive real numbers such that $\Sigma_i \delta_i < \infty$. Suppose that the function g is such that for every number $i \in \omega$ and every $k \in \omega$, $g(i, k+1) > q(g(i, k), \delta_i)$. Then find the function $f \in \omega^\omega$ such that $f(n) > g(n, n)$. I claim that the capacity η is preserved by the forcing $PT_{f,g}$.

In order to show this, assume that $S \Vdash \dot{O} \subset Y$ is a set of η-mass $< \varepsilon$. Fix a countable topology basis $\mathcal{O}$ of the Polish space Y closed under finite unions. Using the continuous reading of names, thinning out the tree S I may assume that there is a function $h : S \to \mathcal{O}$ such that $s \in S$ implies $\eta(h(s)) < \varepsilon$, $t \subset s \to h(t) \subset h(s)$, and $S \Vdash \dot{O} = \bigcup_i h(\dot{x}_{gen} \restriction i)$.

First, a couple of definitions. If $R \subset S$ is a (possibly finite) tree and $t \in R$ is a nonterminal node then write a_t^R for the set of immediate successors of the node t in the tree R. Let $t \in R$. I will say that R is 1-*smaller* than S if for every non-terminal node $s \in R$ with $t \subseteq s$, $\phi_s(a_s^R) \geq \phi_s(a_s^S) - 1$. For every tree $R \subset S$ and every node t let $A(R, t) = \bigcup \{h(s) : s \in R, t \subseteq s\}$. The following claim is key.

Claim 4.4.18. *For every node $t \in S$ and every number $k \geq |t|$ there is a 1-smaller tree $R \subset S$ of height k such that $\eta(A(R, t)) < \varepsilon + \Sigma_{|t|<i<k}\delta_i$.*

Proof. This is proved by induction on $k - |t|$. If $k - |t| = 0$ this is obvious since $P(t, k) = h(t)$. For the inductive step, let $t \in S$ be a node and $k > |t|$ a number. By the induction hypothesis, for every node $s \in a_t^S$, there is a tree R_s such that $\eta(A(R_s, s)) < \varepsilon + \Sigma_{|s| \leq i < k}\delta_i$. Now use the properties of the function g to find a set $b \subset a_t^S$ such that $\phi_t(b) \geq \phi_t(a_t^S) - 1$ and the set $\bigcup_{s \in b} A(R_s, s)$ has η-mass $\varepsilon + \Sigma_{|t|<i<k}\delta_i + \delta_{|t|}$. But then, the tree $R = \bigcup_{s \in b} R_s$ has the required properties! $\square$

Let $\delta > 0$. I will produce an infinite 1-smaller tree $R \subset S$ with $R \in PT_{f.g}$ and a node $t \in R$ such that $\eta(A(R, t)) < \varepsilon + \delta$. This will clearly suffice as $R \restriction t \Vdash \dot{O} \subset A(R, 0)$. Find a number $l \in \omega$ such that $\Sigma_{i>l}\delta_i < \delta$ and let $t \in S$ be a node at level l. Use the claim to find the 1-smaller finite trees $R_k \subset S$ of height k for every number k, and then use a compactness argument to find a 1-smaller infinite tree $R \subset S$ such that every level of R is included in infinitely trees in the collection $R_k : k \in \omega$. Then R is an $\dot{\phi}$-fat tree, and the continuity of the capacity η in increasing unions shows that $\eta(A(R, t)) \leq \sup_k \eta(A(R_k, t)) \leq \varepsilon + \delta$ as required. $\square$

4.4.4 Diagonalizing an F_σ-ideal

Suppose that J is an F_σ-ideal on ω. I will produce a proper and bounding forcing which adds no splitting reals, and it adds an infinite set $b \subset \omega$ such that $\forall c \in J \cap V \, |b \cap c| < \aleph_0$. This has been achieved earlier by Laflamme through a combinatorial construction, even though his paper does not show that his forcing adds no splitting reals.

By a theorem of Mazur [50], there is a lower semicontinuous submeasure ψ on ω such that $J = \{c \subset \omega : \psi(c) < \infty\}$. Now construct a Hausdorff submeasure on the Cantor space in the following way. Let $U = \{u_n : n \in \omega\}$ where $u_n = \{x \in 2^\omega : x(n) = 1\}$, let $\mathrm{diam}(u_n) = 2^{-n}$, and for a set $a \subset U$ let $w(a) = \phi\{n \in \omega : u_n \in a\}$. Consider the Hausdorff submeasure ϕ obtained from these objects and the forcing $P_{I_{\sigma\phi}}$.

The forcing is proper, bounding and adds no splitting reals by Theorem 4.4.8, since the weight function w is clearly lower semicontinuous. Consider the set $B = \{x \in 2^\omega : \exists^\infty n \ x(n) = 1\}$. I will show that $B \notin I_{\sigma\phi}$ and B forces the generic subset of ω to have finite intersection with all the ground model J-small subsets of ω.

Suppose that $A_n : n \in \omega$ are sets of finite ϕ-mass; I must produce a point $x \in B \setminus \bigcup_n A_n$. I may assume that $\phi(A_n) \leq n$. By induction on $n \in \omega$ construct infinite sets $c_n \subset \omega$ of infinite ψ-mass so that

- $\omega = c_0 \supset c_1 \supset \ldots$
- writing $k_n = \min(c_n)$ it is the case that $0 = k_0 \in k_1 \in \ldots$ and for every infinite subset $d \subset c_{n+1}$ the characteristic function of the set $\{k_i : i \leq n\} \cup d$ is not in A_n.

Once this is achieved then the characteristic function of the infinite set $\{k_n : n \in \omega\}$ belongs to $B \setminus \bigcup_n A_n$, proving that $B \notin I_{\sigma\phi}$. The induction step is easy. If $c_n \subset \omega$ has been constructed then choose a number $j > \max\{k_i : i \leq n\}$ and find a set $a_n \subset \omega \setminus j$ of ψ-mass $\leq n+1$ such that $A_n \subset \bigcup_{i \in a} u_i$. This is possible by the definition of the Hausdorff submeasure. The set $c_{n+1} = (c_n \setminus a_n) \setminus j$ will work as desired.

By absoluteness, $B \Vdash \dot{x}_{gen}$ is a characteristic function of an infinite set. Let $c \in J$ be any set. I will argue that $A = \{x \in B : \{n \in c : x(n) = 1\}$ is finite$\} \in I_{\sigma\phi}$. This happens since for every number $k \in \omega$ the set A is covered by $\bigcup_{n \in c \setminus k} u_n$ which is a union of weight $\leq \phi(c)$ of sets of diameter $< 2^{-k}$. Thus $\phi(A) \leq \phi(c)$, $A \in I_{\sigma\phi}$, and $B \Vdash \{n \in c : \dot{x}_{gen}(n) = 1\}$ is finite as desired.

4.5 Pavement submeasures

In mathematical practice, a number of submeasures on Polish spaces comes from the following construction.

Definition 4.5.1. *Let X be a Polish space, U a countable set of its Borel subsets, and $w : U \to \mathbb{R}^+$ a function. The associated* pavement submeasure *ϕ is defined by $\phi(A) = \inf\{\Sigma_n w(u_n) : u_n \in U, A \subset \bigcup_n u_n\}$. The sets in the collection U are referred to as* pavers *and w is a* weight *function. The associated ideal is $I_\phi = \{A \subset X : \phi(A) = 0\}$.*

Note that every presentation of a pavement ideal is a pavement ideal, and so every pavement forcing has a pavement presentation on every uncountable Polish space. The topology on the space X can be increased to make the pavers clopen without changing the forcing P_{I_ϕ} by [40], 13.A. Furthermore, every Polish space is a continuous bijective image of a closed subset of the Baire space. Combining these two tricks it is clear that every pavement forcing has a presentation on the Baire space such that the pavers are open sets.

4.5.1 General theorems

Theorem 4.5.2. *[18] Let X be a Polish space and ϕ a pavement submeasure generated by a countable set of Borel pavers. The forcing P_{I_ϕ} is proper.*

Proof. The proof again uses a determined infinite game. For notational simplicity assume that the underlying space is ω^ω, the pavers are open, and fix the weight function $w : U \to \mathbb{R}^+$ generating the submeasure. Suppose that P is a partial order and $\dot{x}$ is a P-name for an infinite sequence of natural numbers. Consider the game $G(\phi, P, \dot{x})$ of length ω between Player I and Player II played in the following fashion. In the beginning Player I indicates an initial condition p_{ini} and Player II responds with a real number $\varepsilon > 0$. After that, Player I produces one-by-one open dense subsets $\{D_n : n \in \omega\}$ of the poset P, and dynamically on a fixed schedule a Borel set $A \subset X$ of ϕ-mass $\leq \varepsilon$. Player II plays one by one decreasing conditions $p_{ini} \geq p_0 \geq p_1 \geq \ldots$ so that $p_n \in D_n$ and p_n decides the n-th number of the sequence $\dot{x}$. He is allowed to hesitate for any number of rounds before placing his next move. Player II wins if, writing g for the filter he obtained, it is the case that $\dot{x}/g \notin A$.

To make this precise, at round $k \in \omega$ Player I indicates a finite set $a_k \subset U$ so that $a_0 \subset a_1 \subset \ldots$, $\Sigma_{u \in a_k} w(u) \leq \varepsilon$ and for all numbers $j \in k \in \omega$, $\Sigma_{u \in a_k \setminus a_j} w(u) \leq 2^{-j}$. In the end the set $A \subset X$ is recovered as $\bigcup \bigcup_k a_k$. It is clear that given any set B of mass $< \varepsilon$ Player I can play so that $B \subset A$ for his resulting set A.

Lemma 4.5.3. *The following are equivalent:*

1. *$P \Vdash \dot{x}$ is not contained in any ground model Borel ϕ-null set;*
2. *Player II has a winning strategy in the game $G(\phi, P, \dot{x})$.*

Granted this lemma, the whole treatment transfers from the previous section. One direction of the lemma is easy. If there is a condition $p \in P$ such that $p \Vdash \dot{x} \in \dot{B}$ for some ground model coded Borel ϕ-null set B, then Player I can easily win by indicating $p_{ini} = p$, and after Player II responds with a number ε, Player I dynamically produces a suitable superset A of ϕ-mass $< \varepsilon$ of the set B, and mentioning all the open dense sets necessary to make sure that the result of the game falls into the set $B \subset A$.

For the other direction of the lemma note that the payoff set of the game is G_δ in the tree of all possible plays and therefore the game is determined – Fact 1.4.2. Thus it will be enough to obtain a contradiction from the assumption that $P \Vdash \dot{x}$ is not contained in any ground model coded Borel ϕ-null set and yet Player I has a winning strategy σ. A small claim will be used repeatedly:

Claim 4.5.4. *For every condition $p \in P$ there is a real number $\varepsilon(p) > 0$ such that for every set $a \subset U$ with $\Sigma_{u \in a} w(u) \leq \varepsilon(p)$ there is a condition $q \leq p$ forcing $\dot{x} \notin \bigcup a$.*

Proof. Suppose this fails and for every real number $\varepsilon > 0$ find a set $a_\varepsilon \subset U$ with $\Sigma_{u \in a_\varepsilon} w(a_\varepsilon) \leq \varepsilon$ such that $p \Vdash \dot{x} \in \bigcup a_\varepsilon$. But then $p \Vdash \dot{x} \in \bigcap_\varepsilon \bigcup a_\varepsilon$ and the latter set is in the ideal I_ϕ, contradicting the properties of the name $\dot{x}$. □

Player II will find a winning counterplay against the strategy σ in the following fashion.

The first move of Player II will be the real number $\varepsilon = \varepsilon(p_{ini}) \in M$. The rest of the counterplay will be built by induction, Player II's moves denoted by p_n, played at rounds i_n. The initial segment of the play ending after the round i_{n-1} will be denoted by τ_n, and for notational convenience let $p_{-1} = p_{ini}$ and $\tau_0 = \langle p_{ini} \rangle$. The following induction hypotheses will be satisfied:

- $\varepsilon(p_n) \geq 2^{-i_n}$;
- the condition $p_n \in M$ is in the sets D_n, it decides the n-th bit of the sequence $\dot{x}$ and it forces $\dot{x} \notin \bigcup a_{i_n}$. Here the symbols D_n and a_{i_n} refer to Player I's moves in the play τ_{n+1}.

This will certainly be sufficient. Let $\tau = \bigcup_n \tau_n$ and argue that Player II has won. Since the pavers are open, it is enough to show that for every number n and every paver $u \in a_{i_n}$ it is the case that $0_{t_n} \not\subset u$, where $t_n \in \omega^n$ is the sequence of the first n numbers forced onto $\dot{x}$ by the condition p_n. But this is clear from the second inductive item! Note that this argument uses just the second item of the induction hypothesis, the first item just helps keep the induction going.

To perform the induction, suppose the play τ_{n-1} has been constructed. Let b be the collection of all sets in U which the strategy σ will put into the sets $a_k : k \in \omega$ in the infinite play extending τ in which Player II adds no nontrivial moves after the round i_n. Let $B = \bigcup(b \setminus a_k) \subset X$. This is a set of ϕ-mass at most $2^{-i_{n-1}} \leq \varepsilon_{p_{n-1}}$ so there must be a condition $q \leq p_{n-1}$ which forces $\dot{x} \notin \dot{B}$. Find a condition $p_n \leq q$ in the open dense set D_n, deciding the value of the next natural number on the sequence $\dot{x}$. The condition p_n will be the next move of Player II, and it will be played at round i_n such that $2^{-i_n} < \varepsilon(p_n)$. The induction hypotheses continue to hold. $\qquad\qquad\qquad\qquad\qquad\qquad\qquad\qquad\qquad\qquad\qquad\qquad\qquad\quad\square$

The following theorem gathers some useful consequences of the previous proof.

Theorem 4.5.5. *Suppose that ϕ is a pavement measure on a Polish space X. Then*

1. *Whenever $B \in P_{I_\phi}$ is a condition and $D \subset P_{I_\phi}$ is an open dense set then there is a countable set $D' \subset D$ consisting of subsets of B such that $\phi(\bigcup D') = \phi(B)$.*
2. *Whenever $B \in P_{I_\phi}$ is a condition and $\dot{n}$ is a name for a natural number smaller than some fixed $k \in \omega$ then there is a condition $C \subset B$ deciding the value of the name $\dot{n}$ with $\phi(C) \geq 1/k \cdot \phi(B)$.*
3. *The forcing P_{I_ϕ} is $< \omega_1$-proper, and in fact for every countable continuous tower $\vec{M}$ of countable elementary submodels of a large enough structure and every condition $B \in \vec{M}(0)$ the set $\{x \in B : x$ is generic for every model on the tower $\vec{M}\} \subset B$ has the same mass as B.*
4. *If P_{I_ϕ} is nowhere c.c.c. then every Borel set $B \subset X$ can be divided into perfectly many mutually disjoint Borel subsets of the same ϕ-mass.*
5. *If P_{I_ϕ} is nowhere c.c.c. then for every countable collection $\{B_n : n \in \omega\}$ of Borel sets there is a refinement $\{C_n : n \in \omega\}$ such that $C_n \subset B_n$ is a Borel set of the same ϕ-mass as B_n and the sets $C_n : n \in \omega$ are mutually disjoint.*

Proof. For the first item let $B \in P_{I_\phi}$ be a condition, D be an open dense set, and M be a countable elementary submodel of a large enough structure. Let $C = \{x \in B : x$ is M-generic$\}$. The proof of Theorem 4.5.2 shows that $\phi(C) = \phi(B)$; moreover $C \subset \bigcup \{A \in D \cap M : A \subset B\}$ as desired.

For the second item, let $B \in P_{I_\phi}$ be a condition and $\dot{n}$ be a name for a number smaller than some $k \in \omega$. Let M be a countable elementary submodel of a large enough structure, let $C = \{x \in B : x$ is M-generic$\}$, and $C_l = \{x \in C : \dot{n}/x = l\}$ for $l \in k$; so $C = \bigcup_{l \in k} C_l$. A review of the definitions shows that $C_l \Vdash \dot{n} = \check{l}$ if $\phi(C_l) > 0$. As in the previous paragraph, $\phi(B) = \phi(C)$ and by the subadditivity of the submeasure ϕ one of the sets $C_l : l \in k$ must have ϕ-mass not less than $\phi(C)/k = \phi(B)/k$.

(3) is proved by induction on the length of the tower $\vec{M}$. The successor step of the induction is a trivial application of the previous theorem. Now suppose that $\vec{M}$ is a tower of limit length $\alpha \in \omega_1$ and (3) has been verified for all shorter towers. Choose a condition $B \in \vec{M}_0$; I must prove that $\phi(C) = \phi(B)$ where $C = \{x \in B : x$ is $\vec{M}$-generic$\}$. Let $a \subset U$ be a set such that $\Sigma_{u \in a} w(u) < \phi(B)$; I must find a point $x \in C \setminus \bigcup a$. Choose an increasing sequence $\beta_m : m \in \omega$ of ordinals converging to α, an enumeration $a = \{u_n : n \in \omega\}$, and an enumeration $D_m : m \in \omega$ of open

dense subsets of the poset P_{I_ϕ} in $\bigcup \vec{M}$ such that $D_m \in \vec{M}(\beta_m)$. By induction on $m \in \omega$ construct an increasing sequence of numbers $n_m \in \omega$ and descneding chain of conditions $B_m \in P_{I_\phi}$ so that

- $n_0 = 0, B_0 = B$;
- $B_{m+1} \subset \{x \in B : x$ is $\vec{M} \upharpoonright \beta_m\text{-generic}\}$, $B_{m+1} \in D_m$, and $B_{m+1} \in \vec{M}(\beta_m + 1)$;
- $B_m \cap \bigcup_{n \in n_m} a_n = 0$ and $\phi(B_m) > \Sigma_{n \notin n_m} w(a_n)$.

Once this has been done, the intersection $\bigcap_m B_m$ must be nonempty, since the filter generated by the conditions $B_m : m \in \omega$ is $\bigcup \vec{M}$-generic and Proposition 2.1.2 holds in the modle $\vec{M}$. Any point $x \in B$ in this intersection will be $\vec{M}$-generic by the second item, and that point will fall out of the set $\bigcup_n a_n$ by the third item. This will conclude the proof of (3).

The induction itself is easy to perform. Suppose that the set B_m and the number n_m have been constructed. The set $\{x \in B_m : x$ is $\vec{M} \upharpoonright (\beta_m)\text{-generic}\}$ has the same ϕ-mass as the set B_m by the transfinite induction hypothesis and it is covered by the set $\bigcup(D_m \cap \vec{M}(\beta_m))$ by genericity. Thus there must be a set $C \in D_m \cap \vec{M}(\beta_m)$ such that $\phi(C' \setminus \bigcup_{n \notin n_m} a_n) \neq 0$, where $C' = B_m \cap C \cap \{x \in X : x$ is $\vec{M} \upharpoonright \beta_{m+1}$-generic$\}$. Find a number n_{m+1} such that $\phi(C' \setminus \bigcup_{n \notin n_m} a_n) > \Sigma_{n \notin n_{m+1}} w(a_n)$ and let $B_{m+1} = C' \setminus \bigcup_{n \in n_{m+1}} a_n$. The induction hypotheses are satisfied.

(4) is now proved similarly to Proposition 3.7.10. I will prove it under the Continuum Hypothesis. However, it is clear that this assumption is irrelevant since the Continuum Hypothesis can be forced with a σ-closed forcing and so the perfect collection of Borel sets I find in the extension must have existed in the ground model already. So assume CH and fix the set B. Use the assumption of nowhere c.c.c. and the proof of Proposition 3.7.10 to find towers $\vec{M}_t : t \in 2^{<\omega}$ such that $t \subset s$ implies $\vec{M}_t \subset \vec{M}_s$ and whenever t, s are incompatible sequences no point $x \in B$ can be both $\vec{M}_t$ and $\vec{M}_s$-generic. Note the use of CH in that proof. For every infinite binary sequence $y \in 2^\omega$ let $\vec{M}_y = \bigcup_{t \subset y} \vec{M}_t$ and let $B_y = \{x \in B : x$ is $\vec{M}_y\text{-generic}\}$. The previous item shows that these sets have the same ϕ-mass as B, and the choice of the system of towers implies that these sets are pairwise disjoint as desired.

For (5), fix the collection $\{B_n : n \in \omega\}$ to be refined and by induction on $n \in \omega$ find Borel sets $B_n^\alpha : \alpha \in \omega_1$ such that $B_n^\alpha \subset B_n$, $\phi(B_n^\alpha) = \phi(B_n)$, for fixed number n they are mutually disjoint, and if $k \in n$ then every set B_n^α has ϕ-positive intersection with just countably many sets $B_k^\beta : \beta \in \omega_1$, with some ordinal bound $\beta_{nk} \in \omega_1$. Once this has been achieved, let $\alpha \in \omega_1$ be an ordinal larger than all the β_{nk}'s and let $C_n = B_n^\alpha \setminus \bigcup_{k \neq n} B_k^\alpha$. The induction itself is easy to perform. Suppose the sets $B_k^\beta : \beta \in \omega_1, k \in n$ have been constructed. To find the sets $B_n^\beta : \beta \in \omega_1$, choose a countable elementary submodel M of a large enough structure containing in particular the set B_n, and let $B'_n = \{x \in B_n : x$ is M-generic$\}$. Now $\phi(B'_n) = \phi(B_n)$ and by genericity $B'_n \Vdash$ if the generic filter meets some set $B^\beta_k : k \in n, \beta \in \omega_1$, then $\beta \in M$. In particular, $\phi(B'_n \cap B^\beta_k) = 0$ for every number $k \in n$ and every ordinal $\beta \notin M$. The induction step is concluded by an application of (4) to the set B'_n. $\square$

Theorem 4.5.6. *Let X be a Polish space and ϕ a pavement submeasure generated by a countable set of Borel pavers.*

1. *I_ϕ satisfies the third dichotomy. In fact, every analytic set has a Borel subset of the same ϕ-mass.*
2. *I_ϕ satisfies the first dichotomy in the Solovay model. In fact, in that model every set has a Borel subset of the same ϕ-mass and ϕ is continuous in increasing wellordered unions of uncountable cofinality.*
3. *(ZF+DC+AD+) I_ϕ satisfies the first dichotomy and ϕ is continuous in increasing wellordered unions of uncountable cofinality.*
4. *In the constructible universe, if the forcing P_{I_ϕ} is nowhere c.c.c. then there is an ϕ-positive coanalytic set without a ϕ-positive Borel subset.*
5. *ϕ is continuous in increasing wellordered unions of uncountable cofinality less than* add(null). *In particular, if* add(null) $> \aleph_1$ *then every coanalytic set has a Borel subset of the same mass.*
6. *ϕ is ZFC-correct.*

Proof. For the first item, consider the partial order Q_{I_ϕ} of I_ϕ-positive analytic sets ordered by inclusion. Proposition 2.1.11 shows that the forcing adds a single point $\dot{x}_{gen}$ which falls out of all ϕ-null sets and it belongs to all sets in the generic filter. Let $A \in Q_{I_\phi}$ be an analytic ϕ-positive set. I will produce a Borel set $B \subset A$ of the same ϕ-mass, and this will complete the proof of the first item.

Use Lemma 4.5.3 to find a winning strategy σ for Player II in the game $G(\phi, Q_{I_\phi}, \dot{x}_{gen})$, let M be a countable elementary submodel of a large enough structure containing ϕ, A, σ, and let $B = \{x \in A : x$ is M-generic$\}$. The set B is Borel by Fact 1.4.8. The set B also has the same ϕ-mass as the set A by the same proof as in Theorem 4.5.2. The proof of the first item is complete.

The second item is proved in the same way as Theorem 4.3.21(2). The proof of the third item is more sophisticated. Suppose that $A \subset X \times 2^\omega$ is a set and $\varepsilon > 0$ is a real number. If $\phi(\text{proj}(A)) > \varepsilon$, I must produce an analytic set $B \subset A$ such that $\phi(\text{proj}(B)) > \varepsilon$ and then refer to Proposition 3.9.19. Fix a Borel bijection $\pi : 2^\omega \to X \times 2^\omega$ and consider the integer pavement game $G(A, \varepsilon)$ in which Player I produces a set $a \subset U$ with $\Sigma_{u \in a} w(u) < \varepsilon$ and Player II produces a binary sequence $y \in 2^\omega$. Player II wins if $\pi(y) \in A$ and the first coordinate of the point $\pi(y)$ is not in the set $\bigcup a$. To complete the description of the game I stipulate that at round n

Player I must indicate a finite set $a_n \subset a$ such that $\Sigma_{u \in a \setminus a_n} w(u) < 2^{-n}$ while Player II can wait for an arbitrary finite number of rounds to put another bit on his sequence y.

Claim 4.5.7. $\phi(\text{proj}(A)) < \varepsilon$ *implies that Player I has a winning strategy in the game* $G(A, \varepsilon)$ *which in turn implies that* $\phi(\text{proj}(A)) \le \varepsilon$.

Given the claim, the third item of the theorem easily follows. As in Proposition 3.9.20, it will be enough to show that if $\phi(\text{proj}(A)) > \varepsilon$ then A has an analytic subset $B \subset A$ with $\phi(\text{proj}(B)) \ge \varepsilon$. To find B, use the claim and a determinacy argument to find a winning strategy σ in the game $G(A, \varepsilon)$ for Player II and let B be the set of all points in $X \times 2^\omega$ the strategy σ can produce against some counterplay. the definition of the set B shows that it is an analytic set. Since the strategy σ is winning, it must be the case that $B \subset A$, and finally, since the strategy σ remains wining in the game $G(B, \varepsilon)$, the claim implies that $\phi(\text{proj}(B)) \ge \varepsilon$ as desired.

Now I turn to the proof of the claim. The first implication is almost trivial; if $\phi(\text{proj}(A)) < \varepsilon$ then Player I can win by producing a set $a \subset U$ such that $\Sigma_{u \in a} w(u) < \varepsilon$ and $\text{proj}(A) \subset \bigcup a$, ignoring the moves of Player II altogether. The second implication is harder. Fix a strategy σ for Player I. For every finite play t of the game according to the strategy σ let $a_t \subset U$ be the set of all elements $u \in U$ the strategy σ places in the set a after the play t if that play is extended by an infinite sequence of trivial moves by Player II; so $\Sigma_{u \in a_t} < 2^{-|t|}$. Also for a number $n \in \omega$ and a bit $b \in 2$ let tnb be the play extending t in which Player II makes just trivial moves except for the last one at round n where he adds the bit b to his sequence. For a play t and a bit $b \in 2$ let $C_{tb} = \bigcap_n \bigcup a_{tnb}$; clearly $\phi(C_{tb}) = 0$. I will show that if the strategy σ is winning then $\text{proj}(A) \subset \bigcup a_0 \cup \bigcup_{tb} C_{tb}$. This will complete the proof since the latter set has mass $\le \varepsilon$. So suppose $x \in \text{proj}(A) \setminus \bigcup a_0 \cup \bigcup_{tb} C_{tb}$ is a point and $y \in 2^\omega$ is a binary sequence such that $\pi(y) \in A$ is a point projecting to x. I will find a winning counterplay against the strategy σ which produces the sequence y. Clearly, it is just necessary to decide the natural numbers $n_m : m \in \omega$ such that Player II plays the bit $y(m)$ at round n_m. Given the numbers $n_k : k \in m$ write t_m for the finite play against the strategy σ in which Player II places $y(k)$ at round n_k and the and n_{m-1} is the last round of the play t_m; $t_0 = 0$. I will construct the numbers $n_m : m \in \omega$ and the plays t_m inductively maintaining the statement $x \notin \bigcup a_{t_m}$ as the induction hypothesis. This is satisfied at $t_0 = 0$. Now suppose the play t_m has been found. Since $x \notin C_{t_m, y(m)}$, there must be a number $n > n_{m-1}$ such that $x \notin a_{t_m ny(m)}$. Clearly, the number $n = n_m$ together with the play $t_{m+1} = t_m ny(m)$ still satisfy the inductive hypothesis. In the end, the union $\bigcup_m t_m$ is the desired winning counterplay against the strategy σ.

The fourth item is proved with the help of Proposition 3.9.22. I must construct the necessary coding device, that is, a Borel function $f : X \to 2^\omega$ such that for every set $A_0 \in I_\phi$ there is a disjoint set $A_1 \in I_\phi$ such that $f''A_1 = 2^\omega$. In order to do this, choose an $\in$-tower $M_n : n \in \omega$ of countable elementary submodels of a large enough structure and find countable collections $C_n = C_n^0 \cup C_n^1 \in M_{n+1}$ consisting of mutually disjoint sets such that whenever $B \in M_n$ is a condition then it has a subset of the same mass in both C_n^0 and C_n^1. This is possible by Theorem 4.5.5. Let $f(x)(n) = 0$ if x belongs to some set in C_n^0, and $f(x)(n) = 1$ otherwise. I must show that the function f has the requested coding property.

Suppose that $a \subset U$ is a set such that $\Sigma_{u \in a} w(u)$ is smaller than the mass of the whole space. I will find a set $A_1 \in I_\phi$ disjoint from $\bigcup a$ such that $f''A_1 = 2^\omega$. Let $D_n : n \in \omega$ enumerate the open dense subsets of P_{I_ϕ} in the models of the tower such that $D_n \in M_n$. By tree induction on $t \in 2^{<\omega}$ find conditions $B_t \in P_{I_\phi}$ so that $B_0 = X$, $B_t \in M_{|t|}$, $t \subset s$ implies $B_s \subset B_t$, and if $n = |t| > 0$ then:

- $B_t \in D_{n-1}$;
- $B_t \in C_n^{t(n-1)}$;
- $\phi(B_t) < 2^{-2n}$;
- there is a finite set $a_t \subset a$ such that $B_t \cap \bigcup a_t = 0$ and $\phi(B_t) > \Sigma_{u \in a \setminus a_t} w(u)$.

Once this is done, for every binary sequence $y \in 2^\omega$ the intersection $\bigcap_n B_{y \upharpoonright n}$ is nonempty, containing some point $x_y \in X$ which is generic for the model $\bigcup_n M_n$ by the first item. Every such point falls out of the set a by the last item, and $f(x_y) = y$ by the second item. Finally, writing $A_1 = \{x_y : y \in 2^\omega\}$ it must be the case that $\phi(A_1) = 0$ by the third item. This will conclude the proof.

To perform the induction, suppose that the set B_t has been obtained, let $n = |t|$ and work in the model M_n. By Theorem 4.5.5, there is a countable collection $D_n' \subset D_n$ consisting of subsets of B_t of ϕ-mass $< 2^{-2n}$ such that $\phi(\bigcup D_n') = \phi(B_t)$. Since $\phi(B_t) > \Sigma_{u \in a \setminus a_t} w(u)$ there must be a set $C \in D_n'$ such that $\phi(C \setminus \bigcup a) > 0$. Then, there must be a finite set $b_t \subset a$ such that writing $C' = C \setminus \bigcup b_t$, $\phi(C') > \Sigma_{u \in a \setminus b_t} w(u)$. Now use the properties of the collections C_{n+1}^0 and C_{n+1}^1 to find sets $B_{t \frown 0}, B_{t \frown 1} \subset C'$ of the same ϕ-mass as C' in the respective collections. The induction proceeds with $a_{t \frown 0} = a_{t \frown 1} = b_t$.

The fifth item is an immediate corollary of the following claim of independent interest, parallel to Claim 4.3.8. Note that every coanalytic set is an increasing union of $\aleph_1$ many Borel sets.

Claim 4.5.8. *Suppose that κ is an uncountable regular cardinal, $\varepsilon, \delta > 0$ are real numbers, and $\langle B_\alpha : \alpha \in \kappa \rangle$ are sets of ϕ-mass $< \varepsilon$. Then there is an infinite set $b \subset \kappa$ such that $\phi(\bigcup_{\alpha \in b} B_\alpha) < \varepsilon + \delta$. If moreover $\mathrm{add}(\mathrm{null}) > \kappa$ then there is such a set b of size κ.*

Proof. I will start with the last sentence. Let U, w be the set of pavers and the weight function generating the pavement submeasure ϕ. For every number n let K_n be the set of all finite subsets $a \subset U$ such that $\Sigma_{u \in a} w(u) < 4^{-n}$. For every ordinal α find a function f_α with domain ω such that for every $n \in \omega$, $f_\alpha(n) \subset U$ is finite, for all $n > 0$ it is the case that $f_\alpha(n) \in K_n$, $B_\alpha \subset \bigcup\bigcup_n f_\alpha(n)$ and $\Sigma\{w(u) : u \in \bigcup_n f_\alpha(n)\} < \varepsilon + \delta/2$. The assumption $\text{add(null)} > \kappa$ implies that there is a function g with domain ω such that for every number $n \in \omega$, $g(n) \subset K_n$ is a set of size $< 2^n$ and for every ordinal α, for all but finitely many numbers $n \in \omega$ it is the case that $f_\alpha(n) \in g(n)$. Use a counting argument to find a number $m \in \omega$ such that $2^{-m} < \delta/2$ and a function h with domain m such that the set $b = \{\alpha \in \kappa : f_\alpha \upharpoonright m = h \wedge \forall n \geq m \, f_\alpha(n) \in g(n)\}$ has size κ. It is now clear that the set $\bigcup_{\alpha \in b} B_\alpha$ is covered by the pavers in the set $\bigcup \text{rng}(h) \cup \bigcup_{n \geq m} \bigcup g(n)$, whose weights sum up to less than $\varepsilon + \delta/2 + \delta/2 = \varepsilon + \delta$.

To argue now for the first part of the claim, find a countable elementary submodel M of a large structure and an M-generic filter g for a c.c.c. forcing such that $M[g] \models \text{add(null)} > \kappa$. By the work in the previous paragraph, in the model $M[g]$ there is an infinite set $b \subset \kappa$ of size κ such that $M[g] \models \phi(\bigcup_{\alpha \in b} B_\alpha) < \varepsilon + \delta$. This set b is still infinite in V, and it has the desired properties.   □

The last item is a small sweet payback for the work done so far. Suppose M is a transitive model of a large fragment of ZFC containing the code for the pavement submeasure ϕ–that is, the weight function and the pavers. I must show that the model M computes the ϕ-mass correctly for analytic sets. Work in the model M for a while. If $B \subset X$ is an analytic set and $\varepsilon > 0$ is a real number, the statement $\phi(B) > \varepsilon$ is equivalent to Player II having a winning strategy σ in the game $G(A, \delta)$ for some real number $\delta > \varepsilon$ and a Borel set $A \subset X \times 2^\omega$ projecting into B. However, the fact that the strategy σ is winning is coanalytic and therefore absolute between M and V. It follows that $M \models \phi(B) > \varepsilon \leftrightarrow V \models \phi(B) > \varepsilon$ as desired.   □

Theorem 4.5.9. *Let X be a Polish space and ϕ a pavement submeasure generated by a countable set of Borel pavers. The forcing P_{I_ϕ} is regularly embeddable into a σ-closed*c.c.c. iteration.*

Proof. For simplicity (and sanity) assume that the underlying Polish space is the Cantor space 2^ω. Let P be the partial order of pairs $p = \langle a_p, b_p \rangle$ such that:

- a_p is a countable partially ordered set of trees $T \subset (\omega \times 2)^{<\omega}$ such that $\text{proj}[T] \notin I_\phi$; the ordering is defined by $S \leq T$ if $\text{proj}[S] \subset \text{proj}[T]$;
- the set a_p is closed under I_ϕ-positive intersections: if $T_0, T_1 \in a_p$ are trees such that $\text{proj}[T_0] \cap \text{proj}[T_1] \notin I_\phi$ then there is a tree $S \in a_p$ such that $\text{proj}[S] = \text{proj}[T_0] \cap \text{proj}[T_1]$;

- the set a_p is closed under restriction: if $T \in a_p$ and $t \in T$ is a node such that proj$[T \restriction t] \notin I_\phi$ then $T \restriction t \in a_p$;
- the set a_p is closed under subtraction of a finite union of pavers: if $T \in a_p$ is a tree and $a \subset U$ is a finite set such that $\phi(\text{proj}[T] \setminus \bigcup a) > 0$ then there is a tree $S \in a_p$ projecting into the set proj$[T] \setminus \bigcup a$;
- b_p is a countable collection of subsets $z \subset a_p$ such that for every tree $T \in a_p$ it is the case that $\phi(\bigcup z \cap \text{proj}[T]) = \phi(\text{proj}[T])$.

The ordering of the set P is by coordinatewise reverse inclusion. This is clearly a σ-closed forcing. If $G \subset P$ is a generic filter then let $Q \in V[G]$ be the set $\{\text{proj}[T] : \exists p \in G \ T \in a_p\}$ ordered by inclusion. The proof will be complete once I show that Q is c.c.c. and if $D \subset P_{I_\phi}$ is a dense set in the ground model then $D \cap Q \subset Q$ is dense.

Claim 4.5.10. *Suppose* $p = \langle a_p, b_p \rangle \in P$ *is a condition and consider the set* $B = \bigcap_{z \in b_p} \bigcup \{\text{proj}[T] : T \in z\}$. *Then*

1. *for every tree* $T \in a_p$ *it is the case that* $\phi(\text{proj}[T]) = \phi(\text{proj}[T] \cap B)$;
2. *for every* I_ϕ-*positive analytic set* $C \subset B$ *there is a condition* $q = \langle a_q, b_p \rangle \le p$ *such that the set* a_q *includes some tree* T *such that* $C = \text{proj}[T]$.

The claim immediately implies that if $D \subset P_I$ is an open dense set then $P \Vdash \check{D} \cap Q$ is dense. To see this, for every condition $p \in P$ and every tree $S \in a_p$ consider the set $B \subset X$ as in the claim and look at the intersection $B \cap \text{proj}[S]$. It is an ϕ-positive analytic set and therefore contains an ϕ-positive Borel subset and even one from the open dense set D. Choose such a set $C \subset B$ and use the Claim to find a condition $q \le p$ which includes some tree T such that $\text{proj}[T] = B \cap \text{proj}[S]$. The condition q forces that there is an element of the open dense set D below the condition $\text{proj}[S] \in Q$. Finally use a genericity argument to see that $D \cap Q$ is forced to be dense.

For the c.c.c. of the forcing Q in the P-extension, suppose that $p \Vdash \dot{D} \subset Q$ is an open dense set. Let M be a countable elementary submodel of a large enough structure, let $q \le p$ be a the coordinatewise union of some M-generic filter $g \subset P \cap M$ and let $z \subset a_q$ be the set $\{\text{proj}[S] : S \in a_q, q \Vdash \text{proj}[S] \in \dot{D}\}$. I will show that for every tree $T \in a_q$ it is the case that $\phi(\text{proj}[T] \cap \bigcup \{\text{proj}[S] : S \in z\}) = \phi(\text{proj}[T])$. This shows that the pair $r = \langle a_q, b_q \cup \{z\} \rangle \le q$ is a condition in the poset P. It is immediate that $r \Vdash \{\text{proj}[S] : S \in z\} \subset \dot{Q}$ is a countable predense set, proving the c.c.c. of the forcing Q.

For every tree $T \in a_q$ let $\psi(T) = \phi(\text{proj}[T] \cap \bigcup \{\text{proj}[S] : S \in z\})$. I must show that $\psi(T) = \phi(\text{proj}[T])$. First, an elementary argument shows that $q \Vdash \psi(T) = \sup\{\phi(\text{proj}[T] \cap \bigcup E) : E \subset \dot{D} \text{ countable}\}$. Suppose for contradiction that for some tree $T \in a_q$ it is indeed the case that $\psi(T) < \phi(\mathrm{proj}[T])$. Let $a \subset U$ be a set such that $\Sigma\{w(u) : u \in a\} < \phi(\mathrm{proj}[T])$ and $\mathrm{proj}[T] \cap \bigcup\{\mathrm{proj}[S] : s \in z\} \subset \bigcup a$. Let B be the set described in the claim. The set $\mathrm{proj}[T] \cap B \setminus \bigcup a$ is analytic and ϕ-positive, therefore by Theorem 4.5.6, the claim, and a genericity argument there is a condition $r \leq q$ and a tree $S \in a_r$ such that $\mathrm{proj}[S] \subset \mathrm{proj}[T] \cap B \setminus \bigcup a$ and $r \Vdash \mathrm{proj}[S] \in \dot{D}$. Let $a' \subset a$ be a finite set such that $\phi(\mathrm{proj}[S]) > \Sigma\{w(u) : u \in a \setminus a'\}$. Use the closure property of the set a_q to find a tree $T' \in a_q$ such that $\mathrm{proj}[T'] = \mathrm{proj}[T] \setminus \bigcup a'$. Now $\psi(T') \leq \Sigma\{w(u) : u \in a \setminus a'\} < \phi(\mathrm{proj}[S])$ and this contradicts the second sentence of this paragraph for the tree T'.

All that remains to do is to prove the claim. Let $p = \langle a_p, b_p \rangle \in P$ be a condition, let $B = \bigcap_{z \in b_p} \bigcup\{\mathrm{proj}[T] : T \in z\}$ and let $T \in a_p$. I must first argue that $\phi(\mathrm{proj}[T]) = \phi(\mathrm{proj}[T] \cap B)$. Suppose $a \subset U$ is a set such that $\Sigma_{u \in a} w(u) < \phi(\mathrm{proj}[T])$; I must produce a point $x \in \mathrm{proj}[T] \cap B \setminus \bigcup a$. By induction on $n \in \omega$ build trees $T_n \in a_p$ and nodes $t_m^n \in T_m : m \in n$ and finite sets $a_n \subset a$ so that

- $T = T_0$, $\mathrm{proj}[T] \supset \mathrm{proj}[T_1] \supset \mathrm{proj}[T_2] \ldots$
- the nodes $t_m^n \in T_m$ form an increasing sequence, and $\mathrm{proj}[T_n] \subset \mathrm{proj}[T_m \restriction t_m^n]$;
- with some fixed enumeration $b_p = \{z_n : n \in \omega\}$ there is a tree $R \in z_n$ such that $\mathrm{proj}[T_{n+1}] \subset \mathrm{proj}[R]$;
- $\mathrm{proj}[T_n] \cap \bigcup a_n = 0$ and $\phi(\mathrm{proj}[T_n]) > \Sigma_{u \in a \setminus a_n} w(u)$.

Once this has been done, it is clear that for every number $m \in \omega$, the union $\bigcup t_m^n$ forms a branch through the tree T_m, all of these branches project into the same point x and $x \in \mathrm{proj}[T] \cap B \setminus \bigcup a$. The induction itself is easy. Suppose that the trees $T_m : m \in n$ and nodes $t_m^n \in T_m$ are known. Find a tree $S \in a_p$ projecting into the set $\bigcap_{m \in n} \mathrm{proj}[T_m \restriction t_m^n]$. Since $\phi(\mathrm{proj}[S] \cap \bigcup z_n) = \phi \mathrm{proj}[S])$, the set $\mathrm{proj}[S] \cap (\bigcup_{R \in z_n} \mathrm{proj}[R]) \setminus \bigcup a$ has positive ϕ-mass. By the countable subadditivity of the submeasure ϕ, there must be a tree $R \in z_p$ and nodes $t_m^{n+1} \in T_m$ extending the nodes t_m^n such that the set $\mathrm{proj}[R] \cap \bigcap_{m \in n} \mathrm{proj}[T_m \restriction t_m^{n+1}] \setminus \bigcup a$ still has positive ϕ-mass, say $\varepsilon > 0$. There must be a finite set $a_{n+1} \subset a$ such that $\Sigma_{u \in a \setminus a_{n+1}} w(u) < \varepsilon$. Let $T_n \in a_p$ be any tree projecting into the set $\mathrm{proj}[R] \cap \bigcap_{m \in n} \mathrm{proj}[T_m \restriction t_m^{n+1}] \setminus \bigcup a_{n+1}$ and let $t_n^{n+1} = 0$. The induction hypotheses continue to hold.

The second item of the claim now immediately follows. If $C \subset B$ is any analytic ϕ-positive set and T is any tree projecting into it, just close the set $a_p \cup \{T\}$ under the operations in the definition of the forcing P to get a set a_q. The trees in a_q either belong to a_p or else project into subsets of the set $C \subset B$, and therefore the pair $q = \langle a_q, b_p \rangle$ is a condition in the forcing P. □

The Fubini properties common to all the pavement forcings are quite easily identified.

Theorem 4.5.11. *Suppose that J is a σ-ideal on a Polish space Y such that the forcing P_J is proper. The following are equivalent:*

1. $\neg I_\phi \perp J$ *for every pavement submeasure ϕ;*
2. *the forcing P_J has the Sacks property.*

Proof. On one hand, suppose that the forcing P_J does not have the Sacks property. There is a P_J-name for a function $\dot{f} \in \omega^\omega$ such that for every ground model function $h : \omega \to [\omega]^{<\aleph_0}$ such that $|h(n)| = n^2$ it is forced that for infinitely many $n \in \omega$, $\dot{f}(\check{n}) \notin \check{h}(\check{n})$. By Proposition 2.3.1, there is a Borel set $C \in P_J$ and a Borel function $g : B \to \omega^\omega$ such that $B \Vdash \dot{f} = \dot{g}(\dot{y}_{gen})$. Now define a pavement submeasure ϕ on the space X of all functions $h : \omega \to [\omega]^{<\aleph_0}$ such that $|h(n)| = n^2$: set $U = \{u_{nm} : n, m \in \omega\}$, where $u_{nm} = \{h \in X : m \notin h(n)\}$, and $w(u_{nm}) = 1/n^2$.

It is quite clear that $\phi(X) = 1$. If $a \subset U$ is a collection of pavers of total weight < 1 then for no number n there can be more than n^2 many pavers in the set a of the form u_{nm}. As a result, there is a function $h \in X$ such that for all numbers $n, m \in \omega$ such that $u_{nm} \in a$, $m \in h(n)$. Such a function does not belong to any of the pavers in the set a.

It is also clear that for every function $k \in \omega^\omega$, the set $A_k = \{h \in X : \exists^\infty n \; k(n) \notin h(n)\}$ has ϕ-mass zero. Given any real number $\varepsilon > 0$, find a natural number n_0 such that $\Sigma_{n > n_0} 1/n^2 < \varepsilon$ and consider the set $a = \{a_{n,k(n)} : n > n_0\} \subset U$. The sum of the weights of the pavers in this set is $< \varepsilon$ and $A_k \subset \bigcup a$ by the definitions, therefore $\phi(A_f) < \varepsilon$ as required.

To sum up the previous work, writing $D \subset X \times B$ for the Borel set $\{\langle h, y \rangle : \exists^\infty n \; g(y)(n) \in h(n)\}$, it is the case that the horizontal sections of the set D are in the ideal ϕ while the vertical sections of its complement belong to the ideal J. This concludes the proof of (1)→(2).

On the other hand, if the forcing P_J does have the Sacks property then for every pavement submeasure ϕ on a Polish space X it is in fact true that every subset of X in the extension has a Borel superset coded in the ground model of arbitrarily close ϕ-mass. For suppose that the submeasure ϕ is obtained from some set U of pavers and a weight function $w : U \to \mathbb{R}^+$, and suppose that some condition $p \in P_J$ forces $\dot{a} \subset U$ is a set whose sum of weights is less than some fixed $\varepsilon > 0$. It will be enough, for every positive real number $\delta > 0$, to produce a set $b \subset U$ in the ground model whose sum of weights is $< \varepsilon + \delta$, and a condition $q \leq p$ forcing $\dot{a} \subset \check{b}$. It is an elementary matter to find a function $k : \omega \to \mathbb{R}^+$ such that $\Sigma_n n^2 k(n) < \delta$ and P_J-name $\dot{f} : \omega \to [U]^{<\aleph_0}$ such that p forces $\dot{a} = \bigcup \text{rng}(\dot{f})$ and $\Sigma\{w(u) : u \in \dot{f}(n+1)\} < k(n)$ for all $n \in \omega$. Use [2], 2.3.9 again to find a condition $q \leq p$ and a ground model function $g : \omega \to [[U]^{<\aleph_0}]^{<\aleph_0}$ such that for every $n \in \omega$, $g(n)$ consists of at most n^2 many sets and in each of them the pavers sum up to a weight $< k(n)$, and q forces that $\dot{f}(n+1) \in \check{g}(n)$. Moreover, the condition q will decide the value of $\dot{f}(0)$ to be some finite set $c \subset U$. Then clearly q forces $\dot{a} \subset \bigcup \bigcup \bigcup \operatorname{rng}(g)$, and the weights of pavers in the latter set add up to less than $\varepsilon + \delta$ by the construction. This completes the proof. □

Note that the argument shows that among the pavement forcings there is one which has the fewest Fubini-type preservation properties; it is identified in the first paragraph of the proof. This should be compared with the situation for σ-ideals generated by closed sets. Among those, Cohen forcing is the one with the fewest Fubini-type preservation properties.

Finally, there are two theorems regarding the subsets of ω_1 added by pavement forcings.

Theorem 4.5.12. *Suppose that ϕ is a pavement measure on a Polish space X generated by a countable set of Borel pavers. Then P_{I_ϕ} forces that every function $f : \omega_1 \to 2$ in the extension has an infinite ground model subfunction.*

Proof. Fix a condition $B \in P_{I_\phi}$ forcing $\dot{f} : \omega_1 \to 2$. Fix a winning strategy σ for the Nonempty Player In the Borel precipitous game $G(I_\phi)$. Fix $\vec{M}$, a continuous $\in$-tower of countable elementary submodels of a sufficiently large structure of length ω_1 such that the first model $\vec{M}(0)$ contains $X, \phi, \sigma, B, \dot{f}$. For every condition $C \subset B$ write $C(\alpha) = \{x \in C : x \text{ is } M_\beta\text{-generic for all } \beta \le \alpha\}$ and $C(\alpha + 1, b) = \{x \in C(\alpha + 1) : \dot{f}(\vec{M}(\alpha) \cap \omega_1)/x = b\}$. Clearly $C(\alpha + 1) = C(\alpha + 1, 0) \cup C(\alpha + 1, 1)$. If $C \in \vec{M}(0)$ then a bootstrapping argument based on the proof of Theorem 4.5.2 shows that $\phi(C) = \phi(C(\alpha))$. A review of the definitions reveals that $C(\alpha + 1, b) \subset C$ is a Borel set which, if ϕ-positive, forces $\dot{f}(\vec{M}(\alpha) \cap \omega_1) = b$.

There are two distinct cases.

- There is an ordinal $\alpha \in \omega_1$ and a condition C in $P_{I_\phi} \cap \vec{M}(\alpha + 1)$ such that $C \subset B(\alpha)$ and the set of all ordinals β such that $\phi(C(\beta + 1, 1)) < \phi(C)$ is uncountable.
- Otherwise.

To conclude the argument in the first case, apply Claim 4.5.8 to find an infinite set $b \subset \omega_1$ such that the set $\bigcup_{\beta \in b} C(\beta + 1, 1)$ has ϕ-mass smaller than $\phi(C)$ Let $\gamma = \sup(b)$ and consider the set $D = C(\gamma) \setminus \bigcup_{\beta \in b} C(\beta + 1, 1)$. This is an ϕ-positive Borel set and a condition forcing that $\dot{f} \restriction \{\beta_n : n \in \omega\}$ is constantly equal to zero.

In the second case, use the failure of the first case to find an uncountable set $A \subset \omega_1$ such that for every $\alpha \in A$, every ordinal $\alpha \in \beta$ and every condition $C \subset B(\beta)$ in $P_{I_\phi} \cap \vec{M}(\beta + 1)$ it is the case that $\phi(C(\alpha + 1, 1)) = \phi(C)$. Let $\alpha_n : n \in \omega$ be an infinite increasing sequence of elements in the set A. Let $D = \bigcap_n B(\alpha_n + 1, 1)$.

I will show that $\phi(B) = \phi(D)$, so $D \subset B$ is a condition in the poset P_{I_ϕ} forcing $\dot{f} \restriction \{\alpha_n : n \in \omega\}$ is constantly equal to one, completing the argument in the second case and the proof of the theorem. Suppose for contradiction that $\phi(D) < \phi(B)$ and find a set $a \subset U$ such that $\Sigma_{u \in a} w(u) < \phi(B)$ and $D \subset \bigcup a$. By induction on $n \in \omega$ build conditions C_n and sets $a_n \subset a$ so that

- $B = C_0 \supset C_1 \supset \ldots, \ C_n \in \bar{M}(\alpha_n), \ C_{n+1} \subset B(\alpha_n + 1, 1)$ and there are conditions $p_0, q_0, p_1, q_1 \ldots$ such that $C_0 \supset p_0 \supset q_0 \supset C_1 \supset p_1 \supset q_1 \supset C_2 \supset \ldots$ and the sequence $p_0, q_0, p_1, \ldots$ forms a play of the Borel precipitous game $G(I_\phi)$ in which the Nonempty player uses his winning strategy σ;
- $0 = a_0 \subset a_1 \subset \ldots$
- $C_n \cap \bigcup a_n = 0$ and $\phi(C_n) > \Sigma_{u \in a \setminus a_n} w(u)$.

If this construction succeeds then $\bigcap_n C_n$ is a nonempty set by the first item, and any of its elements belongs to the set $D \setminus \bigcup a$, contradicting the choice of the set $a \subset U$.

The initial setup $C_0 = B$ and $a_0 = 0$ satisfies the induction hypothesis. Suppose that C_n, a_n have been constructed. Let $E_n = \{q \in P_{I_\phi} : \exists p \ p_0, q_0, p_1, \ldots q_{n-1}, p, q \text{ is a play utilizing the strategy } \sigma\}$. The set $E_n \in \bar{M}(\alpha_n)$ is dense below C_n and therefore there is a countable subset $F_n \subset E_n$ in the model $\bar{M}(\alpha_n)$ such that $\phi(\bigcup F_n) = \phi(C_n)$. As I am working on the second case, it follows that $\phi(B(\alpha_n + 1) \cap \bigcup F_n) = \phi(C_n)$. By the third item of the induction hypothesis, $\phi(B(\alpha_n + 1) \cap \bigcup F_n \setminus \bigcup(a \setminus a_n)) \neq 0$ and so there must be a set $q_n \in F_n$ such that $\phi(B(\alpha_n + 1) \cap q_n \setminus \bigcup(a \setminus a_n)) = \varepsilon > 0$. Find a finite set $a_{n+1} \subset a$ such that $a_n \subset a_{n+1}$ and $\Sigma_{u \in a \setminus a_{n+1}} < \varepsilon$. Find a set $p_n \in \bar{M}(\alpha_n)$ witnessing the fact that $q_n \in E_n$. Finally, let $C_{n+1} = B(\alpha_n + 1) \cap q_n \setminus \bigcup a_{n-1}$. The induction hypotheses continue to hold. $\qquad\square$

Note that the argument shows that if $\mathrm{add(null)} > \aleph_1$ then there is either an infinite ground model set $b \subset \omega_1$ such that $f \restriction b$ is constantly 0 or an uncountable ground model set $b \subset \omega_1$ such that $f \restriction b$ is constantly 1.

Theorem 4.5.13. *Suppose that ϕ is a pavement submeasure on a Polish space X generated by a countable set of Borel pavers, and $V \subseteq V[H] \subseteq V[G]$ are a ground model, a P_{I_ϕ} extension, and an intermediate extension. Then*

1. *either $V[H] = V[G]$ or $V[H]$ is a c.c.c. extension of V;*
2. *($\mathrm{add(null)} > \aleph_1$) $V[H]$ is generated by a single real.*

I do not know if the assumption $\mathrm{add(null)} > \aleph_1$ can be removed. I also do not know what c.c.c. reals can be added by a pavement forcing. The proof of the theorem is very similar to Theorem 4.1.7 and I omit it.

4.5.2 Laver forcing

Laver defined the *Laver forcing* as the poset of all Laver trees ordered by inclusion. Here an infinite tree $T \subset \omega^{<\omega}$ is Laver if it has a stem $t \in T$ such that every node of the tree T is compatible with it and all longer nodes in the tree t split into infinitely many immediate successors.

Consider the σ-ideal I on ω^ω generated by all sets $A_g = \{f \in \omega^\omega : \exists^\infty n\ f(n) \in g(f \restriction n)\}$ as g varies over all functions from $\omega^{<\omega}$ to ω. The following fact has been proved by a number of people, I think the priority should go to [4]. It shows that the forcing P_I has a dense subset naturally isomorphic to Laver forcing.

Proposition 4.5.14. *Let $A \subset \omega^\omega$ be an analytic set. Either $A \in I$ or A contains all branches of some Laver tree.*

Proof. Suppose $C \subset \omega^\omega \times \omega^\omega$ is a set and consider the following game $G(C)$. In it, Players I and II alternate, Player II starts out with a finite sequence $t_{ini} \in \omega^{<\omega}$ and then at round n, Player I indicates a natural number m_n and Player II responds with a larger number k_n. Moreover, Player II in some rounds mentions another natural number l_i smaller than the index of the round. In the end, let $x = t_{ini}^\frown k_0^\frown k_1^\frown \ldots$ and $y = l_0 l_1 l_2 \ldots$. Player II wins if he obtained infinitely many numbers $l_i : i \in \omega$, and $\langle x, y \rangle \in C$. The following claim is key.

Claim 4.5.15. *Player I has a winning strategy if and only if $\mathrm{proj}(C) \in I$. If Player II has a winning strategy then $\mathrm{proj}(C)$ contains all branches of a Laver tree.*

The proposition immediately follows. If $A \subset \omega^\omega$ is an analytic set then it is a projection of some closed set $C \subset \omega^\omega \times \omega^\omega$. The game $G(C)$ has G_δ payoff and is determined by Fact 1.4.2. A reference to the claim then completes the proof. □

Corollary 4.5.16. *The ideal I is homogeneous.*

Proof. Suppose that $T \subset \omega^{<\omega}$ is a Laver tree. The natural homeomorphism $\pi : \omega^\omega \to [T]$ has the properties required in the definition of homogeneity. □

It turns out that the σ-ideal I is generated by a pavement submeasure. Just let $U = \{O_{t,n} : t \in \omega^{<\omega}, n \in \omega\}$ where $O_{t,n} = \{f \in \omega^\omega : t \subset f, f(|t|) \in n\}$ and let $w(O_{t,n}) = \varepsilon_t$ for some sequence $\varepsilon_t : t \in \omega^{<\omega}$ of positive real numbers with converging sum.

4.5.3 Solovay forcing

The outer Lebesgue measure λ on the Cantor space 2^ω is a pavement submeasure. The forcing P_{I_λ} was first considered by Solovay.

Fact 4.5.17. *I_λ is a homogeneous ideal.*

Proof. This is an immediate corollary of the measure isomorphism theorem. Whenever $B \in P_{I_\lambda}$ is a Borel set of nonzero measure then there is a Borel bijection $\pi : 2^\omega \to B$ which preserves the measure. This means among other things that preimages of λ-null sets are λ-null and the ideal is homogeneous. $\square$

4.5.4 Maharam algebras

A submeasure ϕ on a Boolean algebra is *exhaustive* if for every sequence of pairwise disjoint elements the masses converge to 0. One example of an exhaustive submeasure is the outer Lebesgue measure on Borel sets. Maharam asked whether this is essentially the only example: if X is a Polish space and ϕ an exhaustive submeasure on it, is there a measure with the same null sets as ϕ? Restated in forcing terms, if ϕ is a countably subadditive exhaustive submeasure on 2^ω, is it necessarily true that P_{I_ϕ} is in the forcing sense equivalent to the Solovay forcing? This question remained open for several decades until Talagrand gave a negative answer [75]. It turns out that all countably subadditive exhaustive submeasures on 2^ω are in fact pavement submeasures. This development provides new examples of pavement submeasure forcings. They are all c.c.c. and add splitting reals, and if ϕ is an exhaustive submeasure then P_{I_ϕ} collapses outer Lebesgue measure by a result of Christensen [7] and many other outer regular submeasures by a result of [16].

4.6 Analytic P-ideal forcings

It is possible to associate a forcing with every P-ideal K on natural numbers. This forcing adds an infinite subset of ω with finite intersection with every ground model element of the ideal K; moreover, in the case of an analytic P-ideal this forcing is proper. The resulting forcings are apparently very complex.

Definition 4.6.1. *[18] An ideal K on ω is a P-ideal if for every sequence A_n $(n \in \omega)$ of sets in K there is $A \in K$ such that $A_n \setminus A$ is finite for all n. The associated P-cover σ-ideal I on $\mathcal{P}(\omega)$ is generated by sets $A_x = \{y \subset \omega : x \setminus y$ is infinite$\}$ as x varies through all elements of K.*

I-positive sets are sometimes called *approximations* to K. The family of compact hereditary sets in I plays an important role in the proof of the structure theorem for analytic P-ideals [71]. Note that since K is a P-ideal, the sets A_x with all their subsets form a σ-ideal and so they form a basis for the ideal I consisting of G_δ sets. It is quite obvious that the ideal I does not contain all singletons, for example $\{\omega\} \notin I$. However, the ideal I does contain all singletons when restricted to some interesting Borel sets B, such as $B = K$.

4.6.1 General theorems

Theorem 4.6.2. *[18] If I is a P-cover ideal then the forcing P_I is proper.*

Proof. Fix the analytic P-ideal K on ω which generates the ideal I. Use the classical result of Solecki [71] to find a finite lower semicontinuous submeasure $\mu : \mathcal{P}(\omega) \to \mathbb{R}^+$ such that $K = Exh(\mu)$. That is to say, $\mu(y) = \sup\{\mu(x) : x \subset y$ finite$\}$ for every set $y \subset \omega$, and $K = \{y \subset \omega : \lim_n \mu(y \setminus n) = 0\}$. Note that in fact K is Borel.

Suppose that P is a forcing and $\dot{x}$ is a P-name for a subset of ω. Consider the P-cover game $G(P, \dot{x})$ between Player I and Player II. In it, Player I produces an initial condition $p_{ini} \in P$, one by one open dense sets $D_n \subset P$ and dynamically on a fixed schedule a set $y \subset \omega$, $y \in K$. Player II produces one by one a descending chain $p_{ini} \geq p_0 \geq p_1 \geq \ldots$ of conditions such that $p_n \in D_n$ and p_n decides the statement $\check{n} \in \dot{x}$. He can hesitate for an arbitrary finite number of steps before placing his next move. In the end, let $g \subset P$ be the filter Player II created. Player II wins if $y \subset \dot{x}/g$ modulo a finite set.

To make this precise, I need to specify Player I's schedule for the set y. At round n Player I decides whether $n \in y$ or not and specifies a number $m_n \in \omega$ such that $\mu(y \setminus m_n) \leq 2^{-m_n}$. The latter demand is equivalent to the condition that for every number $k \in \omega$, $\mu(y \cap k \setminus m_n) \leq 2^{-m_n}$. It is quite clear that Player I can produce any given set in the ideal K under this schedule.

As in the previous sections, it will be enough to prove the following lemma.

Lemma 4.6.3. *The following are equivalent.*

- *$P \Vdash \dot{x}$ falls out of all ground model coded Borel I-small sets.*
- *Player II has a winning strategy in the game $G(P, \dot{x})$.*

One direction of the lemma is again trivial. If there is a condition $p \in P$ forcing the set $y \setminus \dot{x}$ to be infinite, then Player I can win by playing on the side an increasing sequence $\langle M_n : n \in \omega \rangle$ of countable elementary submodels of some large structure, enumerating all the open dense subsets of P in $M = \bigcup_n M_n$, producing $p = p_{ini}$ and the set y, and playing so that Player II's filter g is M-generic. In the end, $M[g] \models y \setminus \dot{x}/g$ is infinite by the forcing theorem, and so $y \setminus \dot{x}/g$ is infinite and Player I won.

The opposite direction is harder. Suppose that the first item of the lemma is satisfied. A small claim will be used repeatedly.

Claim 4.6.4. *For every condition $p \in P$ there are numbers $m(p)$ and $k(p)$ such that for every set $y \in K$ of submeasure $\leq 2^{-m(p)}$ there is a condition $q \leq p$ forcing $\check{y} \setminus \dot{x} \subset \check{k}(p)$.*

Proof. Suppose that this fails for some p. By induction on $n \in \omega$ find sets $y_n \in K$ and increasing numbers k_n such that:

- $\mu(y_n) \le 2^{-n}$ and $\mu(\bigcup_{m \in n} y_m) \setminus k_n \le 2^{-n}$;
- $y_n \cap k_n = 0$;
- $p \Vdash \check{y}_n \setminus \dot{x} \ne 0$.

To start, let $k_0 = 0$. To find the set y_n and the number k_{n+1} once the sets $y_m : m \in n$ and the number k_n are known, use the failure of the claim at p, $-n$ and k_n to find a set $y_n \in K$ such that $y_n \cap k_n = 0$, $\mu(y_n) \le 2^{-n}$ and $p \Vdash \check{y}_n \setminus \dot{x} \ne 0$. Then $z = \bigcup_{m \in n+1} y_n \in K$ and therefore there is a number $k_{n+1} \in \omega$ such that $\mu(z \setminus k_{n+1}) \le 2^{-n-1}$. This concludes the inductive step.

In the end, let $y = \bigcup_n y_n$. It is not difficult to verify from the first induction hypothesis that $\mu(y \setminus k_n) \le 2^{-n} + \Sigma_{m \ge n} 2^{-m}$ and therefore $y \in K$. The last two induction hypotheses then show that $p \Vdash \check{y} \setminus \dot{x}$ is infinite, contradiction. □

The payoff set of the game $G(P, \dot{x}, I)$ is Borel and the game is therefore determined by Fact 1.4.2. To conclude the proof of the lemma, it will be enough to derive a contradiction from the assumption that Player I has a winning strategy σ. To find Player II's counterplay, let M be a countable elementary submodel of a large enough structure and let $p = p_{ini} \in P$ be Player I's initial condition. Let $m(p), k(p)$ be the numbers from the claim. The idea now is to construct a counterplay such that the resulting filter $g \subset P \cap M$ is M-generic and $y \setminus \dot{x}/g \subset \max\{k(p), m_{m(p)}\}$. In order to do that, find Player II's moves $p_n \in P \cap M$ played at rounds i_n in such a way that:

- the condition $p_n \in M$ belongs to the n-th open dense set Player I played, to the n-th open dense subset of P in the model M under some fixed enumeration, and it decides the statement $\check{n} \in \dot{x}$;
- $p_n \Vdash \dot{x} \cap \check{y} \cap i_n \subset \max\{k(p), m_{m(p)}\}$; note that the set $y \cap i_n$ is known at round i_n;
- the number $m_{m(p_n)}$ is known at round i_n, and $i_n > k(p_n), m_{m(p_n)}$.

The second induction hypothesis then immediately implies that Player II won the resulting play of the game, obtaining the desired contradiction. To construct p_0, i_0, let $y \in K$ be the set the strategy σ produces if Player II forever hesitates to place a nontrivial move in the play. By the claim, there is a condition $q \le p$ forcing $\check{y} \setminus \dot{x} \subset \max\{m_{m(p)}, k(p)\}$. Let $p_0 \le q$, $p_0 \in M$ be a condition in the first open dense subset of the poset P in the model M and in the first open dense set Player I played, deciding the statement $0 \in \dot{x}$. Let i_0 be a sufficiently large number so that the last induction hypothesis is satisfied. The induction step is similar. Going through the same motions as in the previous sections will then conclude the proof of the theorem. □

Theorem 4.6.5. *Suppose that I is a P-cover ideal on $\mathcal{P}(\omega)$.*

1. *I satisfies the third dichotomy.*
2. *In the choiceless Solovay model, I satisfies the first dichotomy.*
3. *(ZFC+DC+AD+) I satisfies the first dichotomy and it is closed under wellordered unions.*

Example 4.6.6. Laver forcing. Let K be the collection of sets $x \subset \omega \times \omega$ with finite vertical sections. It is not difficult to see that $P_I \restriction K$ is isomorphic to the poset P_J where J is a σ-ideal of nondominating subsets of ω^ω. It has been known for some time that P_J is in the forcing sense equivalent to the Laver forcing [4].

Example 4.6.7. The optimal amoeba forcing for measure. Let K be the collection of sets $x \subset 2^{<\omega}$ such that the set $B_x = \{r \in 2^\omega :$ for infinitely many numbers $n \in \omega$, $r \restriction n \in x\} \subset 2^\omega$ is Lebesgue null. It is well-known and not difficult to verify that K is an analytic P-ideal [71]. The poset $P_I \restriction K$ adds a Lebesgue null set containing all ground model coded Lebesgue null sets. It is not the same as the standard amoeba forcing for measure, in particular it is not c.c.c. Note that the same procedure will work with the Maharam submeasures in place of the Lebesgue measure.

Example 4.6.8. Every quotient forcing associated with a pavement submeasure is isomorphic to $P_I \restriction B$ for a suitable P-cover ideal I and Borel I-positive set B. To see this, let U, w be the weight generating the null ideal J, and let X be the underlying space. Let $K = \{a \subset U : w(a)$ is finite$\}$; so this is a typical F_σ P-ideal on the set $\mathcal{P}(U)$. Let I be the associated P-cover ideal on $\mathcal{PP}(U)$. Consider the function $\pi : X \to \mathcal{P}(U)$ defined by $\pi(r) = \{u \in U : r \notin u\}$ and the set $B = \mathrm{rng}(\pi) \subset \mathcal{P}(U)$. I claim that B is an I-positive Borel set and the bijection $\pi : X \to B$ moves the ideal J to the ideal I below B. If $A \subset X$ is a set in the ideal J, for every $n \in \omega$ find a set $a_n \subset U$ such that $w(a_n) \leq 2^{-n}$ and $A \subset \bigcup a_n$, and set $b = \bigcup_n a_n \subset U$. Clearly, $b \in K$ and the image $\pi''A$ is included in the I-small set $\{c \subset U : b \setminus c$ is infinite$\}$. On the other hand, if $A \subset \mathcal{PP}U$ is a set in the ideal I, find a set $b \subset U$ of finite weight such that $A \subset \{c \subset U : b \setminus c$ is infinite$\}$ and note that the preimage $\pi^{-1}A$ is J-small since it is covered by the union of every cofinite subset of b.

The dependence on Solecki's result and the determinacy of the associated game make it difficult to extend the result to the case of P-cover ideals generated by undefinable P-ideals. It is not difficult to observe that if K is the complement of a Ramsey ultrafilter F, I' is the P-cover ideal derived from K and I is the ideal generated by I' and $\{F\}$ then P_I is in the forcing sense equivalent to the standard c.c.c. poset Q diagonalizing the Ramsey ultrafilter F, since it adds a diagonalizing real and such a real is Q-generic by the Mathias criterion for Q-genericity.

The posets P_I associated with an analytic P-ideal K are strongly inhomogeneous, and some singletons such as $\{\omega\}$ are positive in the ideal I. The P-ideal K itself is a condition in the forcing P_I and below this condition the poset has much more reasonable properties. Note that it adds an element of the analytic P-ideal K which modulo finite includes all ground model elements of K.

Lemma 4.6.9. *The ideal I is homogeneous below K.*

Proof. Recall that an ideal $I \restriction K$ is homogeneous if and only if for every Borel I-positive set $B \subset K$ there is a function $f : K \to B$ such that f-preimages of I-small sets are I-small [83], Definition 2.3.1. In this case, every function f mapping a set $y \in K$ to a set $x \subset B$ which covers y modulo a finite set will clearly work. $\square$

It is not immediately clear if the poset $P_I \restriction K$ is homogeneous per se.

4.7 Other examples

The forcings described in the previous subsections all share a common feature: the ideals associated with them satisfy the first dichotomy, and therefore all the intermediate models inside their respective generic extensions are c.c.c. extensions of the ground model. There are many definable partial orders which do not have this property. Obvious examples of such a behavior are the iterations or side by side products. There are also less obvious but still very common examples of definable partial orders which add a minimal real but can be decomposed into σ-closed*c.c.c. iterations, and therefore do not satisfy the first dichotomy. Some of them are connected to very familiar objects.

4.7.1 The E_0 equivalence

Definition 4.7.1. E_0 *is the equivalence relation on 2^ω defined by xE_0y if and only if $x\Delta y$ is finite.*

E_0 is the minimal non-smooth equivalence relation, as the Glimm–Effros dichotomy shows:

Fact 4.7.2. *[47] Suppose that F is a Borel equivalence relation on some Polish space X. Then:*

1. *either there is a Borel function $g : X \to \mathbb{R}$ such that xFy if and only if $g(x) = g(y)$; the equivalence F is then said to be* smooth;
2. *or there is a topological embedding $h : 2^\omega \to X$ such that rE_0s if and only if $h(r)Fh(s)$.*

Now let I be the σ-ideal on 2^ω generated by Borel partial E_0-transversals, that is, Borel sets which meet every E_0-equivalence class in at most one element. This coincides with the ideal of Borel sets B such that the equivalence $E_0 \restriction B$ is smooth.

Theorem 4.7.3. *The forcing P_I is proper and preserves Baire category.*

Proof. By Corollary 3.5.4 it is enough to show that whenever M is a countable elementary submodel of a large enough structure then the forcing $P_I \cap M$ forces its generic point to fall out of all ground model I-small sets.

Suppose for contradiction that some condition $B \in P_I \cap M$ forces the generic point $\dot{x}_{gen}$ into a Borel partial E_0-transversal A. The following is a key claim.

Claim 4.7.4. *There is a finite binary sequence s such that $(B+s) \cap B \notin I$.*

Here the set $B+s$ consists of all sequences obtained from elements x in the set B by adding the finite sequence s coordinatewise to x modulo 2. If the claim failed then $\bigcup_{s \in 2^{<\omega}} (B+s) \cap B \in I$ and the set $B \setminus \bigcup_s (B+s)$ is a Borel partial E_0-transversal, therefore the set B is I-small, contradiction.

Fix the binary sequence $s \in 2^{<\omega}$ as in the claim. Note that the operation $C \mapsto C-s$ is an automorphism of the forcing $P_I \cap M$. Thus $(B+s) \cap B \Vdash (\dot{x}_{gen} - s)$ is also a P_I-generic point, and since both $\dot{x}_{gen}$ and $\dot{x}_{gen} - s$ both meet the condition B they must both belong to the set $\dot{A}$. This contradicts the assumption that the set A is an E_0-selector. $\qquad\square$

A combinatorially simple dense subset of the forcing P_I has been isolated in [83]. A perfect binary tree T is called an E_0-tree if for every splitnode $t \in T$ the next two splitnodes s_0, s_1 extending t have the same length and T below s_0 is equal to T below s_1.

Fact 4.7.5. *[83], 2.3.29. Every analytic set is either in the ideal I or it contains all branches of some E_0-tree.*

Corollary 4.7.6. *The ideal I is homogeneous.*

Proof. The natural continuous bijection between 2^ω and a given E_0-tree preserves the E_0 equivalence and therefore the membership of sets in the E_0 ideal. $\qquad\square$

The presentation of the forcing P_I as the poset of E_0-trees ordered by inclusion makes it possible to prove many theorems about its forcing properties using the standard fusion arguments. Let me just mention at this point that the fact that the forcing P_I is bounding can be abstractly derived form the Glimm–Effros dichotomy in the following way. No matter what Polish topology t one chooses on the set 2^ω, the equivalence E_0 is still not going to be smooth, and the dichotomy shows that every I-positive Borel set B will contain a set C such that there is a t-continuous map $\pi : 2^\omega \to C$ reducing the E_0-equivalence on 2^ω to E_0 on the set C. But then the set C must be I-positive and t-compact. A reference to Theorem 3.3.2 concludes the argument.

Lemma 4.7.7. P_I *naturally decomposes into a strategically σ-closed*c.c.c. iteration.*

Proof. Let P be the poset of Borel E_0-invariant I-positive sets. If $H \subset P$ is a generic filter, in $V[H]$ let J be the ideal of Borel sets which are disjoint from some set in H, and consider the forcing P_J. I will show that P is σ-closed, $P \Vdash P_J$ is c.c.c., the $V[H]$-generic point $x \in 2^\omega$ for P_J is in fact V-generic for P_I, and the filter H can be recovered from it as the collection of all Borel E_0-invariant sets which contain it. This will complete the proof.

Recall that the forcing P is *strategically closed* if Player II has a winning strategy in the descending chain game, a game in which Players I and II alternate to obtain a descending chain $p_0 \geq q_0 \geq p_1 \geq q_1 \ldots$ of conditions in P and Player II wins if the chain has a lower bound. Player II will win the game by playing conditions $q_n = E_0$-saturation of some E_0-tree $T_n \subset 2^{<\omega}$ such that the trees $T_n : n \in \omega$ form a fusion sequence – the n-th splitting level of the tree T_n is still the n-th splitting level of the tree T_{n+1}. Then the E_0-saturation of the tree $T = \bigcap_n T_n$ will be the required lower bound of the chain obtained in the play. In order to show that Player II can maintain his commitments, suppose that at round n he played $q_n = E_0$-saturation of a tree T_n and Player I answered with some set $p_{n+1} \subset p_n$. Choose a sequence t in the $(n+1)$-th splitting level of the tree T_n. The countable sompleteness of the ideal I shows that the set $p_{n+1} \cap [T_n \restriction t]$ is I-positive, containing all branches of some E_0-tree S. T_{n+1} will then be the smallest E_0-tree containing S and the $(n+1)$-th splitting level of the tree T_n as subsets. the invariance of the set p_{n+1} under the E_0 equivalence shows that the E_0-closure of the set $[T_{n+1}]$ is included in p_{n+1} as required. $\square$

4.7.2 The E_1 equivalence

Definition 4.7.8. E_1 *is the equivalence relation on $(2^\omega)^\omega$ defined by $\vec{x} \, E_1 \, \vec{y}$ if the sequences $\vec{x}, \vec{y}$ agree on all but finitely many entries.*

Fact 4.7.9. *[35] Suppose that F is a hypersmooth Borel equivalence relation. Then either F is Borel reducible to E_0 or E_1 is reducible to F by a continuous injection.*

Here, a *hypersmooth* equivalence relation is one which is an increasing union of countably many smooth equivalence relations. In particular, E_1 is hypersmooth, as are all of its restrictions to Borel subsets of $(2^\omega)^\omega$. So let I be the σ-ideal of Borel sets $B \subset (2^\omega)^\omega$ such that $E_1 \restriction B$ is reducible to E_0. What are the properties of the forcing P_I? The previous Fact shows that it is a homogeneous notion of forcing: if $B \in P_I$ is a Borel set then the equivalence relation $E_1 \restriction B$ is not reducible to E_0 and therefore there is a continuous injective reduction $f : (2^\omega)^\omega \to B$ of E_1 to $E_1 \restriction B$. Clearly, the reduction f extends to an isomorphism of the forcing P_I to $P_I \restriction \mathrm{rng}(f)$. So perhaps we have an interesting new forcing at hand? In fact it turns out otherwise:

Proposition 4.7.10. *The forcing P_I is not proper.*

Proof. I will find a representation of the ideal I as a decreasing nonstabilizing intersection of σ-ideals and then use Proposition 2.2.6. Let I_n be the σ-ideal on $(2^\omega)^\omega$ generated by I and all sets $B \subset (2^\omega)^\omega$ for which there exists a Borel function $g : (2^\omega)^{\omega \setminus n} \to (2^\omega)^n$ such that for all sequences $\vec{x} \in B$, $\vec{x} \restriction n = g(\vec{x} \restriction (\omega \setminus n))$.

It is immediate that the ideals $I_n : n \in \omega$ form an inclusion descending sequence and $I \subset \bigcup_n I_n$. To see $\bigcup_n I_n \subset I$, suppose that a Borel set B belongs to all ideals I_n. For each number n I can then find a Borel partition $B = C_n \cup \bigcup_m D_m^n$ such that $C_n \in I$ and for each natural number m there is a Borel function g_m^n. Let $C = \bigcup_n C_n \subset B$. This is a set in the σ-ideal I and the proof will be complete once I show that the remainder set $B \setminus C$ is in the ideal I as well. Note that the equivalence relation E_1 has only countable equivalence classes on $B \setminus C$: whenever $\vec{x} \in B \setminus C$ is a sequence then $[\vec{x}]_{E_1} \cap B \setminus C \subset \{\vec{y} : \exists n \in \omega \ \vec{y} \restriction (\omega \setminus n) = \vec{x} \restriction (\omega \setminus n) \wedge \exists m \ \vec{y} \restriction n = g_m^n(\vec{y} \restriction (\omega \setminus n))\}$. The equivalence relation E_1 cannot be Borel reduced to a Borel equivalence relation with countable classes, therefore $B \setminus C \in I$ as desired.

The last point is to show that the sequence $I_n : n \in \omega$ does not stabilize below any Borel set. In order to do that, I must show that for every number $n \in \omega$ and every Borel set $B \in P_I$ there is a Borel set $C \subset B$ in $I_n \setminus I$. Use Fact 4.7.9 to find a continuous injection $f : (2^\omega)^\omega \to B$ reducing E_1-equivalence to $E_1 \restriction B$. The range $\mathrm{rng}(f) \subset B$ is compact, and so is the set $C' \subset (2^\omega)^{\omega \setminus n}$ which is the projection of $\mathrm{rng}(f)$ to $(2^\omega)^{\omega \setminus n}$. Thus, the function $g : C' \to (2^\omega)^n$ which assigns to each sequence $\vec{x} \in C'$ the lexicographically least sequence $\vec{y}$ such that $\vec{x} ^\frown \vec{y} \in \mathrm{rng}(f)$, is continuous. Writing $C = \{\vec{x} \in \mathrm{rng}(f) : \vec{x} \restriction n = g(\vec{x} \restriction (\omega \setminus n))$, it is clear that $C \subset B$ is compact set in I_n. To see that $C \notin I$, note that the equivalence relation E_1 reduces to $E_1 \restriction \mathrm{rng}(f)$ via the function f, and $E_1 \restriction \mathrm{rng}(f)$ reduces to $E_1 \restriction C$ via the function $h : \mathrm{rng}(f) \to C$ defined by $h(\vec{x}) = (\vec{x} \restriction (\omega \setminus n))^\frown g(\vec{x} \restriction (\omega \setminus n))$. $\square$

4.7.3 The E_2 equivalence

Let J be the summable ideal on ω; $J = \{a \subset \omega : \Sigma_{n \in a} \frac{1}{n+1} < \infty\}$. Let E_2 be the equivalence on 2^ω defined by $x E_2 y \leftrightarrow \{n \in \omega : x(n) \neq y(n)\} \in J$. The equivalence E_2 is not reducible to a countable Borel equivalence relation and is in fact minimal such. In order to state this precisely, I need a couple of definitions and a dichotomy result due to Hjorth. The presentation owes much to Kanovey's [34].

Definition 4.7.11. *For points* $x, y \in 2^\omega$ *let* $d(x, y) = \Sigma\{\frac{1}{n+1} : x(n) \neq y(n)\}$. *A set* $B \subset 2^\omega$ *is* grainy *if there is a real number* $\varepsilon > 0$ *such that for every finite sequence* $x_m : m \in n$ *of points in* B, *if* $\forall m \in n$ $d(x_m, x_{m+1}) < \varepsilon$ *then* $d(x_0, x_n) \leq 1$.

Note that d is a metric on each E_2-equivalence class, while points from different classes have infinite distance.

Fact 4.7.12. *[34] Let* I *be the* σ-*ideal on* 2^ω *generated by Borel grainy sets.*

1. *An analytic set* $A \subset 2^\omega$ *is* I-*positive if and only if there is a continuous* E_2-*preserving injection* $f : 2^\omega \to A$. *An analytic set* $A \subset 2^\omega$ *is* I-*small if and only if there is a Borel function* $f : B \to 2^\omega$ *with domain* $B \supset A$ *which reduces* $E_2 \upharpoonright B$ *to a countable Borel equivalence relation on* 2^ω.
2. *Moreover, whenever* $A \subset 2^\omega \times 2^\omega$ *is an analytic set whose projection is* I-*positive, then there is a continuous* E_2-*preserving injection* $f : 2^\omega \to \text{proj}(A)$ *and there is a continuous function* $g : \text{rng}(f) \to 2^\omega$ *uniformizing the set* A *on* $\text{rng}(f)$.
3. *The ideal* I *is* $\mathbf{\Pi}^1_1$ *on* $\mathbf{\Sigma}^1_1$. *Every analytic grainy set is a subset of a Borel one.*

Note that the second item shows that the ideal I satisfies the third dichotomy, and if the forcing P_I is proper, then it is homogeneous and bounding. Kanovei proved the properness of the forcing. I will include a proof of a stronger theorem.

Theorem 4.7.13. *The forcing* P_I *is proper and preserves Baire category.*

Proof. For analytic sets $A, B \subset 2^\omega$ write AE_2B if for every point $x \in A$ there is $y \in B$ such that xE_2y and vice versa, for every $y \in B$ there is $x \in A$ such that xE_2y. If moreover $\varepsilon > 0$ is a real number, write $AE_2^\varepsilon B$ if for every $x \in A$ there is $y \in B$ such that $d(x, y) < \varepsilon$ and vice versa. Note that $AE_2^\varepsilon B$ implies AE_2B and the latter relation is an equivalence, while the former is not. A small claim will be useful.

Claim 4.7.14.

1. *If* $A, B \subset 2^\omega$ *are analytic sets and* AE_2B *then* $A \in I \leftrightarrow B \in I$.
2. *If* $A \subset 2^\omega$ *is an* I *positive analytic set and* $\varepsilon > 0$ *is a real number then there is a finite collection* $A_m : m \leq n$ *of analytic* I-*positive subsets of* A *such that* $\forall m \in n A_m E_2^\varepsilon A_{m+1}$ *while* $\forall x \in A_0 \forall y \in A_n$ $d(x, y) > 1$.

To prove the first item, suppose $A \notin I$, and let $C \subset 2^\omega \times 2^\omega$ be the set $\{\langle x, y \rangle : x \in A \wedge y \in B \wedge xE_2y\}$. This is an analytic set with $\text{proj}(C) = A$, and so by the second item of Fact 4.7.12 there is an E_2-preserving continuous injection $f : 2^\omega \to A$ and a continuous function $g : \mathrm{rng}(f) \to 2^\omega$ uniformizing the set C on $\mathrm{rng}(f)$. Then $g \circ f : 2^\omega \to B$ is a continuous E_2-preserving function witnessing the I-positivity of the set B.

For the second item, for a natural number $n \in \omega$, a finite set $a \subset \omega$ such that $\Sigma_{n \in a} \frac{1}{n+1} > 1$, and a function $f \in 2^a$ let $B_{n,f} = \{x \in A : f \subset x$ and there is an ε-path of length $n+1$ through the set A starting with x such that $1 - f \subset$ its endpoint$\}$. There must be a pair n, f such that $B_{n,f} \notin I$. To see this, suppose for contradiction that $B_{n,a} \in I$ for all pairs n, a, find a Borel set $B'_{n,a} \in I$ covering it, and observe that the analytic set $A \setminus \bigcup_{n,f} B'_{n,f}$ is grainy, therefore I-small. This provides a decomposition of the set A into countably many I-small sets, contradicting the positivity of the set A. Now fix the pair n, f such that $B_{n,f} \notin A$, and for $m \leq n$ let $A_m = \{x : x$ is on m-th position of an ε path through the set A of length $n+1$ such that $f \subset$ the first endpoint and $1 - f \subset$ the other endpoint$\}$. Clearly, $A_0 = B_n \notin I$, $A_m E_2^\varepsilon A_{m+1}$, and by the first item the analytic sets A_m are all I-positive as required.

In order to prove the properness of P_I, I will deal with the poset Q_I of analytic I-positive sets, since $P_I \subset Q_I$ is dense by Fact 4.7.12 (1). Suppose that M is a countable elementary submodel of a large enough structure and let $A \subset 2^\omega$ be a Borel grainy set. I must show that the forcing $Q_I \cap M$ forces its generic real to fall out of the set A. For contradiction assume that some condition $B \in Q_I \cap M$ forces $\dot{x}_{gen} \in \dot{A}$. Let $\varepsilon > 0$ be a real number witnessing the grainy property of the set A, and use the claim in the model M to find a finite collection of sets $B_m : m \in n$ as in the second item of the claim. Now consider the forcing R consisting of n-tuples $C_m : m \in n$ such that $C_m \in Q_I \cap M$, $C_m \subset B_m$ and $C_m E_2^\varepsilon C_{m+1}$, ordered by coordinatewise inclusion. For each number $m \in n$ let $\dot{G}_m$ be the R-name for the filter on the poset $Q_I \cap M$ generated by the m-th coordinates of conditions in the generic filter.

Claim 4.7.15. *R forces each filter $\dot{G}_m : m \in n$ to be generic for $Q_I \cap M$.*

For the ease of notation I will prove it just for $m = 0$. Suppose $D \subset Q_I \cap M$ be an open dense set, and $C_m : m \in n$ be a condition in the forcing R. I will find a stronger condition $C'_m : m \in n$ in R such that $C' \in D$. This will certainly be sufficient. Just choose any set $C'_0 \subset C_0$ in the set D and then define by induction on $m \in n$ $C'_{m+1} = \{x \in C_{m+1} : \exists y \in C'_m \ d(x, y) < \varepsilon\}$. These are analytic sets, and since $C'_0 \notin I$ and $C'_m E_2^\varepsilon C'_{m+1}$, all of the sets C'_m are I-positive and form the desired condition in the forcing R.

Now let $H \subset R$ be a generic filter and for every $m \in n$ write $x_m \in 2^\omega$ for the generic real of the m-th generic filter on Q_I. It is immediate that $d(x_m, x_{m+1}) < \varepsilon$ for every $m \in n$, while $d(x_0, x_n) > 1$. On the other hand, it follows from the forcing theorem that $x_m \in A$ for all $m \in n$. However, this contradicts the grainy property of the set A. $\qquad \square$

It is now known that the forcing E_2 is proper, bounding, preserves Baire category and outer Lebesgue measure, together with host of other properties. These results will appear in forthcoming work.

4.7.4 Silver forcing

Another variation of the E_0-ideal is the *Silver ideal*. Let G be the graph on 2^ω connecting two binary sequences if and only if they differ in exactly one place. Let I be the σ-ideal of sets σ-generated by Borel G-independent sets. To facilitate the expressions below, for a partial function $f : \omega \to 2$ write $2^\omega \restriction f$ for the set $\{g \in 2^\omega : f \subset g\}$. A typical I-positive set is then $2^\omega \restriction f$ for a function f mapping a coinfinite subset of ω to 2: it is not difficult to see that every Borel G-independent set has a meager intersection with the space $2^\omega \restriction f$ equipped with the natural topology.

Fact 4.7.16. *[83], 2.3.37. Every analytic subset of 2^ω is either in the ideal I or else it contains a subset of the form $2^\omega \restriction f$ for some function f mapping a coinfinite subset of ω to 2.*

Corollary 4.7.17. *The ideal I is homogeneous.*

Proof. Suppose that f is a function mapping a cofinite subset of ω to 2. The natural continuous bijection between 2^ω and $2^\omega \restriction f$ preserves edges of the graph G and therefore the ideal I. $\square$

Corollary 4.7.18. *The forcing P_I has a dense subset naturally isomorphic to Silver forcing.*

Proof. Recall that Silver forcing is the poset of functions $f : \omega \to 2$ with coinfinite domain ordered by inclusion. It is clear that the map $f \mapsto 2^\omega \restriction f$ is an isomorphism of Silver forcing and a dense subset of the forcing P_I. $\square$

4.7.5 Other Borel graphs

Kechris, Solecki, and Todorcevic [38] defined a Borel graph G_0 on the Cantor set 2^ω in the following way. First choose a dense set $D \subset 2^{<\omega}$ such that every level of $2^{<\omega}$ contains exactly one element of D. For every sequence $s \in D$ define the function $\text{flip}_s : O_s \to O_s$ by setting $\text{flip}_s(x) =$ the binary sequence identical to x except that $x(|s|) = 1 - \text{flip}_s(x)(|s|)$. Let then G_0 be the graph on 2^ω which is the union of the graphs of all the functions $\text{flip}_s : s \in D$.

It is not difficult to check that G_0 is locally countable acyclic Borel graph with uncountable Borel chromatic number, and its connectedness components are exactly the E_0-equivalence classes. In fact, Ben Miller proved that it is minimal with this property in a very strong sense:

Fact 4.7.19. *Suppose that G is a locally countable acyclic Borel graph with uncountable Borel chromatic number on a Polish space X and E is a Borel countable equivalence relation on X with $G \subset E$. Then there is a continuous injection $\pi : 2^\omega \to X$ such that for every $x, y \in 2^\omega$ it is the case that $xE_0y \leftrightarrow \pi(x)E\pi(y)$ and $xG_0y \leftrightarrow \pi(x)G\pi(y)$.*

Let I be the σ-ideal generated by the Borel G_0-independent sets. The previous Fact shows immediately that compact sets are dense in P_I and the ideal I is homogeneous: whenever $B \subset 2^\omega$ is an I-positive set, there will be a Borel injection $\pi : 2^\omega \to B$ as in the Fact, its range will be an I-positive compact subset of B, and π-preimages of I-small sets are small. So perhaps this is another example of a homogeneous proper bounding forcing? It turns out quite otherwise:

Theorem 4.7.20. *The forcing P_I is not proper.*

Proof. Compared to the graph considered in the previous subsection, the graph G_0 is much less homogeneous, as the following claim shows.

Claim 4.7.21. *(Ben Miller) Suppose that $s \in D$ is a sequence and $B \subset O_s$ is a Borel I-positive set. Then there is a compact I-positive subset $C \subset B$ such that $\mathrm{flip}_s''C$ is G_0-independent.*

To see how the theorem follows from the claim, for every sequence $s \in D$ find a maximal antichain $A_s \subset P_I \upharpoonright O_s$ in the set $\{C \subset O_s : \mathrm{flip}_s''C \in I\}$. It will be enough to show that for every condition $B \in P_I$ there is a sequence $s \in D$ such that B is compatible with uncountably many elements of the antichain A_s. This in fact shows that the forcing P_I adds an ω-sequence of ground model reals that cannot be covered by a ground model countable set.

Suppose for contradiction that $B_0 \in P_I$ is a condition compatible with only countably many elements of every antichain $A_s : s \in D$. For every $s \in D$ and every condition $C \in A_s$ compatible with B_0 remove the set $\mathrm{flip}_s''C$ from B_0, obtaining a Borel I-positive set $B_1 \subset B_0$. There must be a sequence $s \in D$ such that $B_2 = \{x \in O_s \cap B_1 : \mathrm{flip}_s(x) \in B_1\} \notin I$. (If there was not such a sequence $s \in D$, removing all the countably many I-small sets $\{x \in O_s \cap B_1 : \mathrm{flip}_s(x) \in B_1\}$ from the set B_1 one ends up with a G_0-independent set, contradicting the positivity of the set B_1.) Now use the claim to find a set $C \in A_s$ compatible with B_2, and consider any point $x \in C \cap B_2$. Clearly, $\mathrm{flip}_s(x) \in B_1$ by the definition of the set B_2, but on the other hand, the set C is compatible with B_2, therefore with B, and therefore $\mathrm{flip}_s''C \cap B_1 = 0$ by the definition of the set B_1. Contradiction!

To prove the claim, without loss of generality asssume that $B \subset O_{s \frown 0}$ and consider the graph $G = G_0 \upharpoonright B$ and the equivalence relation $E = E_0 \upharpoonright B$. Fact 4.7.19 yields a continuous injection $\pi : 2^\omega \to B$ such that for every $x, y \in 2^\omega$, $xG_0y \leftrightarrow \pi(x)G_0\pi(y)$ and $xE_0y \leftrightarrow \pi(x)E_0\pi(y)$. I claim that the image $\pi''2^\omega$ works. Since π reduces the graph G_0 to the graph $G_0 \upharpoonright C$, it must be the case that $C \notin I$. Now suppose that $x, y \in 2^\omega$ and assume that $\text{flip}_s(\pi(x))G_0\text{flip}_s(\pi(y))$. Since $\text{flip}_s \subset G_0$, this means that $\pi(x)$ and $\pi(y)$ belong to the same connected component of the graph G_0, in other words, $\pi(x)E_0\pi(y)$. Since the injection π preserves E_0, this means that xE_0y and there is a finite path p connecting x and y in G_0. Then $\pi''p$ is a finite path in the set B connecting $\pi(x)$ and $\pi(y)$. However, the sequence $\langle \pi(x), \text{flip}_s(\pi(x)), \text{flip}_s(\pi(y)), \pi(y) \rangle$ is a path outside of the set B which connects these two points, contradicting the acyclicity of the graph G_0. □

4.7.6 Weakly wandering sets

Definition 4.7.22. *Let G be a group acting on some Polish space X. A set $A \subset X$ is* weakly wandering *if there is an infinite set $H \subset G$ such that the sets $g \cdot A : g \in H$ are mutually disjoint.*

Let G be a countable group acting in a Borel way on some Polish space X and consider the σ-ideal I generated by Borel weakly wandering sets. There are many actions for which this ideal is nontrivial; one such an example is the shift action of $\mathbb{Z}$ on the space $2^{\mathbb{Z}}$, where every Borel weakly wandering set must be meager.

Theorem 4.7.23. *The forcing P_I is proper and preserves Baire category.*

Proof. Let M be a countable elementary submodel of a large structure. The forcing $P_I \cap M$ adds a single point $\dot{x}$ which belongs to all sets in the generic filter. By Corollary 3.5.4, it will be enough to show that $P_I \cap M$ forces $\dot{x}$ to fall out of all ground model coded Borel sets in the ideal I.

Suppose for contradiction that $B \in P_I \cap M$ is a condition and $A \in I$ is a weakly wandering set such that B forces $\dot{x} \in \dot{A}$. Let $H \subset G$ be an infinite set witnessing the fact that A is weakly wandering.

Claim 4.7.24. *There are distinct elements $g, h \in H$ such that $h^{-1}gB \cap B \notin I$.*

If this failed then the set B would decompose into the countably many sets $\{h^{-1}gB \cap B : g \neq h \in H\}$ and $B \setminus \bigcup_{g \neq h \in H} h^{-1}gB$. The former sets are in the ideal I by the assumption, and the latter set is weakly wandering as witnessed by the set $H \subset G$. Thus $B \in I$, a contradiction.

Fix elements $g, h \in H$ as in the claim. Note that $g, h \in M$, the map $\pi : P_I \to P_I$, $\pi(C) = \{h^{-1}g(x) : x \in C\}$ is an automorphism of the partial order P_I, and its restriction to $P_I \cap M$ is an automorphism of the partial order $P_I \cap M$. Thus if $K \subset P_I \cap M$ is a generic filter containing the condition $h^{-1}g \cap B$, the filter $\pi''K \subset P_I \cap M$ is a generic filter as well. Clearly $h^{-1}g(\dot{x}/K) = \dot{x}/\pi''K$ and $g(\dot{x}/K) = h(\dot{x}/\pi''K)$. However, both filters K and $\pi''K$ contain the condition B, so $\dot{x}/K$ and $\dot{x}/\pi''K$ are both elements of the set A and so their images under the action by g and h should be distinct, contradiction. $\square$

Note that as long as dichotomies for this ideal are missing it is not clear whether the forcing P_I is homogeneous or whether or how it depends on the dynamical system in question.

4.7.7 Mathias forcing

Adrian Mathias introduced the forcing P of all pairs $p = \langle s_p, a_p \rangle$ where $s_p \subset \omega$ is finite and $a_p \subset \omega$ is infinite, with the ordering $q \leq p$ if $s_p \subset s_q$, $a_q \subset a_p$, and $s_q \setminus s_p \subset a_p$. The forcing is designed to add an infinite set $\dot{x}_{gen} \subset \omega$ which is the union of all conditions in the generic filter. The following Fact records the basic information about this partial order.

Fact 4.7.25.

1. *P is proper.*
2. *(Direct decision) If $p \in P$ is a condition and $\dot{b}$ a name for a bit then there is a definite value $c \in 2$ and a condition $q \leq p$ with $s_p = s_q$ such that $p \Vdash \dot{b} = \check{c}$.*
3. *(Geometric property) If M is a transitive model of a large fragment of ZFC, $x \subset \omega$ is Mathias generic for M and $y \subset \omega$ is an infinite set up to finitely many elements included in x then y is Mathias generic for M as well.*
4. *(Generic decomposition) P naturally decomposes into a two-step iteration of the forcing $\mathcal{P}(\omega)$ mod fin and the Prikry-type forcing $P(U)$ where U is the ultrafilter on ω obtained from the first stage of the iteration.*

Let I be the σ-ideal of sets $A \subset \mathcal{P}(\omega)$ which are nowhere dense in the algebra $\mathcal{P}(\omega)$ mod fin. Note that the ordering of $\mathcal{P}(\omega)$ mod fin is σ-closed and therefore these sets indeed do form a σ-ideal. The following proposition immediately implies that the forcing P_I has a dense subset isomorphic to the Mathias forcing P.

Proposition 4.7.26. *Suppose $A \subset \mathcal{P}(\omega)$ is a universally Baire set. Then $A \notin I$ if and only for some finite set $s \subset \omega$ and some infinite set $a \subset \omega$ the set A contains all sets $x \subset \omega$ such that $s \subset x$ and $x \setminus s \subset a$.*

Proof. Suppose that $A \subset \mathcal{P}(\omega)$ is not in the ideal I and let $c \subset \omega$ be an infinite set under which it is dense. Let M be a countable elementary submodel of a large enough structure containing a and let $b \subset \omega$ be a Mathias-generic set below the condition $\langle 0, c \rangle$. Since the set A is dense below c there must be a set $a \in A$ which is almost included in the set b. By the geometric property of the Mathias forcing – Fact 4.7.25 – the set a is Mathias generic over the model M, and since the set A is universally Baire, it must be the case that $M[a] \models a \in A$. By the forcing theorem there must be a condition $p = \langle s, d \rangle \in M$ in the generic filter determined by y such that $p \Vdash \dot{x}_{gen} \in \dot{A}$. Consider the collection of all sets $x \subset a$ with $s \subset a$; I claim this is a subset of the set A as desired. And indeed, every element $x \subset \omega$ of this collection is a subset of the set a and by the geometric property it is Mathias generic over the model M; it meets the condition p and by the forcing theorem $M[x] \models x \in A$. A wellfoundedness argument shows that $x \in A$ as needed.

On the other hand, if $s \subset \omega$ is a finite set and $a \subset \omega$ is infinite then the collection $\{x \subset \omega : x \text{ is infinite and } s \subset x, x \setminus s \subset a\}$ is dense in $\mathcal{P}(\omega)$ mod fin below the set a. The proposition follows. $\qquad\square$

5

Operations

5.1 The countable support iteration

Suppose that I is a σ-ideal on a Polish space X such that the forcing P_I is proper. If $\alpha \in \omega_1$ is a countable ordinal then it is natural to look for a σ-ideal I^α on the Polish space X^α equipped with the product topology such that the forcing with P_{I^α} is in the forcing sense equivalent to the countable support iteration of the forcing P_I of length α. The iterations of arbitrary uncountable length will then be realized using a natural limit of the ideals I^α, and it is treated in Section 5.5.

The natural candidate for the ideal I^α is described in the following definition.

Definition 5.1.1. *Let I be a σ-ideal on a Polish space X and $\alpha \in \omega_1$ be a countable ordinal. A set $A \subset X^\alpha$ is in the ideal I^α if Player I has a winning strategy in the iteration game $G(A)$ of length α. In round $\beta \in \alpha$ of the game, Player I indicates a set $B_\beta \subset X$ in the ideal I and Player II responds with a point $x_\beta \in X \setminus B_\beta$. Player II wins if the sequence $\langle x_\beta : \beta \in \alpha \rangle$ of his answers belongs to the set A. The ideal I^α is referred to as the* transfinite Fubini iteration *of I.*

In the following sections I am going to show that I^α is the correct ideal to consider, but several loose ends have to be tied before precise theorems can be stated. Of a particular concern will be the fact that the membership of Borel sets in the ideal I^α may depend on more than just the membership of Borel sets in I. A related issue is the determinacy of the game $G(A)$ for various payoff sets $A \subset X^\alpha$.

Of course, the basic stepping stone is the suitable definability of the σ-ideal I so that the notion of iteration of P_I is well-defined. This is straightforward to define if suitable large cardinals are present, however in the ZFC case it is necessary to resort to several ugly patches. The following two definitions record the situation.

225

Definition 5.1.2. *(LC) A σ-ideal I on a Polish space X is* iterable *if*

1. *the set of all (codes for) analytic sets in the ideal I is universally Baire;*
2. *the ideal satisfies the second universally Baire dichotomy 3.9.8;*
3. *the forcing P_I is proper in all forcing extensions.*

Note that the third item is equivalent to the properness of the forcing in some extension satisfying CH by Corollary 2.2.9, in particular it is equivalent to the properness of the forcing under CH. Note also that all proofs in the previous chapter yield the second (even the first) dichotomy for the concerned ideals. Note also that every σ-ideal I satisfying (1) and (3) can be amended to one satisfying the second dichotomy without affecting the forcing P_I by Proposition 3.9.10. The second dichotomy will be used to simplify the statements of numerous theorems.

Definition 5.1.3. *(without LC) A σ-ideal I on a Polish space X is* iterable *if*

1. *the ideal is ZFC-correct;*
2. *it satisfies the third dichotomy 3.9.21;*
3. *for every transitive countable model M of set theory and every condition $B \in P_I \cap M$ the set $\{x \in B : x$ is M-generic for $P_I\}$ is I-positive.*

Note that the properness proofs in the previous chapter all yield the iterability of the ideals in question, in both LC and non-LC case. One important point connecting the third items in the two definitions is

Proposition 5.1.4. *(LC) Suppose that I is an iterable σ-ideal on a Polish space X. Then for every countable elementary submodel M of a large enough structure, every generic extension $M[g]$ of M and every Borel set $B \in P_I \cap M[g]$ the set $\{x \in B : x$ is M-generic for $P_I\}$ is I-positive.*

Proof. Fix the model M and the filter g and for the sake of simplicity identify the model $M[g]$ with its transitive collapse. Let δ be a Woodin cardinal of $M[g]$, and in V find an $M[g]$-nonstationary tower $M[g]$-generic filter h and the corresponding embedding $j : M[g] \to N$. Now $P_I \cap M[g][h] = P_I^{M[g][h]}$ since the poset P_I is universally Baire and $M[g][h]$ is a generic extension of an elementary submodel of a large structure in V. The absoluteness of the definition also shows that $P_I^N = P_I^{M[g][h]}$ since the models N and $M[g][h]$ have the same reals. Choose an $M[g]$-inaccessible cardinal $\lambda \in \delta$ and look into the model N. Since $N \models j(P_I^{M[g]}) = P_I$ is proper, $N \models C = \{r \in B : r$ is P_I-generic for the model $j''(M[g] \cap V_\lambda)\}$ is I-positive. By the absoluteness of the ideal I between the models N and $M[g][h]$, $M[g][h] \models C$ is I-positive. And by the universally Baire absoluteness for the model $M[g][h]$, the set C is I-positive in V as desired. $\qquad\square$

5.1.1 A topological restatement of the iteration

Definition 5.1.5. *Let I be a σ-ideal on a Polish space X and $\alpha \in \omega_1$ be a countable ordinal. An I, α-tree is a tree $p \subset X^{<\alpha}$ satisfying the following requirements:*

1. *the tree p as well as the sets $p \upharpoonright \beta = \{\vec{x} \in X^{<\beta} : \exists \vec{y} \ \vec{x}^\frown \vec{y} \in p\}$ for all $\beta \in \alpha$ are Borel;*
2. *for every ordinal $\beta \in \alpha - 1$ and every node $t \in p \cap X^\beta$ the set $\{y \in X : \vec{x}^\frown y \in p\}$ of immediate successors is I-positive;*
3. *if $\vec{x}_n : n \in \omega$ is an inclusion increasing sequence of nodes in the tree p whose lengths are bounded below α then $\bigcup_n \vec{x}_n \in p$.*

I will study the forcing P_α of I, α-trees ordered by inclusion and compare it with the countable support iteration R_α of the forcing P_I. Write $\vec{y}_{gen}$ for the R_α-generic α-sequence of points in the space X, and for an tree $p \in P_\alpha$ write $\pi(p)$ be the function with domain α defined by $\pi(p)(\beta) = $ the canonical R_β-name for the Borel set $\{x \in X : \vec{y}_{gen} \upharpoonright \beta^\frown x \in \dot{p}\}$. The key theorem:

Theorem 5.1.6. *Suppose that $\alpha \in \omega_1$ is a countable ordinal and I is an iterable ideal on a Polish space X. Then*

1. *$\pi : P_\alpha \to R_\alpha$ is an isomorphism between P_α and a dense subset of R_α and for every tree $p \in P_\alpha$, it is the case that $\pi(p) \Vdash_{R_\alpha} \vec{y}_{gen} \in [\dot{p}]$;*
2. *the collection $(I^\alpha)^* = \{A \in X^\alpha : A \text{ is analytic and there is no tree } p \in P_\alpha \text{ such that } [p] \subset B\}$ is an iterable σ-ideal, so P_α is a dense subset of $P_{(I^\alpha)^*}$.*

It follows immediately that writing $\vec{x}_{gen}$ for the canonical P_α-generic α-sequence of points in the space X, the isomorphism π sends the P_α-name $\vec{x}_{gen}$ to the R_α-name $\vec{y}_{gen}$.

Proof. This is really just the proof of the preservation of properness under the countable support iteration [64], Section III.3.3 in disguise. The argument proceeds by simultaneous induction on α. Suppose that the induction hypothesis has been verified up to α.

First verify that that the range of the function π is indeed a subset of R_α and for every tree $p \in P_\alpha$ it is the case that $\pi(p) \Vdash \dot{y}_{gen} \in [\dot{p}]$. This is clear for α limit by the definition of the iteration R_α. At the successor stage $\alpha = \beta + 1$ note that $p \upharpoonright \beta \in P_\beta$ and $\pi(p \upharpoonright \beta) \Vdash_{R_\beta} \vec{y}_{gen} \upharpoonright \beta \in [\dot{p} \upharpoonright \beta]$. I just have to check that $\pi(p \upharpoonright \beta) \Vdash_{R_\beta} \{x \in X : (\vec{y}_{gen} \upharpoonright \beta)^\frown x \in [\dot{p}]\} \notin I$. To see this, let M be a countable elementary submodel of a large structure and let $\vec{y} \in X^\beta$ be an M-generic sequence for the poset R_β below the condition $\pi(p \upharpoonright \beta)$. By the induction hypothesis, $M[\vec{y}] \models \vec{y} \in p$ and by an absoluteness argument $\vec{y} \in p$. The set $B_{\vec{y}} = \{x \in X : \vec{y}^\frown x \in [p]\}$ is an I-positive

Borel set with code in the model $M[\vec{y}]$, and by the ZFC-correctness or absoluteness argument, $M[\vec{y}] \models B_{\vec{y}} \notin I$. By the forcing theorem, $M \models \pi(p \restriction \beta) \Vdash B_{\vec{y}_{gen}} \notin I$, by elementarity $\pi(p \restriction \beta) \Vdash B_{\vec{y}_{gen}} \notin I$, and $\pi(p) \in R_\alpha$ as required.

Second, I must show that the range of the function π is dense in R_α. In order to do that, choose a countable elementary submodel M of a large enough structure and by induction on $\beta \in \alpha + 1$ prove the foolowing statement.

(*) For every ordinal $\gamma \in \beta$, every condition $r \in R_\beta \cap M$ and every $p \in P_\gamma$ whose branches are all M-generic sequences for the poset R_γ which meet the condition $r \restriction \gamma$, there is a tree $q \in P_\beta$ whose branches are M-generic sequences for the poset p_β which meet the condition r, and such that $p = q \restriction \beta$. Moreover for every such a tree p, $\pi(p) \le r \restriction \gamma$, and for every such a tree q, $\pi(q) \le r$.

The argument is essentially identical to the proof in [64] and its details are left to the reader. Let me just indicate the situation in which the third item of the iterability definition is needed. It is in the successor stage of the induction. Suppose that $\beta = \beta' + 1$ are ordinals, $r \in R_\beta$ is a condition and $p \in P_{\beta'}$ is a tree whose branches are M-generic sequences for the poset $R_{\beta'}$ which meet the condition $r \restriction \beta'$. Then the tree $q = p \cup \{\vec{x} \in X^\beta : \vec{x} \restriction \beta' \in p$ and $\vec{x}$ is an M-generic sequence meeting the condition $r\}$ is in the poset P_β. and $B \restriction \beta = A$. The only nontrivial observation in the proof of this statement is that for every sequence $\vec{y} \in [p]$ the set $\{x \in r(\beta)/\vec{y} : r$ is $M[\vec{y}]$-generic for the poset $P_I^{M[\vec{y}]}\}$ is I-positive. But this is exactly the contents of the condition (3).

The density of the range of the function π immediately follows. Suppose that $r \in R_\alpha$ is a condition. Let M be a countable elementary submodel of a large enough structure containing the condition r, and use (*) with $\gamma = 0$ and $\beta = \alpha$ to find the tree $p \in P_\alpha$ such that $\pi(p) \le r$.

The closure of the collection $(I^\alpha)^*$ under countable unions has a similar proof. Suppose that $p \in P_\alpha$ is a tree and $[p] = \bigcup_n A_n$ is a countable union of analytic sets. It will be enough to show that one of the sets $A_n : n \in \omega$ contains all branches of some I, α-tree. By a Shoenfield absoluteness argument, $[p] = \bigcup_n A_n$ holds in every forcing extension, so $\pi(p) \Vdash \exists n \in \omega \; \dot{y}_{gen} \in \dot{A}_n$. let $r \le \pi(p)$ be some condition deciding the value of n, let M be a countable elementary submodel of a large structure, and use (*) with $\gamma = 0$ and $\beta = \alpha$ to find a tree $q \in P_\alpha$ whose branches consist of M-generic points for the forcing R_α meeting the condition r. For every branch $\vec{x} \in [q]$ the forcing theorem implies that $M[\vec{x}] \models \vec{x} \in A_n$, by absoluteness $\vec{x} \in A_n$, and so $[q] \subset A_n$ as required.

The ZFC-correctness of the ideal $(I^\alpha)^*$ is left to the reader. $\square$

The ideal $(I^\alpha)^*$ is not quite the same as I^α. Under mild assumptions on the ideal I, the two contain the same analytic sets though. This is the key point in understanding the countable support iteration. The theorems to this effect are proved in the three following sections.

5.1.2 The Solovay model

Life in the Solovay model offers many advantages, even though (because?) choices are somewhat limited there. One of the advantages is a very smooth statement of the countable support iteration dichotomy theorem. This arised in my conversations with Saharon Shelah in February 2002.

Theorem 5.1.7. *(In the choiceless Solovay model) Suppose that I is a σ-ideal closed under wellordered unions on some Polish space X, and α is a countable ordinal. The ideal I^α is closed under wellordered unions and every I^α-positive set contains all branches of some I, α-tree.*

Proof. Let κ be an inaccessible cardinal and $V(\mathbb{R})$ the related Solovay model. Let I be a σ-ideal closed under well-ordered unions. By a standard argument, it is possible to assume that all the real parameters of the definition of the ideal I are in V; I will neglect these parameters in the sequel. Assume that $\alpha \in \kappa$ is an ordinal and $A \subset X^\alpha$ is a set in $V(\mathbb{R})$. Working in the model $V(\mathbb{R})$, I will show that if the set A is not I^α-small, then it contains all branches of some I, α-tree. By a standard homogeneity argument I may assume that $\alpha \in \omega_1^V$ and the set A is definable as $A = \{\vec{x} \in X^\alpha : \phi(\vec{x}, t)\}$ from some parameter t in the ground model. Consider the following strategy σ for Player I in the iteration game: the strategy σ applied to a string $\vec{y}$ of Player II's answers gives the set $\bigcup(I \cap \mathrm{OD}^{V[\vec{y}]})$. Since the ideal I is closed under well-ordered unions, this is a legal strategy for Player I. If the set A is not I^α-small, there must be a sequence in the set A which is a legal counterplay against this strategy. This means that back in V there is a poset P of size $< \kappa$ and a P-name $\dot{x}$ for an α-sequence of points in the Baire space such that $P \Vdash$"for every $\beta \in \check{\alpha}$, $\dot{x}(\beta) \notin \bigcup(I \cap \mathrm{OD}^{V[\dot{x} \restriction \beta]})$; moreover $\mathrm{Coll}(\omega, < \kappa) \Vdash \phi(\dot{x}, \check{t})$". I will show that in this case there is an I, α-tree q such that $[q] \subset A$.

Back in the model $V(\mathbb{R})$. Call a sequence $\vec{y} \in X^{\leq \alpha}$ *P-generic* if there is a V-generic filter $g \subset P$ such that $\vec{y} \subset \dot{x}/g$. For such a sequence $\vec{y}$ and a condition $p \in P$ we will say that p is *$\vec{y}$-good* if there is a V-generic filter $g \subset P$ containing the condition p such that $\vec{y} \subset \dot{x}/g$. An important observation is that for every ordinal $\beta \leq \alpha$ the set $B = \{\vec{y} \in X^\beta : p$ is $\vec{y}$-good$\}$ is Borel by Fact 1.4.8.

The following claim is reminiscent of the classical preservation theorems for countable support iterations. Note that P is a countable set in the model $V(\mathbb{R})$.

Claim 5.1.8. *For every ordinal $\beta \leq \alpha$, for every ordinal $\gamma \in \beta$, every I, γ-tree q whose branches are P-generic, and every Borel function $f : [q] \to P$ such that for*

*every sequence $\vec{y} \in [q]$ the condition $f(\vec{y})$ is $\vec{y}$-good, there is a $1, \beta$-tree r whose
cofinal branches are P-generic such that $q = r \restriction \gamma$ and for every sequence $\vec{y} \in [r]$
the condition $f(\vec{y} \restriction \gamma)$ is $\vec{y}$-good.*

Once the claim has been proved, we will apply it with $\beta = \alpha$ and $\gamma = 0$ to get
an I, α-tree q whose branches are all P-generic. A standard homogeneity argument
shows that $[q] \subset A$ and the dichotomy follows.

The claim is proved by induction on the ordinal β. First the successor step.
Suppose the induction hypothesis holds at β and we want to verify it at $\beta + 1$
for some ordinal $\gamma \leq \beta$, an I, γ-tree q and a Borel function $f : [q] \to P$. Use
the induction hypothesis to get a I, β-tree q' such that $q = q' \restriction \gamma$ and every branch
$\vec{y} \in [q]$ is P-generic and even $f(\vec{y} \restriction \gamma)$-good. Now consider the tree r whose branches
are exactly those sequences $\vec{y} \in X^{\beta+1}$ for which $\vec{y} \restriction \beta \in q'$ and $\vec{y}$ is P-generic and
even $f(\vec{y} \restriction \gamma)$-good. It is clear that $q' = r \restriction \beta$, r is Borel, and so in order to
conclude the induction step I just have to verify that for every sequence $\vec{y} \in [q']$
the set $\{x \in X : \vec{y}^\frown x \in r\}$ is I-positive. But this is clear: this set is nonempty, it is
definable from parameters in the model $V[\vec{y}]$ and all of its elements fall out of all
I-small sets definable from such parameters, by the choice of the forcing P and the
name $\dot{x}$.

For the limit step, suppose that β is a limit of an increasing sequence of ordinals
$\langle \beta_n : n \in \omega \rangle$, $\gamma \in \beta_0$ is an ordinal and q is a I, γ-tree and $f : [q] \to P$ is a Borel function as in the assumption of the claim. Let $\langle O_n : n \in \omega \rangle$ be an enumeration of open
dense subsets of the poset P in the ground model V, and by use the inductive hypoth-
esis on the ordinals β_n repeatedly to construct a sequence $\langle q_n, f_n : n \in \omega \rangle$ such that

- q_n is an I, β_n-tree, $q = q_0 \restriction \gamma$ and $q_m = q_n \restriction \beta_m$ for $m \in n$;
- $f_n : [q_n] \to P$ are Borel functions such that for every sequence $\vec{y} \in [q_n]$ the condi-
  tions $f(\vec{y} \restriction \gamma) \geq f(\vec{y} \restriction \beta_0) \geq \cdots \geq f(\vec{y} \restriction \beta_n)$ form a descending chain in the poset P;
- every sequence $\vec{y} \in [q_n]$ is P-generic and in fact $f_n(\vec{y})$-good.

The construction is very easy to perform: at each number $n \in \omega$ first apply
the induction hypothesis at β_n to get a tree q_n as asserted in the claim, and then
for every sequence $\vec{y} \in B_n$ let $f_n(\vec{y})$ be some condition in the open dense set O_n
smaller than $f_{n-1}(\vec{y} \restriction \beta_{n-1})$ for which $\vec{y}$ is good, say the first condition with this
property in some fixed enumeration of the poset P.

In the end, let $r = \bigcup_n q_n$. It is clear that r is an I, β-tree. Moreover, every
sequence $\vec{y} \in [r]$ is P-generic and even $f(\vec{y} \restriction \gamma)$-good, as witnessed by the
V-generic filter $g \subset P$ obtained from the descending sequence $\langle f_n(\vec{y} \restriction \beta_n) : n \in \omega \rangle$
of conditions in the poset P.

The closure of the ideal I^α under well-ordered unions immediately follows.
Suppose that $\langle A_\beta : \beta \in \gamma \rangle$ is a well-ordered sequence of I^α-small sets. By a

standard argument I may assume that the sequence is definable from elements of the ground model V. Note that then every single set on the sequence is definable from such elements. The previous argument shows that Player I's strategy in the iteration game commanding him to play $\bigcup(I \cap OD^{V[\bar{y}]})$ whenever the play is in the position where Player II created a sequence $\bar{y}$, is winning for all the sets A_β simultaneously, therefore it is winning for their union. $\qquad \square$

5.1.3 Π_1^1 on Σ_1^1 ideals

The main virtue of the Π_1^1 on Σ_1^1 σ-ideals from the point of view of this book is that the countable support iteration dichotomy theorem is true for them in ZFC. This arose in my conversation with Vladimir Kanovey in April 2002.

Theorem 5.1.9. *Suppose that I is an iterable Π_1^1 on Σ_1^1 σ-ideal on a Polish space X and $\alpha \in \omega_1$ is a countable ordinal. Then every I^α-positive analytic set contains all branches of some I, α-tree.*

Proof. In order to simplify the notation, suppose that the ideal I is ligthface Π_1^1 on Σ_1^1, the space X is actually the Baire space ω^ω, and for every real z let U_z denote the $\Pi_1^1(z)$ set $\bigcup(\Sigma_1^1(z) \cap I)$ from Proposition 3.8.6.

Let $\alpha \in \omega_1$ be a countable ordinal, let a be a real number coding the ordinal α in a suitable fashion, and let b be an arbitrary real. Let $U_{a,b}^\alpha = \{\vec{r} \in X^\alpha : \exists \beta \in \alpha \ \vec{r}(\beta) \in U_{a,b,\vec{r} \restriction \beta}\}$. It is clear that $U_{a,b}^\alpha$ is a $\Pi_1^1(a,b)$ set in the ideal I^α. I will show that every $\Sigma_1^1(a,b)$ set of α-sequences of reals is either a subset of $U_{a,b}^\alpha$ (and so is I^α-small) or else it contains all branches of an I, α-tree. The argument proceeds by induction on the ordinal α for all reals a, b simultaneously.

First the successor stage of the induction. Let $A \subset X^{\alpha+1}$ be a $\Sigma_1^1(a,b)$ set which is not a subset of $U_{a,b}^{\alpha+1}$. Subtracting this $\Pi_1^1(a,b)$ set if necessary we may assume that A is a nonempty $\Sigma_1^1(a,b)$ set disjoint from $U_{a,b}^{\alpha+1}$. Consider the set $\bar{A} = \{\vec{x} \in X^\alpha : \exists y \in X \ \vec{x} ^\frown y \in A\} \subset X^\alpha$. It is not difficult to see that $\bar{A}$ is a nonempty $\Sigma_1^1(a,b)$ set disjoint from $U_{a,b}^\alpha$. Therefore, by the induction hypothesis it contains all branches of some I, α-tree $\bar{p}$. Also for every sequence $\vec{x} \in \bar{A}$ the set $\{y \in X : \vec{x} ^\frown y \in A\}$ is a nonempty $\Sigma_1^1(a,b,\vec{x})$ disjoint from the set $U_{a,b,\vec{x}}$, therefore I-positive, and thus contains a Borel I-positive subset. Consider the countable support iteration $(P_I)^{\alpha+1}$ of the poset P_I of length $\alpha + 1$; certainly $(P_I)^{\alpha+1} = (P_I)^\alpha * \dot{P}_I$. Define a condition p in this two step iteration by $p = \langle \bar{p}, \dot{B} \rangle$ where $\dot{B}$ is a name for an arbitrary Borel I-positive subset of the I-positive analytic set $\{y \in X : \vec{x}_{gen}^\frown y \in A\}$. Let M be a countable elementary submodel of a large enough structure containing the condition p, and use Theorem 5.1.6 to obtain an I, α-tree $q \subset X^{\alpha+1}$ all of whose branches are M-generic sequences meeting the condition p. For every sequence $\vec{x} \in [q]$,

$M[\vec{x}] \models \vec{x} \in A$ by the forcing theorem and $V \models \vec{x} \in A$ by the analytic absoluteness between V and $M[\vec{x}]$. Thus q is the desired $I, \alpha+1$-tree with $[q] \subset A$.

The limit step is trickier. Let $\alpha = \bigcup_n \beta_n$ be a limit of an increasing sequence of ordinals, and suppose that $A \subset X^\alpha$ is a $\Sigma_1^1(a, b)$ set, which is not a subset of $U_{a,b}^\alpha$. As in the previous paragraph we may assume that A is a nonempty set actually disjoint from $U_{a,b}^\alpha$. Let T be a tree recursive in a, b that projects into the set A under a suitable coding. For an integer n, a sequence $\vec{x} \in X^{\beta_n}$, and a node $t \in T$ of length n let $A(n, \vec{x}, t)$ denote the set $\{\vec{y} \in X^{\beta_{n+1}-\beta_n} :$ there is a branch through the tree T meeting the node t which gives rise to an α-sequence of which $\vec{x}^\frown\vec{y}$ is an initial segment$\}$. Several observations are in order. First, the set $A(n, \vec{x}, t) \subset X^{\beta_{n+1}-\beta_n}$ is $\Sigma_1^1(\vec{x}, a, b)$ and disjoint from the set $U_{\vec{x},a,b}^{\beta_{n+1}-\beta_n}$. Thus by the induction hypothesis, if the set $A(n, \vec{x}, t)$ is nonempty, then it is $I^{\beta_{n+1}-\beta_n}$-positive and contains all branches of some $I, \beta_{n+1} - \beta_n$-tree. Moreover, if the set $A(n, \vec{x}, t)$ is nonempty and $\vec{y}$ is its element, then there is a node $t' \in T$ of length $n+1$ extending the node t such that the set $A(n+1, \vec{x}^\frown\vec{y}, t')$ is nonempty. I must produce an I, α-tree q such that $[q] \subset A$. The countable support iteration of the forcing P_I of length α is (naturally isomorphic to) the full support iteration of length ω whose n-th iterand is the partial order P_n of Borel $I^{\beta_{n+1}-\beta_n}$-positive subsets of $X^{\beta_{n+1}-\beta_n}$ by the induction hypothesis and the iterability of the ideal I. Write Q_n for the n-step iteration of $P_m : m \in n$ and $\vec{x}_n \in X^{\beta_n}$ for its generic sequence. Define a condition $p = \langle \dot{p}_n : n \in \omega \rangle$ in the iteration of length ω, together with Q_n-names $\langle \dot{t}_n : n \in \omega \rangle$ for nodes in the tree T such that $p \upharpoonright n = \langle \dot{p}_m : m \in n \rangle \Vdash_{Q_n} \dot{t}_0 \subset \dot{t}_1 \ldots$ and $A(n, \vec{x}_n, \dot{t}_n) \neq 0$. This is easy to do by induction: just let $\dot{p}_n$ be a name for some $I, \beta_{n+1} - \beta_n$-tree inside the set $A(n, \vec{x}_n, \dot{t}_n)$, and $\dot{t}_{n+1}$ be a Q_{n+1}-name for a node in the tree $\check{T}$ extending $\dot{t}_n$ such that $p \upharpoonright n+1 \Vdash \langle \vec{x}_{n+1}, \dot{t}_{n+1} \rangle \in \check{T}$. Now let M be a countable elementary submodel of a large enough structure and use Theorem 5.1.6 to find an I, α-tree q all of whose branches are M-generic sequences meeting the condition p. For every sequence $\vec{x} \in [q]$ it is the case that $M[\vec{x}] \models \vec{x} \in \dot{A}$ as witnessed by the branch $\bigcup_n t_n \subset \check{T}$, and by analytic absoluteness $\vec{x} \in A$. Thus $[q] \subset A$ as desired. $\square$

Corollary 5.1.10. *P_{I^α} is in the forcing sense equivalent to the countable support iteration of P_I of length α.*

Corollary 5.1.11. *The ideal I^α is Π_1^1 on Σ_1^1, and every analytic I^α-positive set has a Borel I^α-positive subset.*

Proof. Let a be a real coding the ordinal α, and let b be an arbitrary real. The proof of the lemma shows that a $\Sigma_1^1(a, b)$ subset of X^α belongs to the ideal I^α if and only if it is a subset of the $\Pi_1^1(a, b)$ set $U_{a,b}^\alpha$, which is a uniformly $\Pi_1^1(a, b)$ condition. $\square$

As an interesting aside, note that the set $U_{a,b}^\alpha$ constructed in the previous proof, natural as it may be, is *not* equal to the union of all $\Sigma_1^1(a, b)$ sets in the ideal I^α. This union is actually properly smaller. To see this, consider the simplest case of $I = \text{ctble}$, the ideal of countable sets, $\alpha = 2$ and $a, b = 0$. Let $A \subset X \times X$ be a Σ_1^1 set with the following frequently used universality property: for every Σ_1^1 set $B \subset X \times X$ there is a point $x \in X$ such that the vertical sections A_x and B_x are equal. A moment's thought will show that then there must be actually uncountably many such points. Now let B be the union of all Σ_1^1 subsets of $X \times X$ which belong to the ideal I^2. This is a Σ_1^1 set, and by the definition of the ideal I^2, only countably many of its vertical sections are uncountable. Thus there must be a point $x \in X$ such that the vertical section B_x is countable, and is equal to the section A_x. Now A_x is a countable $\Sigma_1^1(x)$ set of reals. As such it countains only hyperarithmetic-in-x reals, but since the set of all such reals is properly $\Pi_1^1(x)$, it also misses some of them. Let y be a hyperarithmetic-in-r real which does not belong to the set A_x, and consider the pair $\langle x, y \rangle$. Since y is hyperarithmetic in x, $\langle x, y \rangle \in U_{00}^2$. And since $y \notin A_x = B_x$, the pair $\langle x, y \rangle$ does not belong to any Σ_1^1 set in the ideal I^2.

5.1.4 The universally Baire dichotomy

Now I will give a version of the dichotomy, which in the presence of large cardinals is applicable to essentially any definable proper forcing adding a single real.

Theorem 5.1.12. *(LC) Suppose that I is an iterable ideal on a Polish space X. Whenever $\alpha \in \omega_1$ is a countable ordinal and A is an I^α-positive universally Baire set then A contains all branches of some I, α-tree.*

Proof. There are several lines of argument available at this point. The shortest one will consider Proposition 3.9.13, Theorem 5.1.7, and universally Baire absoluteness argument. Essentially for sentimental reasons I will give a proof that uses determinacy of long games with real entries – this was the first proof I had in the spring of 2000. It also shows that inner models of ZF+DC+AD$\mathbb{R}$ containing all reals evaluate the ideal I^α correctly.

Let $A \subset X^\alpha$ be a universally Baire set. The problem is that in general Player I's moves in the iteration game $G(A)$ are rather arbitrary subsets of the space X; thus the game has entries more complex than reals and the determinacy results of Section 1.4.4 cannot be directly applied to it.

Consider a surrogate game $H(A)$. It is again played between Players I and II for α many rounds. This time, at round $\beta \in \alpha$, Player I starts, plays an I-positive Borel set $B_\beta \subset X$, Player II answers with a Borel I-positive subset $C_\beta \subset B_\beta$, and finally Player I chooses a point $x_\beta \in C_\beta$. Player I wins if the α-sequence of his answers belongs to the set A. Under a suitable coding of Borel sets by reals, this is a game of countable length with real entries and universally Baire payoff, and therefore one of the players has a universally Baire winning strategy by Fact 1.4.3. I will show that if Player I has a winning strategy in the game $H(A)$ then he has a winning strategy in $G(A)$, and if Player II has a winning strategy in the game $H(A)$ then the set A contains all branches of some I, α-tree. This will complete the proof. Thus there are two cases.

First, assume that Player I has a universally Baire winning strategy σ in the game $H(A)$; I must describe his winning strategy in the game $G(A)$. As the game $G(A)$ develops, he will simulate a play of the game $H(A)$ according to the strategy σ on the side so that the α-sequences of points in the space X obtained in the two plays are the same and:

• After Player II has completed his moves $\langle x_\gamma : \gamma \in \beta \rangle$ up to an ordinal $\beta \in \alpha$ Player I will have created the first β many rounds of the auxiliary run of the game $H(A)$ in such a way that the points played by Player II on the G and H side coincide.

• At the round β of the game G he will play the set $D_\beta = X \setminus \bigcup \{C \subset X :$ some Player II's challenge on the H-side makes the strategy σ answer with the set $C\}$. The set in this union is universally Baire and dense in the poset P_I, therefore the complement of its union is universally Baire and has no Borel I-positive subset. By the second dichotomy of the ideal I it is an I-small set, and a legal move for Player I on the G-side.

• After Player II responds with a real $r_\beta \notin D_\beta$, Player I will use the definition of the set D_β to find auxiliary moves $B_\beta \supset C_\beta \ni x_\beta$ on the H-side such that the move C_β follows the strategy σ, and adds them to his auxiliary run of the game $H(A)$.

In the end, Player I wins the run of the game $G(A)$ since he wins the auxiliary run of the game $H(A)$.

Second, assume that Player II has a universally Baire winning strategy σ in the game $H(A)$. Write R_β for the countable support iteration of the forcing P_I of length β, adding a generic β-sequence $\vec{y}_{gen} \in X^\beta$. By induction on $\beta \leq \alpha$ build conditions $p_\beta \in R_\beta$ and R_β-names t_β for partial plays of the game $H(A)$ of length β observing the strategy σ which result in $\vec{y}_{gen}$ as a sequence of Player II's answers, such that $\gamma \in \beta$ implies $p_\gamma = p_\beta \restriction \gamma$ and $p_\beta \Vdash t_\gamma = t_\beta \restriction \gamma$. After the induction has been performed, just use Theorem 5.1.6 to find a countable elementary submodel M of a large enough structure containing I, p_α, A, σ and an I, α-tree q whose branches are M-generic sequences meeting the condition p_α. By the forcing theorem applied in the model M, for every sequence $\vec{y} \in [p] \; M[\vec{y}] \models t_\alpha / \vec{y}$ is a play along the strategy

σ which results in the sequence $\vec{y}$. By the universally Baire absoluteness, $t_\alpha/\vec{y}$ is really such a play, and since the strategy σ was winning for Player II, $\vec{y} \in A$. Thus $[q] \subset A$ is the as required.

To perform the induction, note that at limit stage β there is no freedom of choice: $p_\beta = \bigcup_{\gamma \in \beta} p_\gamma$ and t_β is the name for the union of $t_\gamma : \gamma \in \beta$. At the successor stage $\beta = \gamma + 1$ move into a generic extension $V[G_\gamma]$ where $G_\gamma \subset R_\gamma$ is a generic filter meeting the condition p_γ. Let $B_\beta \subset X$ be the Borel I-positive set dictated to Player II as his move after the run t_γ by the strategy γ. Let $D_\beta = \{x \in X : \text{there is Player}$ I's move $C \subset B_\beta$ which makes the strategy σ answer with $x\}$. Since the strategy σ is universally Baire, even the set D_β is. Also, $D_\beta \notin I$, in fact $B_\beta \setminus D_\beta \in I$. To see this, note that the universally Baire set $B_\beta \setminus D_\beta$ does not have a Borel I-positive subset, since such a set could be used as Player I's move in the definition of the set D_β to produce an element of the obviously empty set $D_\beta \cap (B_\beta \setminus D_\beta)$; and use the second dichotomy of the ideal I. Now $D_\beta \subset B_\beta$ is a universally Baire I-positive set, and so by the second dichotomy of the ideal I it has a Borel I-positive subset E_β. Let $p_\beta = p_\gamma^\frown \dot{E}_\beta$, and let t_β be a name for a partial run of the game $H(A)$ following the strategy σ of the form $t_\beta = t_\gamma^\frown \dot{B}_\beta^\frown \dot{C}^\frown \dot{x}$ where $\dot{x}$ is the γ-th generic point. Since $p_\beta \Vdash \dot{x} \in \dot{E}_\beta \subset \dot{D}_\beta$, such a run exists by the definition of the set D_β and universally Baire absoluteness. This completes the inductive step. $\qquad\square$

The dichotomy has corollaries similar to the previous one.

Corollary 5.1.13. *(LC) $P_{I\alpha}$ is in the forcing sense equivalent to the countable support iteration of P_I of length α.*

Corollary 5.1.14. *(LC) Suppose that $A_y : y \in 2^\omega$ is a universally Baire collection of sets in the ideal I^α. Then there is a universally Baire system of strategies $\sigma_y : y \in 2^\omega$ witnessing this fact.*

The following corollary is really a part of the proof of the theorem.

Corollary 5.1.15. *(LC) Suppose that $r \subset X^{<\alpha}$ is a universally Baire tree such that every countable inclusion increasing sequence of nodes in r whose lengths are bounded below α has an upper bound in r and every nonterminal node of r branches into I-positively many immediate successors. Then there is an I, α-tree p such that $p \subset r$.*

5.1.5 Laver forcing and large cardinals

The previous section is the real reason why the large cardinal assumptions figure so prominently in this book. In order to prove that the dichotomy "every I^α-positive analytic set contains all branches of some I, α-tree" holds in some model of set theory, it is enough to move to the $\mathrm{Coll}(\omega, < \kappa)$ extension for some inaccessible cardinal κ and use Theorem 5.1.7. However, in order to show that the dichotomy holds in every forcing extension (and that is how the dichotomy is used in the main applications in Section 6.1), it is necessary to resort to either an absoluteness argument or a determinacy argument such as the one in the proof of Theorem 5.1.12. This increases the large cardinal strength of the dichotomy for some ideals I. In view of Theorem 5.1.9 and Proposition 3.8.15, the simplest candidate for such an increase in large cardinal strength is the Laver ideal. This section shows that indeed, the validity of the dichotomy for the Laver ideal in all forcing extensions requires large cardinals already for the case $\alpha = 2$.

Theorem 5.1.16. *Let I be the Laver ideal on ω^ω. The following are equivalent:*

1. *for every set x, $x^\#$ exists;*
2. *in every generic extension, every analytic subset of $(\omega^\omega)^2$ is either in the ideal I^2 or it contains an I, 2-tree.*

Recall from Section 4.5.2 that the Laver ideal I is generated by the sets $A_g = \{f \in \omega^\omega : \text{for infinitely many } n \in \omega, f(n) \in g(f \upharpoonright n)\}$ where $g : \omega^{<\omega} \to \omega$. Such functions g will be referred to as *predictors* and if $f \in A_g$ then we will say that the predictor g *captures* the function f.

Proof. Both directions are mildly challenging. For (1)→(2), I will need the following claim.

Claim 5.1.17. *Suppose that analytic determinacy holds. Every Σ^1_2 set $A \subset \omega^\omega$ is either in the Laver ideal or it contains all branches of a Laver tree.*

Proof. Write $A = \mathrm{proj}(B)$ for some coanalytic set $B \subset \omega^\omega \times 2^\omega$. Consider the game G in which Player II starts with finite sequences $t \in \omega^{<\omega}$ and $s \in 2^{<\omega}$ of the same length, and after that, Player I and II alternate, Player I indicates a number n_i and Player II responds with a number $m_i > n_i$ and a bit b_i. In the end, write $x = t^\frown m_0^\frown m_1^\frown \cdots \in \omega^\omega$ and $y = s^\frown b_0^\frown b_1^\frown \cdots \in 2^\omega$. Player II wins if $\langle x, y \rangle \in B$. This is a game with analytic payoff for Player I and therefore it is determined. The claim will follow once I show that if Player I has a winning strategy then $A \in I$, and if Player II has a winning strategy then A contains all branches of a Laver tree. This is a quite standard argument though, essentially literally contained in [4]. $\qquad\square$

Now suppose that sharps exist, move to an arbitrary generic extension and fix an analytic set $A \subset (\omega^\omega)^2$. Consider the set $A' = \{x \in \omega^\omega : A_x \notin I\}$. The Laver ideal is ZFC-correct by the last item of Theorem 4.5.6 and the construction in Section 4.5.2. By Proposition 2.1.23, the set $A' \subset \omega^\omega$ is Σ^1_2. Analytic determinacy holds and therefore the set A' is either in the ideal I or it contains all branches of some Laver tree $T \subset \omega^{<\omega}$. In the former case clearly $A \in I^2$. In the latter case, let me first argue that $T \Vdash A_{\dot{x}_{gen}} \notin I$. To see this, let M be a countable elementary submodel of a large structure and $x \in T$ be an M-generic point for the Laver forcing. Then $A_x \notin I$ and since the Laver ideal is ZFC-correct, $M[x] \models A_x \notin I$. The forcing theorem and an elementarity argument then show that $T \Vdash A_{\dot{x}_{gen}} \notin I$. Let $C \subset (\omega^\omega)^2 \times \omega^\omega$ be a closed set projecting to A. By the continuous reading of names for Laver forcing in the Laver extension, $T \Vdash$there is a Laver tree $\dot{S} \subset \omega$ and a continuous function $\dot{f} : S \to \omega^\omega$ such that $\forall y \in [S] \ \langle \dot{x}_{gen}, y, f(y) \rangle \in C$. Let M be a countable elementary submodel of a large structure, find a Laver tree $T' \subset T$ consisting of M-generic points for the Laver forcing, and for each point $x \in [T']$ consider the Laver tree $\dot{S}/x \subset \omega^{<\omega}$. By a wellfoundedness absoluteness argument in the model $M[x]$, for every branch $y \in [\dot{S}/x]$ it is the case that $\langle x, y, (\dot{f}/x)(y) \rangle \in C$ and therefore $\langle x, y \rangle \in A$. Thus, the set A contains all the branches of the $I, 2$-tree p defined by $\langle x, y \rangle \in p \leftrightarrow x \in T \wedge y \in \dot{S}/x$.

To show $(2) \to (1)$, I will argue for the contrapositive. Since there is a set a whose sharp does not exist, by the covering theorem there must be a cardinal $\kappa > |a|$ of cofinality ω whose successor is computed correctly in $L[a]$. I will prove that in a suitable generic extension collapsing κ and preserving its successor, the dichotomy fails. This follows from three quite independent observations, interesting in their own right.

Claim 5.1.18. *There is a Borel set $B \subset 2^\omega \times \omega^\omega$ such that the set $\{x \in 2^\omega : B_x \notin I\}$ is Π_1^1-complete.*

This follows from the proof of Proposition 3.8.15 and the fact that Laver forcing adds a dominating real.

Claim 5.1.19. *Suppose that $x \in 2^\omega$ and $\omega^\omega \cap L[x]$ is a dominating set. Then there is a $\Pi_1^1(x)$ subset of ω^ω which is I-positive and contains no Laver subtree.*

Proof. Choose an infinite binary sequence $x \in 2^\omega$ and consider the set A of all functions $z \in \omega^\omega$ such that there exists a predictor $g \in L_{\omega_1^{\dot{z}}}[x]$ which captures the function z.

First, the set A is $\Pi_1^1(x)$: $z \in A$ if and only if there exists a linear ordering o recursive in z which is a wellordering and such that every model of $V = L[x]$ whose ordinals are isomorphic to o contains a predictor capturing z.

Second, if the reals of $L[x]$ are dominating, then the set A is I-positive. For choose an arbitrary predictor g. Fix a recursive partition $\{C_n : n \in \omega\}$ of ω into infinite sets, and using the assumption, find a predictor $g' \in L[x]$ such that for every sequence $t \in \omega^{<\omega}$, $g(t) \in g'(t)$ and moreover $g'(t)$ contains some element of the set C_0 not in $g(t)$. Now it is easy to construct a function $z \in \omega^\omega$ such that for all numbers n $z(n) \notin g(z \restriction n)$, $z(n) \in g'(z \restriction n) \leftrightarrow z(n) \in C_0$ and there are infinitely many such numbers n, and the set $\{m > 0 : \operatorname{rng}(z) \cap C_m \neq 0\}$ is an arbitrary subset of $\omega \setminus \{0\}$, in particular a set coding an ordinal α such that $g' \in L_\alpha[x]$. Thus the function z avoids the predictor g and belongs to the set A as witnessed by the predictor g'.

Finally, the set A does not contain a Laver-perfect subset. Suppose T is a Laver tree. Let M be a countable elementary submodel of a large enough structure containing x and T, and let $z \in T$ be an M-generic Laver real. Clearly, $\omega_1^z \in \omega_1^M = \omega_1^{M[G]}$ and so $L_{\omega_1^z}[x] \subset M$. Thus the function z avoids all the predictors in $L_{\omega_1^z}[x]$ and as such it is a branch of the tree T which does not belong to the set A, proving the claim. $\square$

The set A deserves a further comment. Its variations for the countable ideal and the bounded ideal are the largest $\Pi_1^1(x)$ sets without a perfect (superperfect, respectively) subset. However, there is no largest $\Pi_1^1(x)$ set without a Laver-perfect subset. For suppose Y is such a set. Then for every tree $T \subset \omega^{<\omega}$ recursive in the real x, T is wellfounded iff the $\Delta_1^1(x)$ set B_T is not in the ideal I (where the set B is defined in Claim 5.1.18) iff $B_T \setminus Y \neq 0$ (since Borel I-positive sets have Laver subtrees). However, this gives a $\Sigma_1^1(x)$ definition of the properly $\Pi_1^1(x)$ set of wellfounded trees recursive in x, contradiction.

Claim 5.1.20. *Suppose that there is a set whose sharp does not exist. Then there is a forcing extension in which there exists a real x such that $\omega^\omega \cap L[x]$ is dominating.*

Proof. This is a rather standard covering lemma argument. Let a be a set whose sharp does not exist. Let $\kappa > |a|$ be a cardinal which is a limit of successive cardinals $\kappa_n : n \in \omega$; the model $L[a]$ computes the successor cardinal κ^+ correctly. Collapsing the minimum of the cardinals $\kappa_n : n \in \omega$ to $\aleph_0$ I may assume that in fact $\kappa = \aleph_\omega$. Let $\{C_n : n \in \omega\}$ be a partition of ω into infinite sets. Consider the partial order P consisting of all trees T whose nodes are partial functions from ω to κ such that (i) every function $t \in T$ has domain some $m \in \omega$ and maps $C_n \cap m$ into κ_n in an order-preserving way for all $n \in \omega$, and (ii) for all $t \in T$, all $n \in \omega$ and all $\alpha \in \kappa_n$ there is $s \in T$ such that $t \subset s$ and $\alpha \in \max s''C_n$. The set P is ordered by inclusion. It is quite clear that the forcing P adds a function $\dot{f}_{gen} : \omega \to \kappa$ such that $\dot{f}_{gen} \restriction C_n$ is an increasing cofinal sequence in κ_n, and so $P \Vdash |\check{\kappa}| = \aleph_0$. Let $\dot{x}$ be a name for the real in the generic extension which codes a wellordering on ω of ordertype $\check{\kappa}$, together with the function $\dot{f}_{gen}$ and the set a. I will show that the reals of $L[x]$ are dominating in the generic extension.

Well, assume that $T \in P$ and $T \Vdash \dot{g} \in \omega^\omega$ is a function. A standard argument can be employed to give a fusion sequence $T = T_0 \supset T_1 \supset T_2 \supset \ldots$ together with sets $A_0, A_1, \ldots$ so that

- $A_0 = \{0\}$, for each number n there is k_n such that A_n is exactly the set of all nodes of the tree T_n with domain k_n;
- whenever $n > 0$ then the set A_n has size $\aleph_n$ and for every node $t \in A_n$ the tree $T_n \upharpoonright t$ decides the value of $\dot{f}_{gen}(\dot{g}(\check{n}+1)$-th element of the set $\check{C}_{n+1})$;
- for every number n, every node $t \in A_n$, every $m \in n$ and every $\alpha \in \omega_n$ there is a node $s \in A_{n+1}$ such that $\alpha \in \max s'' C_m$.

Let $S = \bigcap_n T_n$. The third item shows that $S \in P$ and the second item provides for a function $h \in \prod_n \omega_n$ such that $S \Vdash \forall n > 1$ $\dot{f}_{gen}(\dot{g}(\check{n})$-th element of the set $\check{C}_n) \in \check{h}(n)$. By the covering lemma, there is a set $b \in L[a]$ of size $\aleph_1$ covering the range of the function h. Let $\dot{e} \in \omega^\omega$ be a function in the extension defined by $\dot{e}(\check{n}) =$ the least number m such that $\sup(\check{b} \cap \omega_n) \in \dot{f}_{gen}(m$-th element of the set $C_n)$, for $n > 1$. Then clearly the condition S forces $\dot{e} \in L[\dot{x}]$ and for all $n > 1$, $\dot{e}(n) > \dot{g}(n)$ as desired, proving the claim. □

The theorem now easily follows. Suppose that there is a set whose sharp does not exist. Then there is a forcing extension $V[G]$ containing a point $x \in 2^\omega$ such that $\omega^\omega \cap L[x]$ is dominating. Work in $V[G]$. Let $A \subset \omega^\omega$ be a bad $\Pi^1_1(x)$ set from Claim 5.1.19 and let $B \subset 2^\omega \times \omega^\omega$ be a bad Borel set from Claim 5.1.18. Since the set $\bar{B} = \{x \in R : B_x \notin I\}$ is Π^1_1-complete, there is a Borel function $f : \omega^\omega \to 2^\omega$ reducing A to $\bar{B}$, that is, $y \in A$ iff $f(y) \in \bar{B}$. Let $C \subset \omega^\omega \times \omega^\omega$ be the set of all pairs $\langle y, z \rangle$ such that $\langle f(y), z \rangle \in B$. I claim that the set C witnesses the statement of the theorem.

First of all, the set C is I^2-positive, and in fact Player II has a winning strategy in the iteration game of length 2. If Player I plays some I-small set B_0 in the first round, then II can choose a function $y_0 \in A \setminus B_0$ since the set A is I-positive. In the second round, after Player I's move B_1 Player II can choose a point $z \notin B_1$ such that $\langle f(y), z \rangle \in B$ since the set of all such points is positive by the definitions.

Second, there is no Borel I-perfect subset $D \subset C$. If D were such a subset, then $D \upharpoonright 1 = \{y \in \omega^\omega : \exists z \, \langle y, z \rangle \in D\}$ would be an I-positive analytic subset of the set A. However, analytic I-positive sets have Laver-perfect subsets, while the set A has no such a subset. □

5.2 Side-by-side product

The operation of side-by-side product of partial orders is much harder to handle than that of the countable support iteration. The basic questions remain open in their full generality.

Question 5.2.1. Suppose P, Q are universally Baire proper forcings. Is $P \times Q$ proper?

Question 5.2.2. Suppose I, J are σ-ideals on Polish spaces X, Y such that the forcings P_I, P_J are proper. Find a description of an ideal K on $X \times Y$ such that $P_I \times P_J$ is in the forcing sense equivalent to P_K, without using the forcing relation.

While these two question may sound disjoint, in fact they should be considered together. Regarding the second question, there are several natural candidates for the side-by-side product of ideals:

Definition 5.2.3. *Suppose that I, J are σ-ideals on Polish spaces X, Y respectively. The symbol $I \times J$ stands for the collection of those Borel sets $D \subset X \times Y$ which do not contain an positive Borel rectangle: a set of the form $B \times C$ where $B \subset X, C \subset Y$ are Borel I-positive, resp. J-positive sets. If $I \times J$ is a σ-ideal then I will say that the ideals I, J have the* rectangular Ramsey property, $MRR(I, J)$. *Similarly for a product of countably many ideals.*

The problem with the above definition is that in many cases it does not yield a σ-ideal as investigated in Section 3.11. Nevertheless, if $I \times J$ is a σ-ideal, then clearly $P_I \times P_J$ is naturally isomorphic to a dense subset of $P_{I \times J}$. We will see that this is indeed frequently the case. The investigation of the forcing properties of the product is then an exercise in rectangular Ramsey theory and canonization theorems. There are some natural competing candidates for the definition of the cross-product ideal.

Definition 5.2.4. *Suppose that I, J are σ-ideals on Polish spaces X, Y. The σ-ideal $I \otimes J$ on $X \times Y$ is σ-generated by Borel sets of two kinds: the Borel sets with all vertical sections in the ideal J, and the Borel sets with all horizontal sections in the ideal I.*

The comparison of the cross-products $I \times J$, $I \otimes J$, and $P_I \times P_J$ is quite difficult in general.

Fact 5.2.5. *[83], 2.3.57. Let I be the E_0-ideal on 2^ω. Then $I \times I = I \otimes I$ [65]. Let J be the ideal of countable sets. Then $J \times J \neq J \otimes J$.*

Neither statement is easy to prove. The proof of the first sentence uses effective descriptive set theory. To prove the second sentence, Shelah claims in [65], Section 4 to have a consistent example of an F_σ-set containing an $\aleph_2 \times \aleph_2$ rectangle but no rectangle with perfect sides. Such a set is in the ideal $J \times J$ but not in $J \otimes J$.

5.2.1 Preservation of category basis

There is at least one class of cases in which I have essentially full information.

Theorem 5.2.6. *[86] (LC) Suppose that $I_n : n \in \omega$ is a collection of universally Baire ideals whose quotient forcings are proper and preserve category basis. Then their countable support side-by-side product is a category basis preserving proper forcing and the collection $I_n : n \in \omega$ has the rectangular Ramsey property. If the σ-ideals $P_{I_n} : n \in \omega$ are $\mathbf{\Pi}^1_1$ on $\mathbf{\Sigma}^1_1$ then the large cardinal assumption is not necessary. In the latter case the product ideal is $\mathbf{\Pi}^1_1$ on $\mathbf{\Sigma}^1_1$ as well and it satisfies the third dichotomy.*

The case of uncountable side-by-side product is then handled using the techniques of Section 5.5.

Proof. Given the results of Section 3.10.10, this is just a simple fusion argument. For simplicity I will deal with the product of two forcings P_I and P_J, with σ-ideals I, J on Polish spaces X, Y respectively. I will call a set $B \times C \subset X \times Y$ where $B \in P_I$ and $C \in P_J$ a *block*. Consider the game G on the product poset defined in a manner similar to the category base game: first Player II indicates a block p_{ini} and then the two players alternate, Player I indicating a block p_n and Player II answering with a block $q_n \subset p_n$. Moreover, before some rounds $n_0 \in n_1 \in \ldots$ Player I raises a flag; we agree that $n_0 = 0$ by default. Player I wins if the result of the play, the set $\{\langle x, y \rangle \in p_{ini} : \forall i \in \omega \exists n \ n_i \leq n < n_{i+1} \wedge \langle x, y \rangle \in q_n\}$ contains a block.

Claim 5.2.7. *Player I has a winning strategy in the game G.*

Proof. Use Theorem 3.10.24 to find winning strategies σ, τ for Player I in the category base games associated with the posets P_I, P_J. I will fuse them to obtain a winning strategy η for Player I in the game G. Write $p_{ini} = B_{ini} \times C_{ini}$, $p_n = B_n \times C_n$ and $q_n = B'_n \times C'_n$ for the blocks played during the game G. The strategy η is uniquely given by the following demands on the moves it produces:

- the play $b = B_{ini}, B_0, B'_0, B_1, B'_1, \ldots$ follows the strategy σ. Let $0 = l_0 < l_1 < l_2 < \ldots$ be the natural numbers such that the strategy σ calls stops to rounds in the play b just before playing $B_{l_1}, B_{l_2}, \ldots$
- the play $c = C_{ini}, C_0, C'_{l_1-1}, C_{l_1}, C'_{l_2-1}, C_{l_2}, \ldots$ follows the strategy τ;
- for every number $k \notin \{l_i : i \in \omega\}$ the set C_k is just C'_{k-1}; this implies that moves of Player II in the play c are legal in the category basis game;
- let $0 = m_0 < m_1 < m_2 < \ldots$ be the natural numbers such that the strategy τ raises the flag in the play c just before playing $C_{m_1}, C_{m_2}, \ldots$. Thus $\{m_j : j \in \omega\} \subset$

$\{l_i: i \in \omega\}$. Player II raises the flag in the play of the game G just before the moves indexed by $m_j: j \in \omega$.

It is not difficult to see that if $B \in P_I$ is the result of the play b and $C \in P_J$ is the result of the play c, then the result of the play of the game G observing the above demands contains the block $B \times C$ and therefore Player I won.   □

It now follows immediately that $I \times J$ is a σ-ideal. Let $B \times C \subset X \times Y$ be a Borel rectangle with positive sides, decomposed into a countable union $B \times C = \bigcup_m D_m$ of Borel sets. It is enough to show that one of the sets D_m contains a rectangle with positive sides.

Write $\dot{x}_{gen}, \dot{y}_{gen}$ for the $P_I \times P_J$-generic pair of points. Note that $B \times C \Vdash \dot{x}_{gen} \in B, \dot{y}_{gen} \in C$, since these points are P_I and P_J-generic respectively. By an absoluteness argument, there is a rectangle $B_0 \times C_0 \subset B \times C$ which forces $\langle \dot{x}_{gen}, \dot{y}_{gen} \rangle \in \dot{D}_m$ for some definite number $m \in \omega$. I claim that the set D_m contains a Borel rectangle with positive sides.

To see this, let M be a countable elementary submodel of a large enough structure containing the rectangle $B_0 \times C_0$, the set D_m, as well as a winning strategy σ for Player I in the game G. Enumerate all open dense subsets of the poset $P_I \times P_J$ in the model M as $\{E_i: i \in \omega\}$ and simulate a run x of the game G against the strategy σ with the initial move $B_0 \times C_0$ and such that all its moves are in the model M and during the i-th round Player II chooses only sets from the open dense set E_i. The result of the game contains some rectangle $B_1 \times C_1$. I claim that $B_1 \times C_1 \subset D_m$; this will complete the proof.

Let $x \in B_1, y \in C_1$ be points. It is easy to check that the collection of all rectangles in the model M containing this pair is a filter on the poset $(P_I \times P_J) \cap M$. By the simulation above, this filter is M-generic. By the forcing theorem applied in the model M, $M[x, y] \models \langle x, y \rangle \in D_m$, and by an absoluteness argument $\langle x, y \rangle \in D_m$. So $B_1 \times C_1 \subset D_m$ as desired.

Finally I am ready to argue that $P_I \times P_J$ is a proper forcing preserving category basis. It is clear that the poset $P_I \times P_J$ naturally densely embeds into $P_{I \times J}$, the game G is just the category game under another name, and Player I has a winning strategy in it. A reference to Theorem 3.10.21 concludes the argument.

To handle the ZFC case for $\mathbf{\Pi}_1^1$ on $\mathbf{\Sigma}_1^1$ ideals, first observe that Theorem 3.10.24 provides the winning strategies in the category basis game for Player I in ZFC, and therefore the above argument can be performed without any reference to large cardinals. For the third dichotomy note that the above argument in fact shows that the collection K of analytic subsets of $X \times Y$ containing no Borel rectangle with positive sides is a σ-ideal. By Theorem 3.8.9 this is a $\mathbf{\Pi}_1^1$ on $\mathbf{\Sigma}_1^1$ ideal.

Every analytic set in a $\mathbf{\Pi}_1^1$ on $\mathbf{\Sigma}_1^1$ ideal has a Borel superset in the ideal, therefore $K = I \times J$ and the theorem follows. $\square$

5.2.2 Asymmetric theorems

The asymmetric product theorems (those in which there are different assumptions on the ideals entering the product) offer a veritable labyrinth of possibilities. The reader should look back to Section 3.11 to see that the rectangular Ramsey properties fail in a number of cases. Very many cases remain open though.

Theorem 5.2.8. *(LC) Suppose that I, J are universally Baire ideals on the respective Polish spaces X, Y such that the forcings P_I, P_J are proper and P_I preserves category bases and P_J preserves Baire category. Then the forcing $P_I \times P_J$ is proper, it preserves Baire category, and $\mathrm{MRR}(I, J)$ holds.*

Proof. As in the previous proof, call a Borel rectangle $B \times C \subset X \times Y$ a *block* if $B \notin I$ and $C \notin J$. Consider the game G between Players I and II in which Player II starts and indicates some block p_{ini}, and then the two players alternate, Player I indicating blocks p_n and Player II responding with blocks $q_n \subset p_n$. Player I wins if the result of the game, the set $\{\langle x, y \rangle \in p_{ini} : \exists^\infty n \; \langle x, y \rangle \in q_n\}$, contains a block. I will show that Player I has a winning strategy in this game. The theorem then follows by the same argument as in the previous section.

Use Theorem 3.10.24 to find a winning strategy σ for Player I in the category basis game associated with the forcing P_I. Use Theorem 3.10.21 to find a winning strategy τ for Player I in the category game associated with the forcing P_J. The winning strategy η for Player I in the game G is a fusion of the strategies σ, τ. Writing $p_{ini} = B_{ini} \times C_{ini}$, $p_n = B_n \times C_n$, and $q_n = B'_n \times C'_n$, the strategy η is fully determined by the following demands:

- the sets $B_{ini}, B_0, B'_0, B_1, B'_1, \ldots$ form a play b in the category base game observing the strategy σ;
- writing $0 = n_0 \in n_1 \in n_2 \in \ldots$ for the rounds after which Player I raised the flag in the play b of the category base game, the play $C_{ini}, C_0, C'_{n_1}, C_{n_1+1}, C'_{n_2}, C_{n_2+1} \cdots$ form a play c of the category game observing the strategy τ;
- at rounds indexed by $m \notin \{n_i : i \in \omega\}$ Player I puts $C_{m+1} = C'_m$–this implies that the moves of Player II in the previous item were legal in the category game.

In the end, let $B \in P_I$ be the result of the play b and let $C \in P_J$ be the result of the play c. It is immediate that the result of the play of the game G in which

Player I follows the above demands contains the block $B \times C$ and therefore Player *I* wins. □

Note that the statement parallel to the previous theorem only replacing category with Lebesgue measure is false. Consider the product of Laver and Silver forcing. Laver forcing preserves outer Lebesgue measure and Silver forcing preserves measure basis. However, Silver forcing adds a splitting real and therefore the rectangular property fails as shown in Example 3.11.13. Nevertheless, the product forcing *is* proper, since Laver forcing is embeddable into a σ-closed*σ-centered iteration.

Theorem 5.2.9. *(LC) Suppose that I is a universally Baire σ-ideal on a Polish space X, ϕ a capacity on a Polish space Y and*

1. *P_I preserves Baire category;*
2. *ϕ is either subadditive outer regular stable capacity or a pavement capacity;*
3. *$P_I \Vdash$ every subset of the space Y can be covered by a ground model coded set of arbitrarily close capacity.*

Then $\mathrm{MRR}(I, I_\phi)$ *and the forcing* $P_{I_\phi} \times P_I$ *is proper. Moreover, if the forcing P_I is bounding then so is the cross-product.*

The large cardinal assumption can be eliminated in a large class of cases. I need a winning strategy for Player I in the Baire category game, which is available in ZFC if the forcing is in addition bounding and the ideal is $\mathbf{\Pi}^1_1$ on $\mathbf{\Sigma}^1_1$ by Theorem 3.10.21. I also need the capacity ϕ to remain continuous in increasing unions in the extension, which is implied by ZFC-correctness of the capacity ϕ, or typically it follows from a formal ZFC proof of the continuity.

Proof. First argue that for every set $B \in P_{I_\phi}$ in the ground model and every open dense set $D \subset P^V_{I_\phi}$ in the P_I-extension there is a countable set $D' \subset D$ consisting of subsets of B such that $\phi(B) = \phi(\bigcup D')$. This is a consequence of the second and third items only. Work in the P_I-extension, let M be a countable elementary submodel of a large enough structure, and let $D' = \{C \in D \cap M : C \subset B\}$; I will prove that the set $D' \subset D$ is as required.

Suppose first that ϕ is a pavement submeasure and for contradiction assume that $\phi(B) > \phi(\bigcup D')$. Use the third item to find a ground model set $a \subset U$ of pavers such that $\bigcup D' \subset \bigcup a$ and $\Sigma_{u \in U} w(u) < \phi(B)$, where U is the countable collection of Borel pavers and w the associated weight function used to build the capacity ϕ. Find a set $C \in D$ below the ground model coded ϕ-positive set $B \setminus \bigcup a$. Find a finite set $a' \subset a$ such that $\Sigma_{u \in a \setminus a'} w(u) < \phi(C)$. Use an elementarity argument to show that there is in the model M a set $C' \in D$, $C' \subset B \setminus \bigcup a'$, $\phi(C') > \Sigma_{u \in a \setminus a'} w(u)$. The set $C' \in D'$ is not covered by the set $\bigcup a$, a contradiction.

The proof for the case of subadditive outer regular stable capacity ϕ is similar. Suppose for contradiction that $\phi(B) > \phi(\bigcup D')$. Use the third item to find an open set O in the ground model such that $\bigcup D' \subset O$ and $\phi(O) < \phi(B)$ and find a set $C \in D$ below the ϕ-positive ground model set $B \setminus \tilde{O}$. The basic property of the tilde operation implies $\phi(O \cup C) > \phi(O)$. Use the continuity of the capacity ϕ in increasing unions to find a basic open set $P \subset O$ such that $\phi(P \cup C) > \phi(O)$. Use an elementarity argument to find a set $C' \in M \cap D$ such that $\phi(P \cup C') > \phi(O)$. Then the set $C' \in D'$ is not covered by the set O, a contradiction.

In both cases note that in the P_I-extension, since the function ϕ remains a capacity, for every condition $B \in P_{I_\phi}^V$ and every open dense set $D \subset P_{I_\phi}^V$ and every $\varepsilon > 0$ there is a finite collection $D' \subset D$ consisting of subsets of B such that $\phi(B) < \phi(\bigcup D') + \varepsilon$. To restate this in the context of the product forcing, for every condition $\langle A, B \rangle \in P_I \times P_{I_\phi}$ and every open dense set $E \subset P_I \times P_{I_\phi}$ and every real number $\varepsilon > 0$ there is a condition $A' \subset A$ and a finite collection D' of subsets of B such that $\phi(B) < \phi(\bigcup D') + \varepsilon$ and for every $B' \in C'$ it is the case that $\langle A', B' \rangle \in E$.

The proofs of properness and the rectangular Ramsey property are now handled in one scoop. Suppose A, B are Borel I, I_ϕ-positive subsets of the spaces X and Y respectively, let $f : A \times B \to \omega$ be a Borel function, and let M be a countable elementary submodel of a large enough structure. I will find Borel I, I_ϕ-positive respectively sets $A_\omega \subset A$ and $B_\omega \subset B$ such that the rectangle $A_\omega \times B_\omega$ consists of M-generic points and the function f is constant on it. To do this, work in the model M and find an I-positive Borel set $A_0 \subset A$ and a ϕ-positive compact set $B_0 \subset B$ such that the pair $\langle A_0, B_0 \rangle$ forces in the product forcing the f-value of the generic pair to be some fixed number $n \in \omega$, and find a winning strategy σ for Player I in the category game associated with the forcing P_I. Enumerate the open dense subsets of the product forcing in the model M by $E_i : i \in \omega$, let $q_i : i \in \omega$ be a decreasing sequence of rational numbers smaller than $\phi(B_0)$ which converges to a nonzero number, and construct Borel I-positive sets $A_i \subset X, A_i \in M$ and Borel I_ϕ-positive sets $B_i \subset Y, B_i \in M$ such that

- there are sets $A_i' : i \in \omega$ such that the sequence $p_{ini} = A_0, A_1', A_1, A_2', A_2 \ldots$ is a play according to the strategy σ, in particular $A_i \subset A_i'$;
- $B_0 \supset B_1 \supset \ldots$ are compact sets and for every number $i \in \omega$ it is the case that $\phi(B_i) > q_i$ and there is a finite collection $D_i \in M$ such that $B_i = \bigcup D_i$ and for every set $C \in D_i$, $\langle A_i, C \rangle \in \bigcap_{j \in i} E_j$.

The previous paragraph shows that the inductive step of the construction can be performed. In the end, let $A_\omega = \{x \in A_0 : \exists^\infty n \ x \in A_n\}$ and $B_\omega = \bigcap_n B_n$. The set A_ω is the result of the play following the strategy σ constructed in the previous paragraph, so it is I-positive. The set B_ω has capacity $\geq \inf_i q_i$ since it is an intersection of a decreasing collection of compact sets of capacity at least that. Every point $\langle x, y \rangle \in A_\omega \times B_\omega$ is M-generic for the product poset as the construction immediately shows. By the forcing theorem then, for every point $\langle x, y \rangle \in A_\omega \times B_\omega$ it is the case that $M[x, y] \models f(x, y) = n$ and by an absoluteness argument $f(x, y) = n$. This concludes the proof of properness and the rectangular Ramsey property.

Finally, the above proof shows that the forcing $P_{I_\phi}^V$ is bounding in the P_I-extension. Thus if the forcing P_I is bounding then so is the cross-product.

$\square$

Corollary 5.2.10. *(LC) If I is a σ-ideal on a space X such that P_I preserves outer Lebesgue basis and ϕ is either a strongly subadditive stable capacity or a pavement capacity on a Polish space Y then $\mathrm{MRR}(I, I_\phi)$ and the forcing $P_{I_\phi} \times P_I$ is proper. In fact the product forcing is bounding and in the case of a strongly subadditive capacity it preserves outer Lebesgue measure.*

Proof. The first step in the argument is the verification of the last item of the assumptions of the theorem. I will consider the case of a strongly subadditive capacity ϕ, the case of pavement capacities is similar. Suppose in the P_I extension, $O \subset Y$ is an open set of ϕ-mass $< \varepsilon$. For a given real number $\delta > 0$ I must find a ground model coded open set P of mass $< \varepsilon + \delta$ such that $O \subset P$. Let $\mathcal{O} \in V$ be a countable basis of the space Y closed under finite intersections and unions and consider the tree T of all finite sequences t of basic open sets such that $\phi(t(0)) < \varepsilon$, and $n \in m \in \mathrm{dom}(t)$ implies $t(n) \subset t(m)$ and $\phi(t(m)) < \phi(t(n)) + \delta \cdot 2^{-\frac{n^2(n+1)}{2}}$. There is a path p through the tree T such that $O = \bigcup \mathrm{rng}(p)$. Since the forcing P_I preserves outer Lebesgue measure basis, by [2], 2.3.12 there is a subtree $S \subset T$ in the ground model such that every node of S has at most 2^n immediate successors and p is a path through the tree S. Let P be the union of all open sets that occur in the range of some sequence in the tree S. A simple exercise in strong subadditivity shows that $\phi(P) < \varepsilon + \delta$, and certainly $O \subset P$.

Towards the proof of preservation of outer Lebesgue measure λ on 2^ω, suppose that ϕ is a strongly subadditive capacity and $B \times C \subset X \times Y$ is a rectangle with I- and ϕ-positive sides respectively, $\varepsilon > 0$ is a real number and $D \subset (B \times C) \times 2^\omega$ is a Borel set with vertical sections of Lebesgue measure $\leq \varepsilon$. I must prove that the set $Z = \{z \in 2^\omega : B \times C \Vdash \check{z} \in \dot{D}_{\dot{x}_{gen}, \dot{y}_{gen}}\}$ has outer Lebesgue measure $\leq \varepsilon$. Use Choquet's theorem 4.3.6(2) to find a measure $\mu \leq \phi$ on the space Y which assigns a nonzero mass to the set C. For every point $x \in B$ let $E_x = \{z \in 2^\omega : \mu(\{y \in C : z \notin D_{x,y}\}) = 0\}$. The usual Fubini theorem shows that the set $\lambda(E_x) \leq \varepsilon$. Choose a positive real number $\delta > 0$. Since the forcing P_I preserves outer Lebesgue measure basis, there is an I-positive set $B_1 \subset B$ and an open set $O \subset 2^\omega$ of outer Lebesgue measure $< \varepsilon + \delta$ such that $\forall x \in B_1$ $E_x \subset O$. I will be done if I prove that $Z \subset O$. Suppose $z \notin O$. To show that $z \notin Z$, find a positive real number $\eta > 0$ such that the set $B_2 = \{x \in B_1 : \mu(\{y \in C : z \notin D_{x,y}\}) > \eta\}$ is still I-positive. Since the forcing P_I preserves outer Lebesgue measure basis, there is a Borel I-positive set $B_3 \subset B_2$ and a set $P \supset C$ such that $\mu(P) < \mu(C) - \eta$ and for every point $x \in B_3$, $\{y \in C : z \in D_{x,y}\} \subset P$. Now consider the rectangle $B_3 \times (C \setminus P)$. The set $C \setminus P$ has μ-measure at least η and therefor it is ϕ-positive. Also, for no points $x \in B_3$ and $y \in C \setminus P$ it is the case that $z \in D_{x,y}$ and so $B_3 \times (C \setminus P) \Vdash \check{z} \notin D_{\dot{x}_{gen}, \dot{y}_{gen}}$ as required. $\qquad\square$

Corollary 5.2.11. *If ϕ is a Ramsey capacity and either strongly subadditive stable capacity or a pavement capacity and I is the σ-compact ideal on the Baire space then* MRR(I, I_ϕ) *and the forcing $P_I \times P_{I_\phi}$ is proper.*

This now immediately follows from the last item of Theorem 4.3.13 and the fact that Miller forcing has the Laver property.

Theorem 5.2.12. *(LC+CH) Suppose that I is a universally Baire σ-ideal such that P_I is a proper forcing. The following are equivalent:*

1. P_I *is bounding and does not add a splitting real.*
2. *Let J be the Laver ideal. Then* MRR(I, J) *and the forcing $P_I \times P_J$ is proper.*
3. *Let K be the Mathias ideal. Then* MRR(I, K) *and the forcing $P_I \times P_K$ is proper.*

This follows immediately from Example 3.11.13 and the fact that both Laver forcing and Mathias forcing embed into a σ-closed*σ-centered iteration. Note that the forcing P_I remains proper in the σ-closed extension by Corollary 2.2.9, and after forcing with P_I the σ-centered forcing remains c.c.c.

5.3 Unions of σ-ideals

One operation on σ-ideals that does not seem to have a convenient translation into the standard forcing terms is the union. Suppose that I, J are σ-ideals on the same Polish space X and let K be the σ-ideal generated by their union. Thus the forcing P_K adds a point in the space X that simultaneously falls out of all ground model I-small and J-small sets, a feat that may not be achievable by any combination of iteration and product of the forcings P_I, P_J.

Theorem 5.3.1. *Suppose that ϕ is an outer regular subadditive stable capacity and ψ is a pavement submeasure on a Polish space X. Let K be the σ-ideal generated by the union $I_\phi \cup I_\psi$. Then*

1. *the forcing P_K is proper;*
2. *the ideal K satisfies the third dichotomy;*
3. *in the Solovay model or under ZF+DC+AD+, the ideal I is closed under wellordered unions and satisfies the first dichotomy.*

I must rush to assure the reader that none of the results in the above items seem to follow from some more general results. Perhaps surprisingly, I do not know in general if the properness is preserved under unions, if the third dichotomy is preserved under unions, or if the closure under wellordered unions is preserved under unions. The theorem can be generalized much further though, as will be immediately clear from the proof. It is possible to combine any countable number of pavement submeasures, a finite number of strongly subadditive capacities, a countable number of ideals generated by closed sets, and a single Hausdorff submeasure, and the items (1–3) above will still be satisfied.

Regarding the particular situation described in the theorem, it is clear that $K = I_\eta$, where $\eta = \phi \wedge \psi$ is the meet of the submeasures ϕ and ψ defined in Definition 4.3.14. Proposition 4.3.15 then shows that if ψ is a Ramsey capacity (such as is the case in Section 4.3.5) then also η is a capacity and the forcing P_K is bounding. If in addition the capacity ϕ is Ramsey then even η is Ramsey and the forcing P_K does not add splitting reals. The items (2) and (3) above can be improved to include the submeasure η in its statements, for example (2) can be strengthened to every analytic set having a Borel subset of the same η-mass.

Proof. First fix some relevant objects. Let $\mathcal{O}$ be some countable basis of the Polish space X closed under finite unions, let U be a countable set of Borel subsets of the space X and $w : U \to \mathbb{R}^+$ be a weight function generating the submeasure ψ. Finally, fix some Borel bijection $f : 2^\omega \to X$.

Suppose that P is a forcing adding an element $\dot{x}$ of the Polish space X. Consider the game G between Players I and II in which Player I starts with a condition $p_{ini} \in P$ and Player II answers with a positive real number $\varepsilon_{ini} > 0$. After that, at round n Player I indicates an open dense set $D_n \subset P$ and a basic open set O_n and a finite set $a_n \subset U$ so that

- $\phi(O_n) < \varepsilon_{ini}$, $\Sigma_{u \in a_n} w(u) < \varepsilon_{ini}$;
- $O_0 \subset O_1 \subset \ldots$ and $\phi(O_{n+1}) < \phi(O_n) + 2^{-n}$;
- $a_0 \subset a_1 \subset \ldots$ and $\Sigma_{u \in a_{n+1} \setminus a_n} < 2^{-n}$.

Player II counters with a descending chain of conditions $p_{ini} \geq p_0 \geq p_1 \ldots$ such that $p_n \in D_n$ and p_n decides the n-th bit of $f^{-1}x$. The important point is though that Player II does not need to play the condition p_n at round n – he can postpone it for an arbitrary finite number of steps. In the end, let $g \subset P$ be the filter generated by the conditions Player II played and let $x = \dot{x}/g$. Player II wins if $x \notin \bigcup_n O_n \cup \bigcup_n \bigcup a_n$.

As in Theorems 4.3.17 and 4.3.21, it is enough to show the following:

Lemma 5.3.2. *Player II has a winning strategy in the game G if and only if $P \Vdash \dot{x}$ falls out of all K-small sets coded in the ground model.*

The left-to-right direction is easy. If some condition $p \in P$ forces the point $\dot{x}$ to fall into some K-small set $A \subset X$, Player I will win by placing $p_{ini} = p$, and once Player II answers with ε_{ini}, Player I proceeds with the construction of some open set $O \subset X$ and a set $a \subset U$ such that $\phi(O), \Sigma_{u \in a} w(u) < \varepsilon$ and $p \Vdash \dot{x} \in \dot{O} \cup \bigcup \check{a}$. Player I also chooses his open dense sets in such a way that in the end the filter g Player II obtained is M-generic for some countable elementary submodel M of a large enough structure. By the forcing theorem applied in the model M, $M[g] \models \dot{x}/g \in O \cup \bigcup a$, by Borel absoluteness $\dot{x}/g \in O \cup \bigcup a$, and Player I wins.

The proof of the right-to-left direction relies on the fact that the payoff set of the game G is Borel in the tree of all possible plays, and therefore G is determined. Thus it is enough to reach a contradiction from the assumption that Player I has a winning strategy σ while $P \Vdash \dot{x}$ falls out of all K-small sets coded in the ground model. To reach it, I will construct a winning counterplay for Player II against σ. First choose a countable elementary submodel M of a large enough structure; the counterplay will use only moves from the model M and the filter Player II will obtain will be M-generic.

Let p_{ini} be the condition indicated by the strategy σ as its first move. Similarly to, there must be a positive real number $\varepsilon > 0$ such that for every pair B, C of Borel sets of respective ϕ and ψ-masses $< \varepsilon$ there is a condition $q \leq p$ forcing $\dot{x} \notin B \cup C$. This will be the first move in the desired counterplay. Now by induction I will construct finite plays $t_0 \subset t_1 \subset t_2 \ldots$ such that

- $t_0 = \langle p_{ini}, \varepsilon_{ini} \rangle$ and t_{n+1} is an extension of t_n consisting of Player II waiting and at some point playing one more condition p_n, which is the last move of t_{n+1};
- p_n belongs to the n-th open dense subset of the poset P_K in the model M in some fixed enumeration;
- $p_n \Vdash \dot{x} \notin O_m \cup \bigcup a_m$; here and below m is the index of the round at which the condition p_n is played;
- for every pair B, C of Borel subsets of X such that $O_m \subset B, \phi(B) < \phi(O_m) + 2^{-m+1}$ and $\psi(A) < 2^{-m+1}$ there is a condition $q \leq p_n$ such that $q \Vdash \dot{x} \notin \dot{B} \cup \dot{C}$.

This will certainly suffice. In the end, the filter g Player II will have constructed will be M-generic and by the forcing theorem and the third inductive item $M[g] \models$

$\dot{x}/g \notin \bigcup_n O_n \cup \bigcup_n \bigcup a_n$ and by the Borel absoluteness $\dot{x}/g \notin \bigcup_n O_n \cup \bigcup_n \bigcup a_n$ as required.

Suppose the play t_n has been constructed. To find the condition p_n and decide at which round Player II should place it, consider the infinite extension of the play t_n in which Player II just waits forever without placing any further conditions. The strategy σ produces an open set $O = \bigcup_m O_m$ and a set $a = \bigcup_m a_m$. By the third item of the induction hypothesis, there must be a condition $q \le p_{n-1}$ such that $q \Vdash \dot{x} \notin \tilde{O} \cup \bigcup_{m \ge |t_n|} a_m$. Let $p_n \le q$ be some condition in the open dense set D_n. I claim that it is possible for Player II to place the condition p_n at some round m so that the induction hypotheses are satisfied. Clearly, the third item of the induction hypotheses will be satisfied. Now suppose that the last item is not satisfied at any number m. Choose a positive real number $\delta > 0$ such that for all sets B, C of respective ϕ, ψ-masses $< \delta$ there is a condition $r \le p_n$ forcing $\dot{x} \notin \dot{B} \cup \dot{C}$. For every number m with $2^{-m+2} < \delta$ choose Borel sets B_m, C_m witnessing the failure. Look at the sets $B = \bigcap_m B_m \setminus \tilde{O}$ and $C = \bigcup_m C_m$. A review of the definitions shows that $p_n \Vdash \dot{x} \in \dot{B} \cup \dot{C}$. However, this is impossible since just as in Theorem 4.3.17 $\phi(B) = 0$ and $\psi(C) < \delta$. $\qquad\square$

Another similar theorem is the following:

Theorem 5.3.3. *Suppose that I is a translation-invariant σ-ideal on 2^ω generated by closed sets and J is the E_0-ideal. Writing K for the σ-ideal generated by $I \cup J$, P_K is proper and K satisfies the third dichotomy.*

Proof. Suppose that M is a countable elementary submodel. It will be enough to prove that $P_K \cap M \Vdash \dot{x}_{gen}$ falls out of all ground model coded K-small sets. Suppose this fails; then either some condition $B \in P_K \cap M$ forces $\dot{x}_{gen} \in \dot{C}$ for some closed set $C \in I$ or some condition $B \in P_K \cap M$ forces $\dot{x}_{gen} \in \dot{A}$ for some Borel E_0-selector D. The former case is impossible, since similarly to the proof of Theorem 4.1.2 there must be some basic open set O such that $F \cap O = 0$ and $B \cap O \notin K$ and then $B \cap O \Vdash \dot{x}_{gen} \notin \dot{C}$. To reach a contradiction in the latter case, first note that as in Theorem 4.7.3 there must be a nonzero rational $q \in 2^{<\omega}$ such that $B \cap B + q \notin K$. If no such a rational existed, removing the K-small sets $B \cap B + q : q \in \mathbb{Q}$ from the set B I would be left with a Borel selector, contradicting the assumption that $B \notin K$. Now fix a nonzero rational $q \in \mathbb{Q}$ such that $B \cap B + q \notin K$ and let $x \in 2^\omega$ be a $P_I \cap M$-generic point below the condition $B \cap B + q$. By the forcing theorem, $x \in D$. But now, the ideal K is translation-invariant, and so the point $x - q$ is also $P_I \cap M$-generic, it meets the condition B, and by the forcing theorem $x - q \in D$. This contradicts the assumption that D was an E_0-selector.

The proof of the third dichotomy is very similar. Write Q_K for the partial ordering of K-positive analytic sets ordered by inclusion. Since K is generated by Borel sets,

Proposition 2.1.11 shows that Q_K adds a generic point $\dot{x}_{gen}$ which belongs to all sets in the generic filter. Suppose that A is an analytic K-positive set. To find a Borel K-positive subset of A, choose a countable elementary submodel of a large enough structure and consider the Borel set $A' \subset A$ of all M-generic points for Q_K in the set B. I must argue that $A' \notin K$, and for that it will be enough to show that $Q_K \Vdash \dot{x}_{gen}$ does not belong to any ground model coded K-small set. The proof is identical to the previous paragraph, with one important additional point added. As I argue that for every *analytic* set $B \in P_K \cap M$ there is a rational number $q \neq 0$ such that $B \cap B + q \notin K$, if such a rational number did not exist then I can remove from the set B some Borel K-small supersets of the K-small sets $B \cap B + q$ and I will be left with an *analytic* E_0-selector. The additional point comes here: Every analytic E_0-selector has a Borel superset that is an E_0-selector by the first reflection theorem [40], 35.10, and therefore it belongs to the ideal K! $\qquad\square$

Again, it is possible to add various translation invariant capacities and pavements submeasures into the union, but the complexity of arguments grows, and in the absence of any tangible applications I will exercise the liberty of omitting the precise statements and proofs.

A natural question arises: Does the forcing obtained from the union of two ideals inherit any properties from the forcings with the two ideals? The only quotable information I have on this subject is summarized in the following theorem.

Theorem 5.3.4. *(LC) Suppose that J is a universally Baire c.c.c. ergodic σ-ideal on a Polish space Y, $I_n : n \in \omega$ are universally Baire σ-ideals on a Polish space X generated by Borel sets, and let K be the σ-ideal generated by the union $\bigcup_n I_n$. Assume that the forcing P_K is proper. If for every $n \in \omega$ $I_n \not\perp J$ holds, then $K \not\perp J$.*

Proof. Let E be a countable Borel equivalence relation witnessing the ergodicity of the ideal J, and use Proposition 3.7.6 to amend E in such a way that the E-saturations of E-small sets are E-small. Now suppose that $B \in P_K$ and $C \subset P_J$ are Borel sets and $D \subset B \times C$ is a Borel set with all horizontal sections K-small. I must find a J-positive vertical section of its complement.

A standard P_J-uniformization argument as in Proposition 2.3.4 yields a J-positive set $C' \subset C$ and Borel subsets $D_n : n \in \omega$ of the rectangle $B \times C'$ such that $D \cap B \times C' \subset \bigcup_n D_n$ and horizontal sections of the set D_n are I_n-small. Beefing up the sets D_n I may assume that they are invariant under the equivalence E. Now for every number $n \in \omega$ the universally Baire set $B_n = \{x \in B : (C' \setminus D_n)_x \in I_n\}$ contains no I_n-positive Borel subset, since such a subset would contradict the assumption $I_n \not\perp J$. As Proposition 3.9.10 shows, the collection K^* of all universally Baire sets without a Borel K-positive subset forms a σ-ideal, and all the sets $B_n : n \in \omega$ belong to it.

Since $B \notin K^*$, it must be the case that there is a point $x \in B \setminus \bigcup_n B_n$. For this point x, the vertical sections $(C' \setminus D_n)_x : n \in \omega$ are all J-positive, and since they are all E-invariant, the ergodicity of the ideal J implies the must be all J-large in the set $C' \subset Y$. In particular, the set $(C' \setminus D)_x \supset \bigcap_n (C' \setminus D_n)_x$ is J-large in the set C' and therefore J-positive as required. $\square$

Corollary 5.3.5. *(LC) The following properties are preserved by countable unions of definable ideals generated by Borel sets:*

1. *preservation of Baire category;*
2. *preservation of outer Lebesgue measure;*
3. *preservation of category bases;*
4. *preservation of Lebesgue measure bases.*

Proof. All of these properties can be expressed as Fubini properties with an ergodic ideal. The corresponding ergodic ideals, from top to bottom, are the meager ideal, the Lebesgue measure zero ideal, the ideal associated with Hechler forcing, and the ideal associated with amoeba forcing. $\square$

5.4 Illfounded iteration

The practice of illfounded iterations has been hampered by the lack of a suitable definition of a partial order that could serve as the iteration. There is a good reason for this – some forcings simply cannot be illfoundedly iterated no matter how one interprets that term. The approach with σ-ideals immediately finds the suitable definition and detects the surrounding difficulties. Many open questions remain.

This section, as it was the case in the previous sections, handles only the case of illfounded iterations of countable length. The uncountable length iterations must be treated using the methods of Section 5.5.

5.4.1 The general case

The following definition should motivate the development of the treatment of the illfounded iteration in this chapter. If $\langle L, \leq \rangle$ is a linear order and $i \in L$ then I will write $< i, \leq i$, and $> i$ for the sets $\{j \in L : j < i\}, \{j \in L : j \leq i\}$, and $\{j \in L : j > i\}$ respectively.

Definition 5.4.1. *(In the Solovay model) Suppose that I is a σ-ideal on a Polish space X closed under well-ordered unions. The ideals I^L on the spaces X^L are* *defined simultaneously for all countable linear orders $\langle L, \leq \rangle$. Namely, $I^L : L$ a countable linear order is the inclusion-smallest collection of ideals such that:*

1. *Each I^L contains all sets of the form A_C where $C \subset X^{<i} \times X$ is a Borel set with I-small vertical sections for some point $i \in L$ and $A_C = \{\vec{x} \in X^L : \langle \vec{x} \restriction < i, \vec{x}(i) \rangle \in C\}$. The sets A_C are referred to as the generators of the ideal I^L.*
2. *Each I^L is closed under wellordered unions.*
3. *The ideals are closed under fusion. Whenever L is a countable linear order, $i \in L$, and $A_{\vec{x}} : \vec{x} \in\, < i$ is a collection of sets in the ideal $I^{\geq i}$ then the set $\vec{y} \in X^L : \vec{y} \restriction \geq i \in A_{\vec{y} \restriction < i}$ is in the ideal I^L.*

Note that in the Solovay model the Axiom of Choice is absent and therefore the process of closing the collection of generators under the wellordered unions and the operation in (3) can be quite complex. Let I_0^L be the collection of generators and for every ordinal α let I_α^L be the collection of wellordered unions and fusions of sets in $\bigcup_{\beta \in \alpha} I_\beta^L$. The increasing sequence of collections I_α^L will stabilize at some ordinal Ω, where $I_\Omega^L = I^L$. Correspondingly, every set $A \in I^L$ is assigned a rank $\text{rk}(A)$ as the smallest ordinal α such that $A \in I_\alpha^L$. The question how large the ordinal Ω is is closely related to the properness of the forcing P_{I^L}. In all cases where I can find the value of Ω it is in fact equal to 1.

Note that if I is a universally Baire ideal then by a $\text{Coll}(\omega, < \kappa)$ homogeneity argument the ideal $I^L \cap V$ is in the ground model V, and by an abuse of notation I will refer to this ideal by I^L as well. In this section I will show that in a number of cases the ideal I^L is nontrivial, has a natural universally Baire definition, and the forcing P_{I^L} is a proper iteration of the forcing P_I in the sense that it adds a sequence $\vec{x}_{gen} \in X^L$ such that for every point $i \in L$, $\dot{x}_{gen}(i)$ is $V[\vec{x}_{gen} \restriction < i]$-generic for the poset P_I.

I will first state the motivating fact which shows that I am aiming at the smallest possible ideal and a natural generalization of the countable support iteration in the wellfounded case.

Fact 5.4.2. *(LC) Suppose that I is a universally Baire ideal and L is a countable linear ordering.*

1. *I^L is a universally Baire ideal.*
2. *If L is a wellordering and the ideal I satisfies the second universally Baire dichotomy then the Definitions 5.4.1 and 5.1.1 coincide on the universally Baire sets.*
3. *Suppose J is a universally Baire ideal on X^L such that the forcing P_J is a proper iteration of the forcing P_I. Then $I^L \subset J$.*

The definition of the ideal I^L itself raises obvious questions. Why do I use the choiceless Solovay model for the definition? Is the ideal I^L nontrivial at all? The following examples should shed some light at these questions.

Example 5.4.3. (ZFC+CH) There are functions $f_n : (2^\omega)^{\omega \setminus n} \to [2^\omega]^{\aleph_0}$ for $n \in \omega$ such that for every sequence $\vec{x} \in (2^\omega)^\omega$ there is a number n such that $\vec{x}(n) \in f_{n+1}(\vec{x} \restriction \omega \setminus n+1)$. In other words, if one attempts to apply the definition of I^L in a model of ZFC+CH instead of the Solovay model, the result is a trivial ideal already in the simplest illfounded case, $I = $ ctble and $L = \omega^*$. In order to find the functions $f_n : n \in \omega$ fix a wellordering $\prec$ of the Cantor space of ordertype ω_1 and let $f_n(\vec{y}) = \{r \in 2^\omega : r \prec s$ for some s on the sequence $\vec{y}\}$. By a wellfoundedness argument, for every sequence $\vec{x} \in (2^\omega)^\omega$ there are numbers $n \in m \in \omega$ such that $\vec{x}(n) \prec \vec{x}(m)$ and then $\vec{x}(n) \in f_{n+1}(\vec{x} \restriction \omega \setminus n+1)$.

Example 5.4.4. (Hjorth) Suppose that I is a universally Baire ideal on a Polish space X such that the forcing P_I is proper and adds a dominating real. Then the ideal I^{ω^*} is trivial. To outline the proof, I will for simplicity assume that the space X is recursively presented and there is a lightface Δ_1^1 function $f : X \to \omega^\omega$ such that for every function $h \in \omega^\omega$ the set $\{x \in X : f(x) \neq^* h\}$ is in the ideal I.

The key observation is that whenever $y \in X$ is a point and $x \in X$ is a P_I-generic point, then $\omega_1^y < \omega_1^{x,y}$. Note that whenever $T \subset \omega^{<\omega}$ is a tree recursive in y, then T has an infinite branch iff T has an infinite branch in V iff T has an infinite branch modulo finite dominated by $f(x)$ iff for some $n \in \omega$ and some sequence $t \in T \cap \omega^n$, the tree $T' = \{s \in T : s$ is inclusion-compatible with t and dominated by $f(x)$ past $n\}$ is infinite. The last condition gives a $\Delta_1^1(x, y)$ restatement of illfoundedness of trees recursive in y, therefore $\omega_1^y < \omega_1^{x,y}$.

Now consider the set $B \subset X^{\omega^*}$ of all sequences $\vec{x}$ such that there is $n \in \omega$ and an illfounded tree $T \subset \omega^{<\omega}$ recursive in $\vec{x} \restriction (\omega, n)$ such that none of its branches is modulo finite dominated by $f(\vec{x}(n))$. This is an I^{ω^*}-small set by virtue of its definition; I will show that $B = X^{\omega^*}$. Well, if $\vec{x} \in X^{\omega^*} \setminus B$ was a sequence, then the ordinals $\omega_1^{\vec{x} \restriction \omega \setminus n}$ would have to strictly decrease as $n \in \omega$ increases by the observation in the previous paragraph. This is of course impossible.

Example 5.4.5. (LC) If the σ-ideal I is universally Baire and the forcing P_I is proper and preserves Baire category, then for every countable linear order L the ideal I^L is nontrivial. The results of Section 3.5 imply that there is a Cohen forcing name $\dot{x}$ for an element of the space X which falls out of all ground model coded I-small sets. Consider the finite support product P of the Cohen forcing along the linear ordering L, adding Cohen reals $r_i : i \in L$ the resulting sequence $\vec{x} = \langle \dot{x}/r_i : i \in L \rangle \in X^L$. Since r_i is a Cohen real over the model $V[r_j : j < i]$, the point $\vec{x}(i)$ falls out of all I-small Borel sets coded in the model $V[\vec{x}(j) : j < i]$ and so the sequence $\vec{x}$ is forced to fall out of all the generators of the ideal I^L. In the Solovay model, the ideal $J = \{A \subset X^L : \{g \subset P : \vec{x}/g \in A\}$ is meager in the space of all filters on $P\}$ is closed under wellordered unions, I just showed that it contains all the generators of the ideal I^L, and therefore $I^L \subset J$. On the other hand, $X^L \notin J$ and so $X^L \notin I^L$.

Example 5.4.6. (LC) If I is a universally Baire polar ideal then $X^L \notin I^L$. The proof is the same as in the previous example, replacing Cohen forcing with Solovay forcing and the finite support iteration of Cohen forcing with the random algebra with $L \cdot \omega$ many generators.

There is a number of ideals I and illfounded orderings L for which I do not know if I^L is a nontrivial ideal, see Questions 7.4.2 and 7.4.3. The following is the main theorem of this section.

Theorem 5.4.7. (LC) *Suppose that I is a universally Baire σ-ideal on a Polish space X such that the forcing P_I is proper and preserves Baire category. Let L be a countable linear ordering. Then*

1. *P_{I^L} is a proper forcing which is an iteration of the poset P_I along the ordering L. Moreover P_{I^L} preserves Baire category.*
2. *A Borel set $B \subset X^L$ is I^L-positive if and only if the Cohen forcing adds a sequence $\vec{x} \in B$ such that $\vec{x}(i)$ belongs to no Borel I-small set coded in the model $V[\vec{x} \restriction < i]$.*

In fact the second item abstractly follows from the first, but it seems to be impossible to prove the first without the help of the second. Note that the second item yields a description of the ideal which does not refer to the inaccessible cardinal κ at all.

Proof. Let κ be the inaccessible cardinal from which the ideal I^L is defined, let $G \subset \text{Coll}(\omega, < \kappa)$ be a generic filter and work in the model $V[G]$. In the end, I will use a suitable absoluteness argument to transfer the important properties of the forcing P_{I^L} back to the ground model. The key features of the proof are the following two innocent abstract forcing claims.

Claim 5.4.8. *If $K \subset L$ is an initial segment of L then P_{I^K} is in a natural sense a regular subordering of P_{I^L}.*

Proof. If $B \subset X^L$ is a set in the Solovay model I will study the projection $\pi(B) = \{\vec{x} \in X^K : B_{\vec{x}} \notin I^{L \setminus K}\}$ where $B_{\vec{x}} = \{\vec{y} \in X^{L \setminus K} : \vec{x}^\frown \vec{y} \in B\}$. Note that

(*) $B \in I^L \leftrightarrow \pi(B) \in I^K$.

The right-to-left direction of this equivalence immediately follows from the inclusion of the fusion into the definition of the ideals I^L. For the left-to-right direction, argue by an easy induction on the rank of the set $B \subset X^L$.

The argument is now easily finished. If $B \subset X^L$ is a Borel I^L-positive set then by (*) the projection $\pi(B) \subset X^K$ is an I^K-positive set in the Solovay model. The ideal I^K is closed under wellordered unions, and since every set in the Solovay model is a wellordered union of Borel sets, there is a Borel I^K-positive set $C \subset \pi(B)$. (This move is on the formal level the only reason why I considered closing the collection of all generators under wellordered unions.) By (*) again, for every condition $C' \subset C$ in the forcing P_{IK} the set $B' = \{\vec{x} \in B : \vec{x} \restriction K \in C'\}$ is Borel I^L-positive, and $\pi(B') \leq C'$. The claim follows. $\square$

At least in one simple case I must find an explicit form of the remainder P_{IL}/P_{IK}. Let $i \in L$ be a point, $K = \{j \in L : j < i\}$ and $J = K \cup \{i\}$. To find the remainder P_{IJ}/P_{IK}, let $\vec{x} \in X^K$ be a generic sequence for P_{IK} and in the model $V[G][\vec{x}]$ define $Q = \{B \subset X : B$ Borel, I-positive, and there is a universally Baire function $f \in V[G]$, $f: X^K \to \mathcal{B}(X)$, such that $B = f(\vec{x})\}$ ordered by inclusion.

Claim 5.4.9. $P_{IJ} = P_{IK} * Q$.

Proof. This is really an immediate corollary of the previous claim. Consider the map $\eta : P_{IK} * Q \to \mathcal{P}(X^J)$ defined in the following way. If $C \in P_{IK}$ is a condition and $f: X^K \to \mathcal{B}(X)$ is a universally Baire function such that $C \Vdash \dot{f}(\vec{x}_{gen}) \notin I$ then let $\eta(C, f) = \{\vec{x} \in X^J : \vec{x} \restriction K \in C \wedge \vec{x}(i) \in f(\vec{x} \restriction K)\}$.

First of all, the range of η consists of I^J-positive sets. To see this, note that if $C \Vdash \dot{f}(\vec{x}_{gen}) \notin I$ then the set $D = \{\vec{x} \in C : f(\vec{x}) \notin I\}$ must be I^K-positive, in fact modulo I^K equal to C: if its complement in C was I^K-positive, it would contain a positive Borel set which by universally Baire absoluteness would force $f(\vec{x}_{gen}) \in I$. Now $D = \pi(\eta(C, f))$ and so $\pi(\eta(C, f)) \notin I^J$ by the previous claim.

Second, if $B \subset X^J$ is an I^J-positive set in the Solovay model then it has a subset which is in the range of the map η. To see this, thin out the set B to be Borel if necessary, and consider the set $\pi(B) \subset X^K$. This is an I^K-positive set by the previous claim, so it has a Borel I^K-positive subset C. Let $f: X^K \to \mathcal{B}(X)$ be defined by $f(\vec{x}) = \{y \in X : \vec{x}^\frown y \in B\}$. A universally Baire absoluteness argument shows that the pair C, f is in the domain of the function η, and clearly $\eta(C, f) \subset B$. $\square$

Now I come to the key points. Let M be a countable elementary submodel of a large enough structure. I will show that the forcing $P_{IL} \cap M$ forces its generic sequence $\vec{x}_{gen}$ out of every Borel set in the ideal I^L. This will show that the forcing P_{IL} is proper and preserves Baire category by Corollary 3.5.4. In order to do this,

I will show that $P_{IL} \cap M$ forces its generic sequence $\vec{x}_{gen}$ out of every generator of the ideal I^L. This shows that for every generator B the set $\{g \subset P_{IL} \cap M : \vec{x}_{gen}/g \notin B\}$ contains a dense G_δ in the Polish space of all ultrafilters on $P_{IL} \cap M$. After that, an easy induction on the rank of a set $A \subset X^L, A \in I^L$ shows that for every set $A \in I^L$, the set $\{g \subset P_{IL} \cap M : \vec{x}_{gen}/g \in A\}$ contains a dense G_δ set, and therefore $P_{IL} \cap M \Vdash \vec{x}_{gen} \notin \dot{A}$ as desired. The induction uses the closure of the meager ideal under fusion and wellordered unions in the Solovay model.

So choose a point $i \in L$, let $K = \{j \in L : j < i\}$, and choose a Borel set $C \subset X^K \times X$ with I-small vertical sections. I must prove that $P_{IL} \cap M \Vdash \langle \vec{x}_{gen} \restriction K, \vec{x}_{gen}(i) \rangle \notin \dot{C}$. Note that writing $J = \{j \in L : j \leq i\}$ it is the case that $(P_{IK} \cap M) \prec (P_{IJ} \cap M) \prec (P_{IL} \cap M)$ essentially by Claim 5.4.8 since the model M is closed under the projection function described there. Moreover, writing $\vec{y}_{gen}$ for the $P_{IK} \cap M$-generic sequence, it is the case that $(P_{IJ} \cap M) = (P_{IK} \cap M) * (Q \cap M[\vec{y}_{gen}])$ by Claim 5.4.9. Thus it will be enough to show that the forcing $Q \cap M[\vec{y}_{gen}]$ forces $\langle \vec{y}_{gen}, \dot{z} \rangle \notin \dot{C}$, where $\dot{z}$ is the $Q \cap M[\vec{y}_{gen}]$-name for a generic point in the space X. Note that $Q \cap M[\vec{y}_{gen}] \subset P_I$ and $C_{\vec{y}_{gen}} \in I$. Let $\sigma \in M$ be some universally Baire winning strategy for Player II in the category game for the forcing P_I. It is immediate from the definition of the name Q that the set $Q \cap M[\vec{y}_{gen}] \subset P_I$ is closed under the strategy σ, and the same argument as in Corollary 3.5.4 shows that it forces its generic point to fall out of all I-small sets, in particular out of $C_{\vec{y}_{gen}}$, as desired.

Finally, I must wrap up and transfer the whole argument to the ground model. Note that I proved the following to hold in the model $V[G]$.

(**) A Borel set $B \subset X^L$ is I^L-positive if and only if $\phi(B)$ holds, where $\phi(B)$ is the statement that there is a name $\vec{x}_{gen}$ in the Cohen forcing which is forced to fall into B and outside of all generators.

(***) The forcing P_J, where $J = \{B \subset I^L$ Borel, $\neg\phi(B)\}$, is proper and preserves Baire category.

Perhaps I should argue for (**). The left-to-right direction first: if $B \subset X^L$ is a Borel I^L-positive set and M is a countable elementary submodel of a large enough structure then the countable forcing $P_{IL} \cap M \restriction B$ forces its generic sequence into the set B and out of all generators as I just argued. For the opposite direction note that if the Cohen poset forces $\vec{x}_{gen}$ to fall out of all generators, there is a Borel function $f : 2^\omega \to B$ such that f-preimages of generators are meager. Since the meager ideal is closed under well-ordered unions, preimages of any set in the ideal I^L are meager, and therefore no set in I^L can cover the set B.

Note that the property ϕ of (**) is a universally Baire property of a Borel set and therefore absolute among forcing extensions. By the fact that the forcing P_J is proper and category preserving is witnessed by the existence of a universally

Baire winning strategy in the category game for the first player, and this fact is also absolute, and transfers to V. The theorem follows.

As the last step in the proof I must show that the forcing P_{IL} is indeed an iteration of P_I along the linear order L, which is to say that for every index $i \in L$ the point $\vec{x}_{gen}(i)$ is forced to be P_I-generic over the model $V[\vec{x}_{gen} \upharpoonright < i]$. All work necessary for this has already been done. Just use a universally Baire absoluteness argument to transfer Claims 5.4.8 and 5.4.9 to the ground model, and then note that the latter says exactly that the point $\vec{x}_{gen}(i)$ is P_I-generic over the model $V[\vec{x}_{gen} \upharpoonright < i]$: since the forcing $P_{I<i}$ is proper, *every* condition in $P_I \cap V[\vec{x}_{gen} \upharpoonright < i]$ is an image of the sequence $\vec{x}_{gen} \upharpoonright < i$ under a ground model universally Baire (even Borel) function by Proposition 2.3.1.   □

The argument may seem somewhat tortured: first I produce a completely abstract definition of the ideal I^L only to find a quite definite description of the ideal in the above theorem. It may have been indeed easier to start working from the definite description in Theorem 5.4.7(2), and that was the road I took in [83]. The present argument shows how abstract considerations lead to this description. It is also my hope that the reader will compare the argument with the ZFC proofs in the following subsection and see the similarities.

5.4.2 The Π_1^1 on Σ_1^1 case

Considering the development in Section 5.1.3 one may be tempted to expect that the theorems of the previous subsection may be provable in ZFC if the ideal I is Π_1^1 on Σ_1^1. I do not know if this is the case, nevertheless an important class of forcings can indeed be treated in ZFC. Throughout this section I will work under the assumption that the Polish space X is recursively presented, the linear order L is recursive, and the ideal I is lightface Π_1^1 on Σ_1^1. The general case follows by a straightforward relativization.

Definition 5.4.10. *Suppose I is a Π_1^1 on Σ_1^1 σ-ideal on a Polish space X. Suppose that L is a countable linear order. The ideal I_s^L (s for "simple," perhaps a case of wishful thinking) is generated by the sets $A_r = \{\vec{x} \in X^L : \exists i \in L\ x(i) \in \bigcup(I \cap \Sigma_1^1(r, \vec{x} \upharpoonright < i))\}$ as r varies over all elements of the Cantor space 2^ω.*

Note that since the ideal I is Π_1^1 on Σ_1^1 the set $\bigcup(I \cap \Sigma_1^1(y))$ is uniformly $\Pi_1^1(y)$ by Proposition 3.8.6, and therefore the ideal I_s^L is generated by coanalytic sets. Since every coanalytic set is a wellordered union of Borel sets, the ideal I_s^L is a subset of the ideal I^L defined in the previous subsection. I will show that in several important cases these two ideals contain the same Borel sets and the theorems about I^L from the previous section go through in ZFC for I_s^L.

Theorem 5.4.11. *Suppose that I is a $\mathbf{\Pi}_1^1$ on $\mathbf{\Sigma}_1^1$ σ-ideal generated by closed sets and L is a countable linear order. Let $J = I_s^L$. Then*

1. *the ideal J is $\mathbf{\Pi}_1^1$ on $\mathbf{\Sigma}_1^1$ and satisfies the third dichotomy;*
2. *the forcing P_J is a proper category preserving iteration of P_I along L.*

Proof. The first item is a necessary intermediate step to the proof of the second item. The argument uses the Gandy–Harrington forcing of all nonempty $\mathbf{\Sigma}_1^1$ subsets of the space X^L ordered by inclusion. A standard argument close to the proof of Proposition 2.1.11 shows that this forcing adds a generic sequence $\vec{x}_{gen} \in X^L$ which belongs to all sets in the generic filter. Let $A \subset X^L$ be a $\mathbf{\Sigma}_1^1$ set and consider the set $\tilde{A} = \{\vec{x} \in A : \forall i \in L \; \vec{x}(i) \notin \bigcup(I \cap \mathbf{\Sigma}_1^1(\vec{x} \restriction < i))\}$. Since the ideal I is $\mathbf{\Pi}_1^1$ on $\mathbf{\Sigma}_1^1$ this is an analytic set, and there are two cases. Either $\tilde{A} = 0$, in which case clearly $A \in J$; or $\tilde{A} \neq 0$, in which case I will prove that $A \notin J$.

Consider the condition $p = \tilde{A}$ in the Gandy–Harrington forcing P. I will prove that $p \Vdash \vec{x}_{gen} \notin \bigcup(J \cap V)$. However, let me first show how the first item follows from this. By an analytic absoluteness argument, $\tilde{A} \notin J$. To find a Borel J-positive subset $B \subset \tilde{A}$ let M be a countable elementary submodel of a large enough structure and let $B = \{\vec{x} \in \tilde{A} : \vec{x}$ is M-generic for the forcing $P\}$. This is a Borel set, since it is in one-to-one Borel correspondence with the Borel set of all M-generic filters on the countable forcing P, and it is J-positive since $p \Vdash \vec{x}_{gen} \in B \setminus \bigcup(J \cap V)$. This proves the third dichotomy. Finally, to show that J is a $\mathbf{\Pi}_1^1$ on $\mathbf{\Sigma}_1^1$ σ-ideal, note that we just proved that a $\mathbf{\Sigma}_1^1(y)$ set A is in J if and only if $\forall \vec{x} \in A \; \exists i \in L \; \vec{x}(i) \in \bigcup(I \cap \mathbf{\Sigma}_1^1(y, \vec{x} \restriction < i))$ which is a $\mathbf{\Pi}_1^1(y)$ condition.

Towards the proof of $p \Vdash \vec{x}_{gen} \notin \bigcup(J \cap V)$, the key observation is that if $i \in L$ is an arbitrary point then the Gandy–Harrington forcing Q of nonempty $\mathbf{\Sigma}_1^1$ subsets of $X^{<i}$ naturally regularly embeds into P, with the projection function $\pi : P \to Q$ defined by $\pi(q) = \{\vec{y} \in X^{<i} : \exists \vec{x} \in q \; \vec{y} \subset \vec{x}\}$. Now suppose for contradiction that some condition $q \leq p$ forces in P that $\vec{x}_{gen}(i) \in \bigcup(I \cap \mathbf{\Sigma}_1^1(\check{u}, \vec{x}_{gen} \restriction < i)$ for some parameter $u \in 2^\omega$ and a point $i \in L$. Strengthening the condition q if necessary we may find a P-name $\dot{C}$ for a closed $\mathbf{\Sigma}_1^1(\check{u}, \vec{x}_{gen} \restriction < i)$ set in the ideal I such that $q \Vdash \vec{x}_{gen}(i) \in \dot{C}$. Strengthening q further I may assume that $\dot{C}$ is in fact a Q-name where Q is the Gandy–Harrington forcing of nonempty $\mathbf{\Sigma}_1^1$ subsets of $X^{<i}$. Consider the condition $\pi(q) \in Q$ and work in the Q-extension $V[\vec{y}]$ where $\vec{y} \in \pi(q)$ is the Q-generic sequence. The set $D = \{x \in X : \exists \vec{z} \; \vec{y}^\frown x^\frown \vec{z} \in q\}$ is a nonempty $\mathbf{\Sigma}_1^1(\vec{y})$ set which is disjoint from the set $\bigcup(I \cap \mathbf{\Sigma}_1^1(\vec{y}))$ by the definition of the set $\tilde{A}$, and therefore it is I-positive. Since the set $C \subset X$ is closed and in the ideal I, there must be a basic open set $O \subset X$ such that $C \cap O = 0$ and $D \cap O \neq 0$. Return to the ground model and find a condition $r \leq \pi(q)$ in the poset Q which identifies the open set O. An absoluteness argument shows that

the Σ^1_1 set $\{\vec{x} \in X^L : \vec{x} \restriction i \in r \wedge \vec{x}(i) \in O\}$ is nonempty. This set is a condition in P stronger than q which forces $\vec{x}_{gen}(i) \notin \dot{C}$, contradicting the choice of the name $\dot{C}$.

This completes the proof of the first item. The argument for the second item now proceeds along the lines of the previous section. The first observation is that writing $J_{<i} = I^{<i}_s$ and $J_{\geq i} = I^{\geq i}_s$, the forcing $P_{J_{<i}}$ naturally embeds into P_J. To see that note that by the third dichotomy proved above there is no harm in including analytic sets in these forcings, and then the projection function $\pi : P_J \to P_{J_{<i}}$ is defined by $\pi(A) = \{\vec{y} \in X^{<i} : \{\vec{z} \in X^{\geq i} : \vec{y}^\frown \vec{z} \in A\} \notin J_{\geq i}\}$. The important point is that since the σ-ideal $J_{\geq i}$ is $\mathbf{\Pi}^1_1$ on $\mathbf{\Sigma}^1_1$ the set $\pi(A)$ is analytic. The properties of projection are then easily checked for the function π.

To prove that the forcing P_J is proper and preserves category, it is enough to show that for every countable elementary submodel M of a large enough structure, $M \cap P_J \Vdash \vec{x}_{gen} \notin \bigcup(J \cap V)$, where $\vec{x}_{gen}$ is now the $M \cap P_J$-generic sequence in X^L – Corollary 3.5.4. Suppose for contradiction that some condition $p \in M \cap P_J$ forces $\vec{x}_{gen}(i) \in \bigcup(I \cap \Sigma^1_1(\check{u}, \vec{x}_{gen} \restriction < i)$ for some point $i \in L$ and parameters $u \in 2^\omega$. Strengthening the condition p I may assume that there is a name $\dot{C}$ for a closed $\Sigma^1_1(\check{u}, \vec{x}_{gen} \restriction < i)$ set in the ideal I such that $p \Vdash \vec{x}_{gen}(i) \in \dot{C}$, and $\dot{C}$ is in fact a $M \cap P_{J_{<i}}$-name. Note that the forcing $M \cap P_{J_{<i}}$ naturally embeds into $M \cap P_J$ as witnessed by the projection function $\pi \restriction M$. Move to the $M \cap P_{J_{<i}}$-generic extension $V[\vec{y}]$ where $\vec{y} \in \pi(p)$ is the generic sequence. By an absoluteness argument the set $D = \{x \in X : \{\vec{z} \in X^{>i} : \vec{y}^\frown x^\frown \vec{z} \in p\}$ is I-positive, and there must exist a basic open set $O \subset X$ such that $C \cap O = 0$ while $D \cap O \notin I$. Move back to the ground model, let $q \leq \pi(p)$ be a condition in the forcing p which identifies the basic open set O, and consider the set $r = \{\vec{y} \in q : \{x \in O : \{\vec{z} \in X^{>i} : \vec{y}^\frown x^\frown \vec{z} \in p\} \notin J_{>i}\} \notin I\}$. This is an analytic set in the model M. It cannot be $J_{<i}$-small set, because it would then be covered by a coanalytic set $r' \in M \cap J_{<i}$ and the condition $q \setminus r'$ would force $\dot{D} \in I$, a contradiction. But then the set $p' = \{\vec{x} \in p : \vec{x} \restriction i \in r, \vec{x}(i) \in O\} \subset p$ is by the definitions a J-positive analytic subset of p which forces $\vec{x}_{gen}(i) \notin C$, a contradiction.

The very last step in the proof is to show that P_J is indeed an iteration of the forcing P_I, that is, for every $i \in L$ the point $\vec{x}_{gen}(i)$ is forced to be P_I-generic over the model $V[\vec{x}_{gen} \restriction < i]$ where $\vec{x}_{gen}$ now is the name for the P_J-generic sequence in the space X^L. The proof of this is identical to the argument in Theorem 5.4.7 and as such left to the reader. $\square$

A particular special case that has been treated from the combinatorial standpoint [43] is the case of an ideal I σ-generated by a σ-compact collection of compact sets. In this case a more complete information is available.

Theorem 5.4.12. *Suppose that I is a σ-ideal on a Polish space X generated by a σ-compact collection of compact sets and L is a countable linear order. Let $J = I_s^L$. The forcing P_J is bounding.*

Proof. In order to simplify the proof I will assume that the underlying space X is the Cantor space 2^ω, the ordering L is recursive and the ideal I is generated by a recursive increasing sequence of compact collections $F_n : n \in \omega$ of compact sets. $\qquad\square$

I must produce, among other things, a collection of compact subsets of X^L which is dense in the forcing P_J. I will call a nonempty compact set $C \subset X^L$ *solid* if for all basic open sets $O \subset X^L$, all sequences $\vec{x} \in C \cap O$ and all points $i \in L$ the set $\{z \in 2^\omega : \exists \vec{y}\, (\vec{x} \upharpoonright < i)^\frown z^\frown \vec{y} \in C \cap O\}$ is not in the ideal I.

The following two lemmas will complete the proof of the theorem.

Lemma 5.4.13. *Nonempty compact solid sets are J-positive.*

Proof. Fix a nonempty compact solid set $C \subset X^L$. Consider the forcing P of all nonempty basic open subsets of the set C ordered by inclusion. A completeness argument immediately shows that if $G \subset P$ is a generic filter then $\bigcap G$ is a set containing a single element $\vec{x}_{gen} \in C$. I will show that $P \Vdash \dot{x}_{gen} \notin \bigcup(\dot{J} \cap V)$. An analytic absoluteness argument then shows that $C \notin J$.

The key observation is again that for every point $i \in L$ the forcing $P_{<i}$ of all nonempty basic open subsets of the projection $C_{<i}$ of the set C into the space $X^{<i}$ naturally regularly embeds into the poset P, with the projection function $\pi(p) =$ projection of the basic open set p into $C_{<i}$. Note that $\pi(p)$ is indeed a basic open subset of $C_{<i}$ since the set C is compact and zero-dimensional. The other properties of the projection are easily checked for the function π.

Suppose for contradiction that some condition $p \in P$ forces $\vec{x}_{gen}(i) \in \bigcup(I \cap V[\vec{x}_{gen} \upharpoonright < i])$. Strengthening the condition p if necessary I can assume that it also identifies a number $n \in \omega$ such that there is a compact set $K \in F_n \cap Vm[\vec{x}_{gen} \upharpoonright < i]$ with $\vec{x}_{gen}(i) \in K$. Strengthening the condition p even further I may assume that $\dot{K}$ is in fact a $P_{<i}$ name. Let $G \subset P_{<i}$ be a generic filter containing the condition $\pi(p)$ and $\vec{x} \in C^{<i}$ its attendant generic sequence. The set C is solid in the ground model and by a Shoenfield absoluteness argument it is solid even in the generic extension. Thus the set $B = \{z \in X : \exists \vec{y} : \vec{x}^\frown z^\frown \vec{y} \in p\}$ is compact and I-positive, it is not a subset of the set K, and there is some basic open set $O \subset X$ such that $B \cap O \neq 0$ while $K \cap O = 0$. Find a condition $q \leq \pi(p)$ in the generic filter G which identifies this open set O and run back to the ground model. Let $r \leq p$ be the set $\{\vec{x} \in C : \vec{x} \upharpoonright i \in q, \vec{x}(i) \in O, \vec{x} \in p\}$. This is a basic open subset of the set C, it is nonempty since in the generic extension $V[G]$

it is nonempty, and it forces $\vec{x}_{gen}(i) \notin \dot{K}$, contradicting the assumed properties of the condition p. $\qquad\qquad\qquad\qquad\qquad\qquad\qquad\qquad\qquad\qquad\qquad$ □

Lemma 5.4.14. *If $f : (2^\omega)^L \to 2^\omega$ is a partial $\Sigma_1^1(a)$ function then*

1. *either for every sequence $\vec{x} \in \mathrm{dom}(f)$ there is a point $i \in L$ such that $\vec{x}(i)$ belongs to some $\Sigma_1^1(a, \vec{x} \restriction i)$ set $A \in I$;*
2. *or there is a nonempty compact solid set $C \subset A$ such that the function $f \restriction C$ is continuous.*

Proof. Let me simplify the notation by the assumption that $a = 0$; the general case follows by a simple relativization. Consider the set $A = \mathrm{dom}(f)$ and the set $\tilde{A} = \{\vec{x} \in A : \forall i \in L \ \vec{x}(i) \notin \bigcup(\Sigma_1^1(\vec{x} \restriction (-\infty, i)) \cap I)\}$. Note that A is a Σ_1^1 set. The ideal I is $\mathbf{\Pi}_1^1$ on $\mathbf{\Sigma}_1^1$ by Theorems 4.1.8 and 3.8.9; by Proposition 3.8.6 then the set $\bigcup(\Sigma_1^1(y) \cap I)$ is $\Pi_1^1(y)$ uniformly in y and therefore the set $\tilde{A}$ is Σ_1^1. If $\tilde{A} = 0$ then we are in the first case and the proof is complete. So assume $\tilde{A} \neq 0$.

The argument uses a generalization of the Gandy–Harrington forcing. The closed set $C \subset X^L$ will be added by a forcing P consisting of all tuples $p = \langle a_p, n_p, e_p, g_p, h_p \rangle$ where:

- $a_p \subset L$ is a finite set, $n_p \in \omega$ is a number, and $e_p \subset 2^{a_p \times n_p}$ is a collection of sequences. The information carried by this part of the condition is $e_p = \{\vec{x} \restriction a_p \times n_p : \vec{x} \in C\}$.
- g_p is an assignment which attaches to each function $u \in e_p$ a Σ_1^1 set $g_p(u) \subset \tilde{A}$. The information carried here is $\vec{x} \in C \wedge \vec{x} \restriction a_p \times n_p = \vec{u}$ implies $\vec{x} \in g_p(\vec{u})$.
- h_p is an assignment which attaches to each function $u \in e_p$ a finite sequence $h_p(u) \in 2^{<\omega}$. This will carry the information that $\vec{x} \in C \wedge \vec{x} \restriction a_p \times n_p = u$ implies $h_p(u) \subset f(\vec{x})$.
- (The most important condition) There is an assignment $k_p = k : e_p \to X_L$ such that for every function $u \in e_p$ $k(u) \in g_p(u)$, $k(u) \restriction a_p \times n_p = u$, and for every $i \in a_p$ and every two functions $u, v \in e_p$ if $u \restriction (a_p \cap < i) \times n_p = v \restriction (a_p \cap < i) \times n_p$ then $k(u) \restriction < i = k(v) \restriction < i$.

Note that the function k from the last item is not a part of the condition p. The ordering is defined by $q \leq p$ if $a_p \subset a_q, n_p \subset n_q, \{u \restriction a_p \times n_p : u \in e_q\} = e_p$ and for every function $u \in e_q$ it is the case that $g_q(u) \subset g_p(u \restriction a_p \times n_p)$ and $h_p(u \restriction a_p \times n_p) \subset h_q(u)$. Thus $\langle 0, 0, 0, 0, 0 \rangle$ is the largest condition in the poset P.

Let M be a countable elementary submodel and $G \subset P$ an M-generic filter. Consider the set $C = \{\vec{x} \in X^L : \forall p \in G \ \vec{x} \restriction a_p \times n_p \in e_p\}$. I claim that the set C is as required in the second item of the claim. It is clear that the set C is closed, the other properties are less obvious. The following claim is key.

Claim 5.4.15. *If $p \in G$, $u \in e_p$ and $B = g_p(u)$ then $\{\vec{x} \in C : \vec{x} \restriction a_p \times n_p = u\} \subset B$.*

Proof. Let $q \in P$ be a condition, let $T_u : u \in e_q$ be recursive trees projecting to lightface analytic subsets of X^L, and let $u \mapsto l(u) \in T_u : u \in e_q$ be an assignment such that for every function $u \in e_q$, $g_q(u) \subset \mathrm{proj}(T_u \restriction l(u))$ holds. I will prove that for every number $m \in \omega$ there is a strengthening $q' \leq q$ with the same a, n, e, and h coordinate as q and an assignment $u \mapsto l'(u) \in T_u : u \in e_q$ such that for every function $u \in e_q$, $l(u) \subset l'(u)$, $|l'(u)| \geq m$, and $g_{q'}(u) \subset \mathrm{proj}(T_u \restriction l'(u))$. The claim then follows by a straightforward genericity argument.

To find the g coordinate of the condition q', consult an assignment $k : e_q \to X^L$ witnessing the fact that $q \in P$. Note that for every function $u \in e_q$, $k(u) \in g_q(u) \subset \mathrm{proj}(T_u \restriction l(u))$. find a node $l'(u)$ of the tree T_u of length at least m that extends $l(u)$ and such that $k(u) \in \mathrm{proj}(T_u \restriction l'(u))$. Then the map $g_{q'}(u) = g_q(u) \cap \mathrm{proj}(T_u \restriction l'(u))$ will have the desired properties. Note that the same assignment k witnesses the fact that $q' \in P$. □

Now it is clear that $C \subset \tilde{A}$. To show that $f \restriction \tilde{A}$ is a continuous function, suppose that $t \in 2^{<\omega}$ is a finite binary sequence. I must show that the preimage $f^{-1}O_t$ is relatively open in the set C. Let $p \in P$ be any condition, and consider the assignment k_p. For every function $u \in e_p$, find a sequence $h_q(u)$ extending $h_p(u)$ such that $h_q(u) \subset f(k_p(u))$ and $h_q(u)$ is either incompatible with t or extends t. Consider the tuple $q = \langle a_p, n_p, e_p, g_q, h_q \rangle$ where $g_q(u) = \{\vec{x} \in g_p(u) : h_q \subset f(\vec{x})\}$. It is clear that $q \leq p$ is a condition as witnessed by the assignment $k_q = k_p$ and the claim shows that if $q \in G$ then $\forall \vec{x} \in C \; t \subset f(\vec{x}) \leftrightarrow t \subset h_q(\vec{x} \restriction a_p \times n_p)$, therefore the preimage $f^{-1}O_t$ is relatively clopen in the set C. The continuity of the function $f \restriction C$ now follows by a genericity argument.

To show that $C \subset X^L$ is solid, let me re-introduce an observation from Theorem 4.1.8. Call a set $Z \subset 2^{<\omega}$ n-large if no set in the collection F_n of generators for the ideal I contains elements extending every sequence in Z. Since F_n is a compact collection of compact sets, for every I-positive set $Y \subset X$ there is a number $m \in \omega$ such that the set $\{y \restriction m : y \in Y\}$ is n-large.

Claim 5.4.16. *For every condition $p \in P$, every index $i \in L$, every natural number $n \in \omega$ and every function $u \in e_p$ there is a condition $q \leq p$ such that the set $\{t \in 2^{<\omega} : \exists v \in e_q \; u \subset v \text{ and } v \text{ at } i \text{ equals } t\}$ is n-large.*

Proof. Fix p, i, n, u, without loss of generality assume $i \in a_p$, and let $k : e_p \to X^L$ be the assignment witnessing the fact that $p \in P$. Let $\vec{x} = k(u)$ and consider the set $Y = \{y \in X : \exists \vec{z} \in X^{>i} \exists l \; l \text{ is an assignment witnessing that } p \in P \text{ and } l(u) = \vec{x} \restriction < i ^\frown y ^\frown \vec{z}\}$. A review of definitions shows that this is a $\Sigma^1_1(\vec{x} \restriction < i)$ set, and since it is nonempty (containing the point $\vec{x}(i)$), it must be I-positive by the choice of the set

$\tilde{A}$. Find a natural number $m > n_p$ such that the set $Z = \{y \restriction m : y \in Y\}$ is n-large, and for every sequence $t \in Z$ find $y_t \in Y$ extending it and an assignment l_t witnessing $y_t \in Y$. Construct the condition q in the following way:

- $n_q = m$, $a_q = a_p$;
- every sequence $v \in e_p$, $v \neq u$ has a unique extension in e_q, and it is $k(v) \restriction a_q \times n_q$;
- the extensions of u in e_q are exactly the functions $\{l_t \restriction a_q \times n_q : t \in Z\}$;
- g_q, h_q are obtained trivially from g_p, h_p: $g_q(v) = g_p(v \restriction a_p \times n_p)$ for all functions $v \in e_q$, and the same thing for h_q;
- the assignment k_q is defined by $k_q(v) = k_p(v \restriction a_p \times n_p)$ if $u \not\subset v$ and $k_q(v) = l_t(u)$ if $u \subset v$ and v at i is equal to t.

A review of definitions reveals that the condition q is as required. □

The solidity now immediately follows. Suppose $\vec{x} \in C$ is a sequence, $i \in L$ is an index, and $K_n : n \in \omega$ are compact sets in the ideal I, $K_n \in F_n$. To find a sequence $\vec{y} \in x$ such that $\vec{y} \restriction < i = \vec{x} \restriction < i$ and $\vec{y}(i) \notin \bigcup_n K_n$, use the claim and a genericity argument repeatedly to build sequences $\vec{y}_n \in C$ such that $\vec{y}_n \restriction < i$ converges to $\vec{x} \restriction < i$ and $\vec{y}_n(i)$ belongs to some clopen subset of 2^ω disjoint from $\bigcup_{m \in n} K_m$. The sequence $\vec{y} = \lim_n \vec{y}_n$ will then be as required. □

5.5 Directed systems of ideals

It is frequently the case that the forcing job requires adding a more complex object than just a real. There are iterations and products of uncountable length as well as many other constructions. This section explains an operation on ideals helpful in such situations.

Definition 5.5.1. *Let X be a Polish space and L an arbitrary set. A set $B \subset X^L$ is Baire if it is obtained from basic open sets by a repeated application of countable unions, intersections, and complements. Here a basic open set is a set $O \subset X^L$ for which there is an open set $P \subset X$ and an index $i \in L$ such that $O = \{\vec{x} \in X^L : \vec{x}(i) \in P\}$.*

The Baire sets are similar to Borel sets in that they have a natural representation in the generic extension, and I will adopt the same conventions about this representation as in the case of Borel sets. Every Baire set $B \subset X^L$ has a *support*–a countable set $\mathrm{supp}(B) \subset L$ such that the membership of any sequence $\vec{x} \in X^L$ in the set B depends only on $\vec{x} \restriction \mathrm{supp}(B)$. For every countable set $a \supset \mathrm{supp}(B)$ I will write $B \restriction a = \{\vec{x} \in X^a : \exists \vec{y} \in B \ \vec{y} \restriction a = \vec{x}\}$.

Suppose that I is a σ-ideal on the set X^L and consider the forcing P_I of all Baire I-positive sets ordered by inclusion. The following propositions have proofs closely related to those in Chapter 2.

Proposition 5.5.2. P_I *adds a sequence* $\vec{x}_{gen} \in X^L$ *such that for every Baire set* $B \subset X^L$ *in the ground model,* $\vec{x}_{gen} \in B$ *iff* B *is in the generic filter.*

Proposition 5.5.3. P_I *is proper if and only if for every countable elementary submodel* M *of a large enough structure and every set* $B \in P_I \cap M$ *the set* $\{\vec{x} \in B : \vec{x}$ *is* P_I*-generic*$\}$ *is* I*-positive.*

Proposition 5.5.4. *Suppose* P_I *is proper forcing.*

1. *If* $B \in P_I$ *is a condition and* $\dot{y}$ *a* P_I*-name for an infinite binary sequence then there is a condition* $C \subset B$ *and a Borel function* $f : C \upharpoonright \operatorname{supp}(C) \to 2^\omega$ *such that* $C \Vdash \dot{y} = \dot{f}(\vec{x}_{gen} \upharpoonright \operatorname{supp}(C))$.
2. *If* $B \in P_I$ *is a condition,* Y *is a Polish space,* K *is a set, and* $\dot{A}$ *is a name for a Baire subset of* Y^K *then there is a condition* $C \subset B$ *and a Baire set* $D \subset C \times Y^K$ *such that* $C \Vdash \dot{A} = \dot{D}_{\vec{x}_{gen}}$.

Every σ-ideal on Baire sets is obtained as a limit of sorts of a directed system of smaller σ-ideals. While this is an elementary observation, it will be very helpful in the actual construction of interesting σ-ideals on Baire sets.

Definition 5.5.5. *A collection* $\mathfrak{S} = \{I_a : a \in S\}$ *is a directed system of* σ*-ideals if*

1. $S \subset [L]^{\aleph_0}$ *is a cofinal set of countable subsets of* L;
2. I_a *is a* σ*-ideal on the set* X^a;
3. *whenever* $a \subset b$ *are sets in* S *and* $B \subset X^a$, *then* $B \in I_a$ *iff* $\{\vec{x} \in X^b : x \upharpoonright a \in B\} \in I_b$.

The limit $\lim \mathfrak{S}$ *of such a directed system is the collection* I *of Baire sets defined by* $B \in I$ *iff for some (equivalently, every) set* $a \in S$ *with* $\operatorname{supp}(B) \subset a$ *it is the case that* $B \upharpoonright a \in I^a$. *It is not difficult to see that* I *is in fact a* σ*-ideal.*

5.5.1 Uncountable iterations and products

The various iterations and products of uncountable length of partial orders of the form P_J where J is an iterable σ-ideal on a Polish space X can be handled using the technology of this section. Fix an ordinal κ and for a countable set $a \subset \kappa$ let J_a be the ideal on X^a defined just as the Fubini power J^α where $\alpha =$ ordertype of a. It is not difficult to see that the ideals $J_a : a \in [\kappa]^{\aleph_0}$ form a directed system; let J_κ be its limit.

Theorem 5.5.6. *(LC)* P_{J_κ} *is a proper notion of forcing, naturally isomorphic to the countable support iteration of length κ of the forcing P_J. If the ideal J is in addition $\mathbf{\Pi}_1^1$ on $\mathbf{\Sigma}_1^1$, then no large cardinal assumptions are necessary.*

The same construction applies to products. For an iterable σ-ideal K on a Polish space X, an ordinal κ, and a countable set $a \subset \kappa$ let K_a be the collection of those Borel subsets of X^a which do not contain a rectangle $\Pi_{\beta \in a} B_\beta$ where B_β are Borel K-positive sets. It is not difficult to verify that if the collections K_a are in fact σ-ideals then these ideals form a directed system; let K_κ be its limit.

Theorem 5.5.7. *(LC) Suppose that K is an iterable σ-ideal such that the forcing P_K preserves category basis. Then the forcing P_{K_κ} is proper, it is naturally isomorphic to the countable support product of κ many copies of P_K, and it preserves category basis. If the ideal K is in addition $\mathbf{\Pi}_1^1$ on $\mathbf{\Sigma}_1^1$, then on large assumptions are necessary.*

Illfounded iterations can be treated in a similar way. For an iterable σ-ideal J on a Polish space X, a linearly ordered index set L, and a countable set $A \subset L$ let J_a be the σ-ideal on X^a as constructed in Section 5.4. It is not difficult to see that these ideals from a directed system. Let J_L be the limit of this system.

Theorem 5.5.8. *(LC) Suppose that J is an iterable category preserving σ-ideal. Then the forcing P_{J_L} is proper, it is an iteration of the forcing P_J along L, and it preserves Baire category.*

The proofs of the previous propositions are essentially identical to the propositions concerning the countable length iterations and products and as such are left to the reader. Note though that there always is a small additional ingredient in the argument. I do not have a general theorem saying something to the effect "if the forcings associated with the ideals in the directed system are proper then so is the limit." Once the properness is established though, the forcing with the directed system inherits all the Fubini properties from the ideals in the system.

It is possible to develop a very similar theory of countable and uncountable iterations, products, or illfounded iterations for the partial orders of the form P_I where I is a suitably definable σ-ideal of Baire subsets of X^L for some Polish space X and index set L. Since I have not found a single person in the world interested in this a priori promising line of research as yet, I refrain from developing it here.

5.5.2 Shooting a club subset of ω_1

All forcings described in Chapter 4 are $< \omega_1$-proper, and this feature propagates through the countable support iteration [64]. Thus, in the resulting generic extensions, various club-guessing principles will hold. If one wants to violate club guessing, it is necessary to use a more sophisticated approach. Let me demonstrate the approach on the problem of adding a club subset of ω_1 with finite intersection with every ground model set of ordertype ω.

Let $X = 2$ and $L = \omega_1$. Consider the directed system $\Im$ of ideals indexed by countable ordinals. For an ordinal $\alpha \in \omega_1$ let the ideal I_α be generated by the set $\{\vec{x} \in 2^\alpha : \vec{x}$ is not a characteristic function of a closed subset of $\alpha\}$ plus all sets $\{\vec{x} \in 2^\alpha : \{\beta \in \alpha : \vec{x}(\beta) = 1\} \cap b$ is infinite$\}$ as b runs through all subsets of α of ordertype ω. It is not difficult to check that these ideals indeed form a directed system; let $I = \lim \Im$. It is immediate to verify that $P_I \Vdash \vec{x}_{gen}$ is a characteristic function of a club subset of ω_1^V with finite intersection with every ground model set of ordertype ω. The following is a key fact; its proof is essentially a repetition of arguments such as Theorem 4.5.2.

Fact 5.5.9. *[83] 4.2.14. P_I is proper.*

What is the advantage of adding a closed unbounded subset of ω_1 in this seemingly complicated way over the usual finite side condition forcings? It turns out that one has the same tight control over the reals added by the directed system of ideals forcings as one does in the case of the forcings adding a single real. For example, one can show that the forcing P_I above has all the Fubini properties of Laver forcing, in particular, it does not add Cohen or random reals and this feature persists into its countable support iterations. In this way, it is possible to obtain new models of the Ciesielski–Pawlikowski style axioms of Section 6.1 in which various prediction principles on ω_1 do not hold, and therefore these prediction principles do not follow from the axioms. However, in the absence of definite applications I hesitate to develop this subject any further.

6

Applications

6.1 Cardinal invariant inequalities

The contents of this section was the original motivation behind the development of the theory presented in this book. It turns out that for many cardinal invariants $\mathfrak{x}$, $\mathfrak{y}$, if the inequality $\mathfrak{x} < \mathfrak{y}$ can be forced at all then there is an extension $M_{\mathfrak{y}}$ depending only on the invariant $\mathfrak{y}$ which must realize the inequality. Moreover, there is an axiomatization of this phenomenon, a sentence $\phi_{\mathfrak{y}}$ depending only on the invariant $\mathfrak{y}$ such that $M_{\mathfrak{y}} \models \phi_{\mathfrak{y}}$ and in all forcing extensions satisfying the sentence $\phi_{\mathfrak{y}}$ the inequality $\mathfrak{x} < \mathfrak{y}$ holds.

The canonical models $M_{\mathfrak{y}}$ turn out to be natural and well-understood; these are the models such as the iterated Sacks model or the iterated Laver model [46]. Using the methods of the present book, the question whether $M_{\mathfrak{y}} \models \mathfrak{x} < \mathfrak{y}$ translates into a question about Borel or projective sets. The canonical axiomatizations $\phi_{\mathfrak{y}}$ also predate the methods of this book; they have been discovered by Ciesielski and Pawlikowski [9] at least in the case of the iterated Sacks model. What is new in this section is exactly the argument that these models and axiomatizations have the strong absoluteness properties described in the previous paragraph.

In order to be able to state and prove the theorems I must introduce several notions first.

6.1.1 Ciesielski–Pawlikowski axioms

In their book [9], Ciesielski and Pawlikowski isolated a number of closely related axioms which hold in the iterated Sacks model, and used them to study various properties of the model. Here I will use just one of them and add a parameter to it so that I get a family of axioms which hold in various forcing extensions.

Definition 6.1.1. *Let I be a σ-ideal on a Polish space X. The CPA(I) game is played in ω_1 stages between Player I and II in the following fashion. At round $\alpha \in \omega_1$ Player I indicates an I, β_α-tree p_α for some countable ordinal $\beta_\alpha \in \omega_1$, and a Borel function $f_\alpha : [p_\alpha] \to 2^\omega$. Player II responds with a I, β_α-tree $q_\alpha \subset q_\alpha$. Player I wins if in the end, $2^\omega = \bigcup_\alpha f_\alpha''[q_\alpha]$.*

A simple observation shows that if CH holds then Player I has a winning strategy. He simply plays constant functions $f_\alpha : \alpha \in \omega_1$ such that their respective singleton ranges $y_\alpha : \alpha \in \omega_1$ cover the whole Cantor space. It may be the case that this is the only way how Player I can obtain a winning strategy.

Definition 6.1.2. *CPA(I) is the statement "$\mathrm{cov}^*(I) > \aleph_1$ and Player II does not have a winning strategy in the CPA(I) game." Here $\mathrm{cov}^*(I) = \min\{A \subset I : \bigcup A$ contains an I-positive Borel subset$\}$.*

The CPA definition above differs from the one I gave in [83] in that I use the invariant $\mathrm{cov}^*(I)$ in the place of $\mathrm{cov}(I)$. This change is irrelevant in the most important case of homogeneous ideals.

The axiom CPA(I) is really designed to hold in the iterated P_I model.

Proposition 6.1.3. *[9][83] Suppose that I is an iterable σ-ideal on a Polish space X generated by Borel sets. Then CPA(I) holds in the countable support iterated P_I extension.*

The requirement that the ideal I be generated by Borel sets can be replaced by other conditions if enough absoluteness is available.

Proof. Suppose for simplicity that CH holds, write P_γ for the countable support iteration of length γ of the poset P_I and choose a generic filter $G \subset P_{\omega_2}$. I have to show that CPA(I) holds in $V[G]$. The key tools are the $\aleph_2$-c.c. of the poset P_{ω_2} proved in [64], III.4.1 and the representation of the poset P_{ω_2} as found in Section 5.5.

First of all, $\mathrm{cov}^*(I) = \aleph_2$ in $V[G]$. By a chain condition argument every collection of $\aleph_1$ many codes for Borel sets in the ideal I belongs to some model $V[G \cap P_\gamma]$ fo a suitable ordinal $\gamma \in \omega_2$. By the absoluteness demands on the ideal I, the generic points past the ordinal γ added during the iteration fall out of all of these sets.

To complete the proof, I must show that no Player II's strategy for the CPA game is winning in the model $V[G]$. So let $\dot\sigma \in V$ be a P_{ω_2}-name for a strategy for Player II. By a chain condition argument again, there is an ordinal $\gamma \in \omega_2$ such that the partial evaluation $\dot\sigma/(G \cap P_\gamma) = \dot\sigma/G \cap V[G \cap P_\gamma]$ is Player II's strategy in the game in the model $V[G \cap P_\gamma]$. In this model CH holds and so Player I can enumerate all his legal moves in the CPA game as $\{\langle \alpha_\beta, p_\beta, f_\beta \rangle : \beta \in \omega_1\}$ and play them in succession. Player II follows the strategy to produce trees $\{q_\beta : \beta \in \omega_1\}$. I claim that this play of the game is winning in $V[G]$ for Player I, completing the proof. To see this, note that $V[G]$ is a P_{ω_2}-extension of $V[G \cap P_\gamma]$ by the factorization theorem of [2], 1.5.10. Work in the model $V[G \cap P_\gamma]$. If $p \in P_{\omega_2}$ is a condition and $\dot{x}$ is a P_{ω_2}-name for an infinite binary sequence, by Proposition 5.5.4 there is a condition $q \leq p$ in P_{ω_2} and a Borel map $f : X^{\mathrm{dom}(q)} \to 2^\omega$ such that $q \Vdash \dot{x} = f(\vec{x}_{gen} \restriction \mathrm{dom}(q))$. By the choice of Player I's moves there must be an ordinal $\beta \in \omega_1$ such that the unique order-preserving map $\pi : \alpha_\beta \to \mathrm{dom}(q)$ sends the tree p_β to q and the map f_β to f. Let $r \leq q$ be the image of the set q_β under the map π. Then clearly $r \Vdash \dot{x} \in f''_\beta r_\beta$. Since the condition p and the name $\dot{x}$ were arbitrary, it follows that $V[G] \models 2^\omega = \bigcup_\beta f''_\beta[q_\beta]$ and Player I won as desired. $\qquad\square$

Thus typically CPA holds in the iterated Sacks model, CPA(bounded) holds in the iterated Miller model and so on. Other models for the CPA variations are in a short supply. It is for example possible to use a countable support iteration of the forcing constructed in Section 5.5.2 to find a model of CPA(Laver) in which various club prediction principles on ω_1 fail.

It is not at all clear on the first sight what the consequences of an axiom like CPA(*I*) should be. It follows from the work in the following sections that the axioms imply equalities $\mathfrak{x} = \aleph_1$ for large classes of invariants $\mathfrak{x}$. Ciesielski and Pawlikowski [9] studied the special case of $I = \mathtt{ctble}$ and derived a number of other consequences. Many important open questions remain, such as whether any of these axioms imply $\mathfrak{c} = \aleph_2$.

6.1.2 Tame invariants

In order to state the absoluteness theorems in full generality, I need to introduce a syntactical class of cardinal invariants.

Definition 6.1.4. *Let X be a Polish space. A cardinal invariant $\mathfrak{x}$ is tame if it is defined to be the minimum size of a set $A \subset X$ with properties $\phi(A)$ and $\forall x \in X \, \exists y \in A \, \theta(x, y)$ where ϕ quantifies over natural numbers and elements of A only and θ is a projective formula not mentioning the set A. Real parameters are allowed. The set A is called a* witness *for $\mathfrak{x}$.*

Most cardinal invariants studied today are tame. There are exceptions to this rule, and with a little practice it becomes obvious that there is a thick line separating the tame and non-tame invariants. The methods of this section offer no hint on the treatment of non-tame invariants.

Example 6.1.5. $\mathfrak{a}$ is tame. It is defined as the smallest size of a maximal almost disjoint family. To write it in the tame form, let $X = \mathcal{P}(\omega)$, $\phi(A)$ will assert that the elements of A are mutually almost disjoint and $\theta(x, y)$ will assert that $x \cap y$ is infinite.

Example 6.1.6. $\mathfrak{b}$ is tame. It is the smallest possible size of a modulo finite unbounded subset of ω^ω, so $\phi(A) = ``A = A"$ and $\theta(x, y)$ will assert that $x \not< y$ modulo finite.

For a similar reason, invariants like $\mathfrak{s}$ and $\mathfrak{t}$ as well as all covering numbers are tame. The covering numbers can be written in such a way that the formula $\phi(A)$ is trivially true.

Example 6.1.7. (independently proved by Mildenberger) $\mathfrak{h}$ is not tame.

Proof. I will find a model M of ZFC+LC+$\mathfrak{h} > \aleph_1$ and a forcing in it which preserves cardinals, preserves tame invariants and collapses $\mathfrak{h}$ to $\aleph_1$. The model M is the iterated Mathias model and the forcing P is simply the algebra $\mathcal{P}(\omega)$ mod fin. In the model M, P is $\aleph_1$-distributive of size $\mathfrak{c} = \aleph_2$ and therefore does not collapse cardinals or tame invariants and by the results of [66] it forces $\mathfrak{h} = \aleph_1$ as desired. □

Example 6.1.8. (Mildenberger) $\mathfrak{g}$ is not tame.

Proof. In fact I will prove that if $\mathfrak{x}$ is a tame invariant and ZFC+LC proves $\mathfrak{g} \leq \mathfrak{x}$ then ZFC+LC proves $\mathfrak{d} \leq \mathfrak{x}$. To understand the meaning of this, note that ZFC proves $\mathfrak{g} \leq \mathfrak{d}$ by [2] Theorem 4.3.2. It will show that $\mathfrak{g}$ is not tame since $\mathfrak{g} < \mathfrak{d}$ holds in the finite support iterated Cohen model by [2], 4.3.3.

So suppose that $\mathfrak{x}$ is a tame invariant such that ZFC+LC does not prove $\mathfrak{d} \leq \mathfrak{x}$. There must be a countable model M of the consistent theory ZFC+LC+$\mathfrak{x} < \mathfrak{d}$, and by Theorem 6.1.11, in its iterated Miller extension $M[G]$, $\aleph_1 = \mathfrak{x} < \mathfrak{d} = \aleph_2$ holds. By [3] $\mathfrak{g} = \mathfrak{d} = \aleph_2$ holds in the iterated Miller model as well. Thus the model $M[G]$ stands witness to the consistency of the theory ZFC+LC+$\mathfrak{x} < \mathfrak{g}$. □

In the large-cardinal-free context, where not enough absoluteness is available, I will have use for the following weakening of the above definition:

Definition 6.1.9. *Let X be a Polish space. A cardinal invariant $\mathfrak{x}$ is very tame if it is defined as the minimum size of a set A with properties $\phi(A)$ and $\forall x \in \mathbb{R} \ \exists y \in A \ \theta(x, y)$ where $\phi(A) = \forall x_0, x_1 \cdots \in A \ \exists y_0, y_1, \cdots \in A \ \psi(\vec{x}, \vec{y})$ for some arithmetic formula ψ, θ is an analytic formula, and ZFC proves that for every countable set $a \subset \mathbb{R}$ with $\phi(a)$ there is a set $A \supset a$ such that $|A| = \mathfrak{x}$ and A is a witness for $\mathfrak{x}$.*

This notion is slightly more complicated, and due to its last clause, it is not Δ_0 in the definition of the invariant. However, it is still sufficiently intuitive and broad – I do not know of a single tame invariant in use today that would not be very tame.

Example 6.1.10. $\mathfrak{a}$ is very tame.

Proof. Suppose that B is an infinite maximal almost disjoint family, and a is a countable almost disjoint family. I will find a maximal almost disjoint family A extending a of the same cardinality as B. It is not difficult to produce a permutation $\pi : \omega \to \omega$ such that every set in a is modulo finite equal to the π-image of some set in B. Thus the family A can be obtained as the set $\{\pi''x : x \in B\}$ with possible finite modifications to some of its members. $\square$

Some invariants fail to be tame. This book offers no hint for general treatment of non-tame invariants.

6.1.3 Absoluteness theorems

I will provide the archetype result and its ZFC version.

Theorem 6.1.11. *[83] (LC) Suppose that I is an iterable σ-ideal on a Polish space X satisfying the second dichotomy. Suppose that $\underline{x}$ is a tame invariant. If $\underline{x} < \text{cov}^*(I)$ holds in some forcing extension then it holds in every forcing extension satisfying CPA(I).*

In other words, the "optimal" way to prove the consistency of inequality of the form $\underline{x} < \text{cov}^*(I)$ is to study an arbitrary model of CPA(I), such as the iterated P_I extension. It may be however that the easiest way to argue that $\underline{x} < \text{cov}^*(I)$ holds in this extension is to prove that it holds in some other extension and then use the theorem as a transfer tool. The theorem says that the iterated P_I extension will keep the values of as many tame cardinal invariants as possible at $\aleph_1$. The following are common special cases. The ideals are all homogeneous and so $\text{cov}(I) = \text{cov}^*(I)$ in all cases.

- $I =$ the ideal of countable sets. $\text{cov}(I) = \mathfrak{c}$ best increased by countable support Sacks forcing.
- $I =$ the ideal of σ-bounded sets in ω^ω. $\text{cov}(I) = \mathfrak{d}$ best increased by countable support iterated Miller forcing.
- $I =$ the ideal of meager sets. $\text{cov}(I)$ best increased by countable support iterated Cohen forcing. In this case finite support iteration gives a similar effect.

- I = the ideal of Lebesgue null sets. $\mathrm{cov}(I)$ best increased by countable support iterated Solovay forcing. In this case a large measure algebra will give a similar effect.
- I = the c_{min} ideal. $\mathrm{cov}(I)$ = [22] best increased by the countable support iterated c_{min} forcing.
- I = the Laver ideal. $\mathrm{cov}(I) = \mathfrak{b}$ best increased by countable support iteration of the Laver forcing.
- I = sets nowhere dense in $\mathcal{P}(\omega)$ mod fin. $\mathrm{cov}(I) = \mathfrak{h}$ best increased by countable support iteration of Mathias forcing.

Similar theorems can be proved for combinations of several ideals, stating for example

- I = the ideal of compact subsets of the Baire space and J = the Lebesgue null ideal. $\min\{\mathfrak{b}, \mathrm{cov}(\mathrm{null})\}$ best increased by the countable support iteration in which the Miller and Solovay forcings alternate.

There are several attractive corollaries of metamathematical flavor.

Corollary 6.1.12. *(Elimination of forcing) Let I and $\mathfrak{x}$ satisfy the assumptions of the theorem. Then ZFC+LC $\vdash \exists P$ $P \Vdash \mathfrak{x} < \mathrm{cov}^*(I)$ if and only if ZFC+LC+CPA(I) $\vdash \mathfrak{x} < \mathrm{cov}^*(I)$.*

The right-to-left implication is clear, since by Proposition 6.1.3 CPA(I) holds in the iterated P_I extension. For the left-to-right implication assume that $\mathfrak{x} < \mathrm{cov}(I)$ can provably be forced and argue in the theory ZFC+LC+CPA(I). $\mathfrak{x} < \mathrm{cov}^*(I)$ can be forced, and applying Theorem 6.1.11 in the generic extension, $\mathfrak{x} < \mathrm{cov}^*(I)$ holds in every inner model of it satisfying CPA(I), in particular it holds in the ground model. $\mathfrak{x} < \mathrm{cov}^*(I)$ follows.

Corollary 6.1.13. *(Separation at $\aleph_1$ and $\aleph_2$) Let I and $\mathfrak{x}$ satisfy the assumptions of the theorem. Then $\mathfrak{x} < \mathrm{cov}^*(I)$ is consistent with ZFC+LC if and only if $\aleph_1 = \mathfrak{x} < \mathrm{cov}^*(I) = \aleph_2$ is.*

The right-to-left implication is trivial. For the left-to-right implication look at any model of ZFC+LC+$\mathfrak{x} < \mathrm{cov}^*(I)$. There, $\mathfrak{x} < \mathrm{cov}^*(I)$ can be forced by the trivial poset and so an application of Theorem 6.1.11 in it says that $\mathfrak{x} < \mathrm{cov}^*(I)$ will hold in the iterated P_I extension. In that extension though, $\aleph_1 = \mathfrak{x} < \mathrm{cov}^*(I) = \aleph_2$.

Corollary 6.1.14. *(Mutual consistency) Let $I, \mathfrak{x}$ and $\mathfrak{y}$ satisfy the assumptions of the theorem. If each of $\mathfrak{x} < \mathrm{cov}^*(I)$ and $\mathfrak{y} < \mathrm{cov}^*(I)$ can be separately forced, then their conjunction can be forced as well.*

To see this, note that Theorem 6.1.11 says precisely that the iterated P_I extension will satisfy all such forceable inequalities simultaneously. This argument does not give *literal* mutual consistency, the statement "if each of $\mathfrak{x} < \mathrm{cov}^*(I)$ and $\mathfrak{y} < \mathrm{cov}^*(I)$ is consistent, then so is their conjunction." This statement is probably false in general, but I have no counterexample.

Proof. To prove Theorem 6.1.11, let X be a Polish space and $\mathfrak{x}$ be a tame invariant, defined as $\mathfrak{x} =$ the least size of a set $A \subset X$ such that $\phi(A) \wedge \forall x \in X \exists y \in A \; \theta(x, y)$. Consider the sentence $\psi = \exists A \subset X \; \phi(A) \wedge \forall \alpha \in \omega_1 \; \forall p$ an I, α-tree $\forall f : [p] \to X$ Borel $\exists y \in A \; \{\vec{z} \in B : \theta(f(\vec{z}), y)\} \notin I^\alpha$. This may be a mouthful, but note that this is a Σ_1^2 sentence with a universally Baire parameter, the predicate for the ideals $I^\alpha : \alpha \in \omega_1$.

Claim 6.1.15. $\mathfrak{x} < \mathrm{cov}^*(I)$ *implies* ψ.

Proof. Just look at any set $A \subset X$ such that $|A| < \mathrm{cov}^*(I)$ and $\phi(A) \wedge \forall x \in X \exists y \in A \; \theta(x, y)$ holds. I claim that the set A is a witness to the sentence ψ. To see this, choose a countable ordinal $\alpha \in \omega_1$, a Borel set $B \notin I^\alpha$ and a Borel function $f : B \to X$ and for contradiction assume that for every point $y \in A$ the set $C_y = \{\vec{z} \in B : \theta(f(\vec{z}), y)\}$ is in the ideal I^α. Choose winning strategies $\sigma_y : y \in A$ for Player I in the iteration games with the payoff sets C_y, and choose an I, α-tree p such that $[p] \subset B$. By induction on $\beta \in \alpha$ construct sequences $t_\beta \in p$ of length β which are legal answers to all the strategies $\sigma_y : y \in A$ simultaneously. To do that, at limit stage just take unions, and at a successor stage $\beta = \gamma + 1$ note that the sets $\sigma_y(t_\gamma) : y \in A$ are in the ideal I and therefore by the inequality $|A| < \mathrm{cov}^*(I)$ they cannot cover the I-positive Borel set $\{z : t_\gamma^\frown z \in p\}$. Thus it is possible to find a point in the latter set which is a legal response to all the strategies simultaneously.

In the end, the branch $\vec{z}$ of the tree p extending all the nodes $t_\beta : \beta \in \alpha$ belongs to the set B, but at the same time, it belongs to none of the sets $C_y : y \in A$ since it conforms to all the winning strategies $\sigma_y : y \in A$. Contradiction. $\square$

Now work in some forcing extension satisfying CPA(I). Consider a σ-closed poset P forcing CH. In the P extension, ψ holds: it holds in the forcing extension satisfying $\mathfrak{x} < \mathrm{cov}^*(I)$ by the previous claim, and it must then hold in every extension satisfying CH by Σ_1^2 absoluteness 1.4.12. Find a P-name $\dot{A} = \{\dot{y}_\alpha : \alpha \in \omega_1\}$ for a witness to the sentence ψ. Consider the following Player II's strategy in the CPA game: On the side, he is going to create a descending chain of conditions $r_\alpha : \alpha \in \omega_1$ in P such that p_α decides the point $\dot{y}_\alpha$. Also, writing $p_\alpha, \beta_\alpha, f_\alpha$ for α-th move of Player I, the condition r_α is going to identify some point y such that it is forced into the set $\dot{A}$ and the set $\{\vec{z} \in [p_\alpha] : \theta(f(\vec{z}), y)\}$ is not in the ideal I^{β_α}, and Player II will respond with some tree q such that $[q]$ is a subset of this set.

This cannot be a winning strategy for Player II, and therefore there is a winning counterplay by Player I against it. Let $A = \{y_\alpha : \alpha \in \omega_1\} \subset X$ be the set of the points such that $p_\alpha \Vdash \dot{y}_\alpha = \check{y}_\alpha$. A review of the definitions shows that A is a witness for the sentence ψ, in particular A witnesses $\mathfrak{x} = \aleph_1$! $\square$

The time has come to state the ZFC version of Theorem 6.1.11.

Theorem 6.1.16. *Suppose that I is an iterable $\mathbf{\Pi}_1^1$ on $\mathbf{\Sigma}_1^1$ σ-ideal. Suppose that $\mathfrak{x}$ is a very tame cardinal invariant. If $\mathfrak{x} < \mathrm{cov}^*(I)$ holds in some inner model of ZFC containing all ordinals or its generic extension, then $\aleph_1 = \mathfrak{x} < \mathrm{cov}^*(I)$ holds in every generic extension of every inner model of ZFC containing all ordinals whenever this extension satisfies CPA(I).*

The restriction to $\mathbf{\Pi}_1^1$ on $\mathbf{\Sigma}_1^1$ ideals rules out ZFC applications to such common models as the Laver model. There is however an important feature which makes this theorem applicable for several ideals I for which the assumptions of Theorem 6.1.11 are not satisfied. Note that Theorem 6.1.16 does not require the ideal I to satisfy the second dichotomy. This is made possible by the tight proof of the countable support iteration dichotomy for $\mathbf{\Pi}_1^1$ on $\mathbf{\Sigma}_1^1$ ideals. Thus Theorem 6.1.16 is applicable to such ideals as the E_0 ideal or the Silver forcing ideal.

Proof. To prove Theorem 6.1.16, let I be a σ-ideal satisfying the assumptions of the theorem; to simplify the notation assume that it is actually lightface Π_1^1 on Σ_1^1. Let $\mathfrak{x}$ be a very tame invariant defined as in Definition 6.1.9 by $\min\{|A| \subset \mathbb{R} : \phi(A) \wedge \forall x \in \mathbb{R}\, \exists y \in A\ \theta(x, y)\}$. Consider the sentence ψ saying that there is a countable set $a \subset \mathbb{R}$ satisfying ϕ, an ordinal $\alpha \in \omega_1$, a I, α-tree p and a Borel function $f : [p] \to \mathbb{R}$ such that for every countable set $b \supset a$ satisfying ϕ and every real $y \in b$ the set $\{\vec{z} \in [p] : \theta(f(\vec{z}), y)\}$ is in the ideal I^α. I will show that ψ is equivalent to a Σ_2^1 statement, $\psi \to \mathfrak{x} = \mathrm{cov}(I)$ and $\neg\psi \wedge \mathrm{CPA}(I) \to \mathfrak{x} < \mathrm{cov}^*(I)$. Then the theorem follows. Suppose that $\mathfrak{x} < \mathrm{cov}(I)$ holds in some inner model containing all ordinals, as in all generic extensions. Then the inner model must satisfy $\neg\psi$ and by Shoenfield's absoluteness ψ must fail in all inner models containing all ordinals. So in any such a model satisfying CPA(I) $\mathfrak{x} < \mathrm{cov}^*(I)$ must hold. $\square$

Claim 6.1.17. *ψ implies $\mathfrak{x} = \mathrm{cov}(I)$.*

Proof. Suppose that ψ holds as witnessed by some $a \subset X$, $\alpha \in \omega_1$, p and f. By the homogeneity of the invariant $\mathfrak{x}$ it will be enough to show that for every set $A \subset X$ such that $a \subset A$, $\phi(A)$ and $|A| < \mathrm{cov}^*(I)$ there is a point $x \in X$ such that no point $y \in A$ satisfies $\theta(x, y)$. Look at a point $y \in A$. Find a countable set $b \subset A$ such that $\phi(b)$, $a \subset b$ and $y \in b$, and look at the sentence ψ: it says that the set $B_y = \{\vec{z} \in [p] : \theta(f(\vec{r}), y)\}$ is in the ideal I^α. So fix the corresponding Player I's winning strategies σ_y in the iteration games with payoff sets B_y, for all $y \in A$. I will produce a sequence $\vec{z} \in [p]$ which is a legal counterplay against all the strategies $\sigma_y : y \in A$ simultaneously. Then the sequence $\vec{z}$ falls out of all sets $B_y : y \in A$, in other words the point $x = f(\vec{r})$ has the required property. The construction of the sequence $\vec{z} = \langle z_\beta : \beta \in \alpha \rangle$ proceeds by induction. At the stage β, the set $\{s : \exists \vec{t} \ \langle z_\gamma : \gamma \in \beta \rangle \frown s \frown \vec{t} \in B\}$ is Borel and I-positive. On the other hand, the strategies $\{\sigma_y : y \in A\}$ offer only $< \mathrm{cov}^*(I)$ many countable sets to be avoided, which then can be easily done. The limit stages present no difficulty due to the σ-closure condition in the definition of an I-perfect set. $\square$

Claim 6.1.18. $\neg \psi \wedge CPA(I)$ *implies* $\aleph_1 = \mathfrak{x} < \mathrm{cov}(I)$.

Proof. Suppose ψ fails and consider the following Player II's strategy in the CPA(I) game. As the game develops, he will construct an increasing chain $\{a_\beta : \beta \in \omega_1\}$ of countable sets of reals satisfying the property ϕ. Note that the property is preserved under unions of increasing sequences due to its simple syntactical form. When Player I indicates an I, α_β-tree p_β and a Borel function $f : [p_\beta] \rightarrow 2^\omega$, Player II will use the failure of ψ to $\bigcup_{\gamma \in \beta} a_\gamma$, p_β and f to produce a countable set a_β and a point $y_\beta \in a_\beta$ such that the set $\{\vec{z} \in [p_\beta] : \theta(f(\vec{z}), y_\beta)\}$ is not in the corresponding iterated Fubini power of the ideal I. The latter set is analytic and so it must contain all branches of a I, α_β-tree $q_\beta \subset p$. this will be Player II's move. CPA(I) says that this strategy is not winning for Player II and supplies Player I's counterplay which wins, i.e. $\bigcup_{\beta \in \omega_1} f''_\beta[q_\beta] = 2^\omega$. Then the set $A = \bigcup_\beta a_\beta$ that Player I produced on the side during the run of this game is a witness to $\aleph_1 = \mathfrak{x} < \mathrm{cov}^*(I)$: $\phi(A)$ holds since the set A is an increasing union of sets satisfying ϕ, and for every real x there is an ordinal $\beta \in \omega_1$ such that $x \in f''_\beta[q_\beta]$, and by the choice of the tree q_β, there is a real $y \in a_\beta \subset A$ such that $\theta(x, y)$ holds. $\square$

Claim 6.1.19. ψ *can be written in a Σ^1_2 form.*

Proof. Consider the statement $\bar{\psi}$ saying that there is a countable model M of a large fraction of ZFC and its elements $a, x, p, \dot{z}$ such that $M \models a \subset X$ is a countable set satisfying ϕ, x is a countable ordinal, p is a condition in the countable support iteration of P_I-forcing along x and $\dot{b}$ is a name for an infinite binary sequence, such that M is wellfounded and for every countable set $b \supset a$ satisfying ϕ and every point $y \in b$ the set $\{\vec{z} \in Y^x : \vec{z}$ is M-generic and $\theta(\dot{b}/\vec{z}, y)\}$ is I^x small, where Y is the domain Polish space of the σ-ideal I. This is a Σ^1_2 statement since the membership in the ideal I^x for analytic sets is a coanalytic statement by Corollary 5.1.11. To see that $\psi \rightarrow \bar{\psi}$, let a, α, p and f witness the statement of ψ and choose a countable elementary submodel M of large enough structure containing these objects, let

$\dot{b} = \dot{f}(\vec{z}_{gen})$. Then $M, a, p, \dot{b}$ witness the sentence $\bar{\psi}$. Vice versa, if $M, a, x, p, \dot{z}$ witness $\bar{\psi}$, then the ordering x must be isomorphic to some ordinal α and by a properness argument there is a I, α-tree q consisting solely of M-generic sequences meeting the condition p. Then the set B together with the function $f(\vec{z}) = \dot{b}/\dot{z}$ witnesses the sentence ψ. $\qquad\square$

6.2 Duality theorems

The duality heuristic in the science of cardinal invariants is the following statement: if I, J are suitably definable σ-ideals and ZFC proves $\mathrm{cov}(I) \leq \mathrm{non}(J)$ then ZFC also proves $\mathrm{non}(I) \geq \mathrm{cov}(J)$. In fact, every inequality involving invariants add, non, cov, cof of suitably definable invariants can be dualized by switching add with cof and non with cov and reversing the inequality sign, and the duality heuristic will say that an inequality is provable if and only if its dual is. This is really completely false in general, nevertheless it has provided a valuable guidance in many situations. In this section I will provide several theorems that offer formal insight into the heuristic.

Example 6.2.1. Suppose $\leq$ is some simply definable prewellordering on a Polish space X of ordertype ω_1. Let I be a σ-ideal generated by those subsets of X the ranks of whose elements are bounded in ω_1. Clearly ZFC proves that $\mathrm{cov}(I) = \mathrm{non}(I) = \aleph_1$. Now consider the ideal J of countable sets on 2^ω and note that the duality heuristic fails: ZFC proves $\aleph_1 = \mathrm{cov}(I) \leq \mathrm{non}(J) = \aleph_1$, but it certainly does not prove $\mathfrak{c} = \mathrm{cov}(J) \leq \mathrm{non}(I) = \aleph_1$, or at least so we like to think.

In some cases the duality theorems immediately follow from preservation theorems in Section 6.3.

Example 6.2.2. Suppose that ϕ is a pavement submeasure on a Polish space X. ZFC+LC proves $\mathfrak{d} \leq \mathrm{non}^*(I_\phi)$ if and only if ZFC+LC proves $\mathfrak{b} \geq \mathrm{cov}^*(I_\phi)$. To argue for this, I will show that both statements are equivalent to ZFC+LC proving $I_\phi \perp J$ where J is the Miller ideal on ω^ω.

First note that from $I_\phi \perp J$ it abstractly follows that $\mathfrak{d} \leq \mathrm{non}^*(I_\phi)$ and $\mathfrak{b} \geq \mathrm{cov}^*(I_\phi)$. Just find a ϕ-positive Borel set $B \subset X$ and a Borel set $D \subset B \times \omega^\omega$ whose vertical sections are σ-bounded and the horizontal sections of the complement are ϕ-small. A review of the definitions then shows that if $E \subset \omega^\omega$ is a dominating set then the collection $\{(B \times \omega^\omega \setminus D)^y : y \in E\}$ consists of ϕ-null sets and covers the whole set B, and if $F \subset B$ is a ϕ-positive set then for every point $x \in F$ choose a function $f_x \in \omega^\omega$ σ-bounding the section D_x; the set $\{f_x : x \in F\}$ is not σ-bounded. This proves the two cardinal invariant inequalities.

Now suppose that ZFC+LC does not prove $I_\phi \perp J$, find a model M in which this fails, and work in it. As P_J, the Miller forcing, is homogeneous, Proposition 3.2.12 shows that this implies $\neg J \perp \phi$. By Theorem 6.3.13, this feature propagates through the countable support iteration of P_J. In other words, in the iterated Miller extension it will be the case that $\text{non}^*(I) = \aleph_1 < \mathfrak{d}$ and so ZFC+LC does not prove $\mathfrak{d} \leq \text{non}^*(I_\phi)$. For the other inequality note that the ideal J is generated by closed sets, therefore by Theorem 6.3.9 the relation $\neg I_\phi \perp J$ propagates through the countable support iterations of the poset P_{I_ϕ}. In other words, in the iterated P_{I_ϕ}-extension the ground model reals will be still unbounded and $\mathfrak{b} = \aleph_1 < \text{cov}^*(I_\phi)$, and so ZFC+LC does not prove $\mathfrak{b} \geq \text{cov}^*(I_\phi)$.

Such situations are very rare though. In the following subsections I will isolate several theorems in which no preservation theorem is needed.

6.2.1 The countable ideal

Theorem 6.2.3. *[83] Suppose that J is a σ-ideal on a Polish space X generated by vertical sections of some projective set $2^\omega \times X$. If ZFC+LC proves $\text{cov}(J) = \mathfrak{c}$ then ZFC+LC proves $\text{non}(J) \leq \aleph_2$. If the ideal J is generated by vertical sections of an analytic set then the large cardinal assumptions can be dropped.*

Proof. The argument is based on the following combinatorial fact of independent interest. Let I be the ideal of countable subsets of the Cantor space 2^ω.

Lemma 6.2.4. *For every countable ordinal $\alpha \in \omega_1$, $\text{non}(I^\alpha) \leq \aleph_2$.*

Proof. The key tool in the argument is the following

Fact 6.2.5. *(Shelah) For every limit countable ordinal α and every regular cardinal $\lambda > \aleph_1$ there is a stationary set $S \subset \lambda$ consisting of ordinals of cofinality ω and a sequence $\langle C_\delta : \delta \in S \rangle$ of sets such that*

1. *for every ordinal $\delta \in S$ $C_\delta \subset \delta$ is a closed cofinal set of ordertype α;*
2. *for ordinals $\gamma, \delta \in S$ the intersection $C_\gamma \cap C_\delta$ is an initial segment of both of the sets C_γ, C_δ;*
3. *for every closed unbounded set $E \subset \lambda$ there is an ordinal $\delta \in S$ such that $C_\delta \subset E$.*

This fact was announced in [60], p. 136, remark 2.14A. The proof is in [59], available from the Mathematics ArXiv. The reader wishing to stick to published results will have to use a well-known much stronger club-guessing property at $\aleph_3$ [63], and will only get a larger upper bound of $\aleph_3$ for $\text{non}(\text{ctble}^\alpha)$.

So fix a nonzero countable ordinal α, without loss of generality α is limit. The proof of the lemma divides into two branches according to the value of $\mathfrak{c}$. If $\mathfrak{c} \leq \aleph_2$

then we are done since $|(2^\omega)^\alpha| = c \leq \aleph_2$. If $c \geq \aleph_3$ then fix a collection $\{r_\delta : \delta \in \omega_2\}$ of distinct infinite binary sequences and the club guessing sequence $\langle C_\delta : \delta \in S \rangle$ with the set $S \subset \omega_2$ from Fact 6.2.5. For each ordinal $\delta \in S$ define the α-sequence $\vec{r}_\delta \in (2^\omega)^\alpha$ to include the real numbers indexed by the nonaccumulation elements of the set C_δ. The proof will be complete once I show that the set $\{\vec{r}_\delta : \delta \in S\}$ is $\texttt{ctble}^\alpha$-positive.

For the ease of notation regard the sequences $\vec{r}_\delta$ as indexed by the nonaccumulation points of the set C_δ. Let σ be Player I's strategy in the iteration game; I must produce an ordinal $\delta \in S$ such that the sequence $\vec{r}_\delta$ is a valid counterplay against that strategy. To that end, fix a continuous tower $\langle M_\delta : \delta \in \omega_2 \rangle$ of elementary submodels of large enough structure, all of them of size $\aleph_1$ and containing ω_1 as a subset. Let $E = \{\delta \in \omega_2 : M_\delta \cap \omega_2 = \delta\}$. The set E is closed unbounded, so there must be an ordinal $\delta \in S$ such that $C_\delta \subset E$. Then $\vec{r}_\delta$ is the sought sequence. Indeed, for every ordinal γ among the first α nonaccumulation points of the set C_δ the set $C_\delta \cap \gamma$ is in the model M_γ by (2) above. So even the sequence $\vec{r}_\delta \restriction \gamma$ and Player I's challenge $\sigma(\vec{r}_\delta \restriction \gamma)$ at round γ are in the model M_γ. But that challenge is just a countable set, so $\sigma(\vec{r}_\delta \restriction \gamma) \subset M_\gamma$ does not contain the point $r_\gamma \notin M_\gamma$ as desired. $\qquad\square$

It may be of certain interest to give a diametrally different argument for the weaker

Lemma 6.2.6. $\texttt{non}(\texttt{ctble}^\omega) \leq \aleph_{\omega+1}$.

Proof. Recall the basic pcf structure fact:

Fact 6.2.7. *[63], II.1.5. There is an increasing sequence $\langle \kappa_n : n \in \omega \rangle$ of cardinals below $\aleph_\omega$ such that the true cofinality of the product $\prod_n \kappa_n$ modulo finite is $\omega_{\omega+1}$, meaning that there is a modulo finite increasing cofinal sequence $\langle f_\gamma : \gamma \in \omega_{\omega+1} \rangle$ in the product.*

To see that $\texttt{non}(\texttt{ctble}^\omega) \leq \aleph_{\omega+1}$ assume without loss that $c \geq \aleph_\omega$ and fix a sequence $\langle r_\gamma : \gamma \in \omega_\omega \rangle$ of distinct infinite binary sequences. For each function f_γ from the above Fact and every integer n let $\vec{r}_{\gamma n} \in \mathbb{R}^\omega$ be the sequence defined by $\vec{r}_{\gamma n}(m) = r_{f_\gamma(n+m)}$. It will be enough to prove that the set $\{\vec{r}_{\gamma n} : \gamma \in \omega_{\omega+1}, n \in \omega\}$ is $\texttt{ctble}^\omega$-positive. Suppose that σ is Player I's strategy in the iteration game, and let $g \in \prod_n \kappa_n$ be a function defined by $g(n) = \sup\{\gamma \in \kappa_n :$ for some partial play of the game respecting the strategy σ and using just the reals indexed below κ_{n-1} as Player II's answers, the real r_γ belongs to some Player I's move$\}$. Since the functions $f_\gamma : \gamma \in \omega_{\omega+1}$ are cofinal in the product $\prod_n \kappa_n$, there will be an ordinal $\gamma \in \omega_{\omega+1}$ and an integer $n \in \omega$ such that the function f_γ dominates g from n on. The reader can easily check that the sequence $\vec{r}_{\gamma n}$ is a legal sequence of answers against the strategy σ as desired. $\qquad\square$

The remainder of the proof is easy. Suppose that ZFC+LC proves $\text{cov}(J) = \mathfrak{c}$. Then ZFC+LC proves that this equality holds in the iterated Sacks extension. Now argue formally in the theory ZFC+LC. The work of Sections 5.5 and 5.1 shows that there is a countable ordinal $\alpha \in \omega_1$, a Borel I^α-positive Borel set $B \subset (2^\omega)^\alpha$, and a Borel function $f: B \to X$ such that f-preimages of J-small sets are ctble^α-small. The function f just represents a point of the space X which is forced to fall out of all ground model coded J-small sets. Now a separate argument is necessary to prove that ctble^α is a homogeneous ideal and therefore the set B can be taken as the whole space $(2^\omega)^\alpha$. Let $E \subset (2^\omega)^\alpha$ be a ctble^α-positive set from Lemma 6.2.4. A review of the definitions shows that $f''E$ is a J-positive set of size $\leq \aleph_2$! □

The cardinal bound $\aleph_2$ in the theorem is the best possible; it cannot be replaced with $\aleph_1$.

Example 6.2.8. Let J be the σ-ideal on $(2^\omega)^\omega$ σ-generated by the sets $A_g = \{\vec{r} \in \mathbb{R}^\omega : \exists i \in j \in \omega \; \vec{r}(j) = g(\vec{r}(i))\}$ as g varies over all Borel functions from 2^ω to 2^ω. Then ZFC proves $\text{cov}(J) = \mathfrak{c}$ and ZFC+PFA proves $\text{non}(J) = \aleph_2$.

Proof. If $\{g_\gamma : \gamma \in \kappa\}$ are $< \mathfrak{c}$ many Borel functions from 2^ω to 2^ω, by induction on $j \in \omega$ find binary sequences $\vec{r}(i)$ so that $\vec{r}(i)$ is not in the set $\{g_\gamma(\vec{r}(j)) : j \in i, \gamma \in \kappa\}$ of size $< \mathfrak{c}$. Clearly the sequence $\vec{r} \in (2^\omega)^\omega$ is not in any of the sets A_{g_γ} showing that $\text{cov}(J) = \mathfrak{c}$.

Now work in the context of PFA and fix an arbitrary set $A \subset (2^\omega)^\omega$ of size $\aleph_1$. I will show that $A \in J$, and therefore $\text{non}(J) = \aleph_2$. Let $r_\alpha : \alpha \in \omega_1$ be an enumeration of the set X of all the reals occuring on the sequences in the set A. By an application of PFA to the Baumgartner's poset adding a club with finite conditions, there is a club $C \subset \omega_1$ such that for every sequence $\vec{r} \in A$ there are integers $i \in j$ and two successive elements of the club C such that the reals $\vec{r}(i), \vec{r}(j)$ were enumerated between these. Let $f_n : X \to X$, for $n \in \omega$ be functions such that for every ordinal $\alpha \in \omega_1$ the functional values $\{f_n(r_\alpha) : n \in \omega\}$ include every real enumerated before the least element of the club C larger than α. By an application of MA to a c.c.c. coding poset there are Borel functions $\{g_n : n \in \omega\}$ such that for every integer n $f_n \subset g_n$. It is immediately clear that $A \subset \bigcup_n A_{g_n}$. □

6.2.2 The c_{min} ideal and variations

Recall the definition of the c_{min} ideal of Section 4.1.5. The function $c_{min} : [2^\omega]^2 \to 2$ is defined by $c_{min}(x, y) = x\Delta y \mod 2$ and the c_{min} ideal, which I will here denote by I_2, is generated by the c_{min}-monochromatic sets. A natural generalization of this procedure yields partitions $c_n : [2^\omega]^2 \to n$ and σ-ideals I_n on the Cantor space 2^ω

by setting $c_n(x, y) = x \Delta y \mod n$ and defining the ideal I_n as σ-generated by sets $A \subset 2^\omega$ such that $c_n'' A^2 \neq n$.

The basic feature of these ideals from the set theoretic point of view is the following.

Fact 6.2.9. *For every number* $n \in \omega$, $\text{cov}(I_{n+1}) \leq \text{cov}(I_n)$ *and* $\text{cov}(I_n)^{+n} \geq \mathfrak{c}$.

In particular, the invariants $\text{cov}(I_n) : n \in \omega$ form a nonincreasing sequence of cardinals and therefore stabilize at some point. ZFC does not decide the value of the natural number at which this happens. Also, if $\mathfrak{c}$ is a limit cardinal, all of these invariants are equal to $\mathfrak{c}$.

Theorem 6.2.10. *[67] (LC) Suppose that J is a σ-ideal on a Polish space X generated by vertical sections of some projective set $2^\omega \times X$. If ZFC+LC proves $\text{cov}(J) \geq \text{cov}(I_n)$ then ZFC+LC proves $\text{non}(J) \leq \aleph_n$. If the ideal J is generated by vertical sections of an analytic set then the large cardinal assumptions can be dropped.*

Theorem 6.2.11. *[67] (LC) Suppose that J is a σ-ideal on a Polish space X generated by vertical sections of some projective set $2^\omega \times X$. If ZFC+LC proves $\text{cov}(I) \geq \min_n \text{cov}(I_n)$ then ZFC+LC proves $\text{non}(I) \leq \aleph_{\omega_2+1}$. If the ideal J is generated by vertical sections of an analytic set then the large cardinal assumptions can be dropped.*

Proof. The arguments are the same in both cases, and they use the line of thinking from the previous section. The following lemmas are critical.

Lemma 6.2.12. *For every number $n \in \omega$ and every countable ordinal $\alpha \in \omega_1$, $\text{non}(I_n^\alpha) \leq \aleph_{n+1}$.*

Lemma 6.2.13. *If K is a transfinite Fubini product of any countable combination of the ideals $I_n : n \in \omega$ then $\text{non}(K) \leq \aleph_{\omega_2+1}$.*

In order to prove these statements, for every natural number $n \in \omega$ I will consider ideals J_n on $(2^\omega)^{n+1}$ defined in the following way. For every number $k \in n+1$ and every function $f : (2^\omega)^{n+1} \setminus \{k\} \to 2^\omega$ the set $A_f = \{x \in (2^\omega)^{n+1} : x(k) = f(x \upharpoonright (n+1 \setminus \{k\}))\}$ is in the ideal J_n, and the ideal J_n is σ-generated by all sets of this form. The ideal J_n is trivial if $\mathfrak{c} = \aleph_n$, but I will have to study its behavior in more complex situations than that.

For every number $n \in \omega$ let $g : (2^\omega)^{n+1} \to 2^\omega$ be the function defined by $g(x)(m \cdot n + k) = x(k)(m)$ whenever $k \leq n$ and $m \in \omega$. A simple observation now shows that preimages of I_n-small sets are J_n-small. It immediately follows that it will be enough to prove the lemmas for the ideals J_n in place of I_n.

Towards the proof of Lemma 6.2.12 fix a number $n > 0$ and a countable ordinal $\alpha \in \omega_1$, without loss assuming that α is limit. For every number $m \in n$ use Fact 6.2.5 to choose a club guessing sequence $\vec{C}_m = \langle C^m_\delta : \delta \in \omega_{m+2} \rangle$ in ω_{m+2}. I may certainly assume that $\mathfrak{c} \geq \aleph_{n+1}$ and so we can choose a sequence $\vec{s} = \langle s_\delta : \delta \in \omega_{n+1} \rangle$ of pairwise distinct elements of 2^ω. For every n-tuple $\vec{\delta} \in \prod_{m \in n} \omega_{m+2}$ such that the sets $C^m_{\delta(m)}$ have ordertype α, let $\langle \vec{\delta}(m)(\beta) : \beta \in \alpha \rangle$ enumerate the nonaccumulation points of these sets in the increasing order, and let $\vec{r}_{\vec{\delta}} \in (((2^\omega)^n)^\alpha)$ be the α-sequence whose β-th element is the point $\langle s_{\vec{\delta}(m)(\beta)} : m \in n \rangle$ in the space $(2^\omega)^n$. It will be enough to show that the set $\{\vec{r}_{\vec{\delta}} : \vec{\delta} \in \prod_{m \in n} \omega_{m+2}\}$ is J^α_n-positive.

To prove this, for every strategy σ for Player I in the iteration game I need to find an n-tuple $\vec{\delta}$ such that the sequence $\vec{r}_{\vec{\delta}}$ is a legal counterplay against the strategy σ. So fix the strategy σ and by downward induction on $m \in n$ find

- a continuous increasing $\in$-tower $\vec{M}_m = \langle M^m_\delta : \delta \in \omega_{m+2} \rangle$ of elementary submodels of a large enough structure, each of them of size $\aleph_{m+1}$ and such that $M^m_\delta \cap \omega_{m+2} \in \omega_{m+2}$. In particular, we require $\vec{C}_k : k \in n+1, \vec{s}, \sigma \in M^m_0$. Let $E_m = \{\delta \in \omega_{m+2} : M^m_\delta \cap \omega_{m+2} = \delta\}$; this is a closed unbounded subset of ω_{m+2}.
- an ordinal $\delta_m \in \omega_{m+2}$ such that the set $C^m_{\delta_m} \subset \delta_m$ is cofinal of ordertype α and it is a subset of the club E_m. For every two numbers $k \in m \in n$ we demand that $\delta_m \in M^k_0$.

Let $\vec{\delta} = \langle \delta_m : m \in n \rangle$. I claim that the sequence $\vec{r}_{\vec{\delta}}$ is the desired legal counterplay against the strategy σ. So look at an arbitrary round $\beta \in \alpha$ and suppose that the sequence does constitute a legal counterplay up to this point. What happens at round β?

The important observation is that for every integer $k \in n$, the play up to this round is in the model $M^k_{\vec{\delta}(k)(\beta)}$ since it is defined from objects that belong to this model. In particular, one of the parameters in the definition is the set $C^k_{\vec{\delta}(k)} \cap \vec{\delta}(k)(\beta)$ which is in the model by the coherence requirement (2) in Fact 6.2.5.

The strategy σ now indicates functions $\{f_{mk} : k \in n, m \in \omega\}$ such that f_{mk} is a function from $(2^\omega)^{n \setminus \{k\}}$ to 2^ω. We must show that the point $\vec{r}_{\vec{\delta}}(\beta)$ is not contained in the graph of any of these functions. So choose integers $m \in \omega$ and $k \in n$. Consider the set $Y = \{f_{mk}(\vec{u}, \vec{v}) : \vec{u} \in (2^\omega)^k$ is a sequence all of whose entries are on the $\vec{s}$-sequence, indexed by ordinals smaller than ω_{k+1} and $\vec{v} \in (2^\omega)^{n \setminus (k+1)}$ is a sequence all of whose entries are on the $\vec{s}$-sequence, indexed by ordinals in the set $\bigcup_{k+2 \in l \in n+2} C^l_{\vec{\delta}(l)}\}$. This set is of size $< \aleph_{k+2}$ and it belongs to the model $M^k_{\vec{\delta}(k)(\beta)}$. Thus $Y \subset M^k_{\vec{\delta}(k)(\beta)}$, in particular $s_{\vec{\delta}(k)(\beta)} \notin Y$, which by the definition of the set Y

means that the point $\vec{r}_{\vec{s}}(\beta) = \langle s_{\vec{\delta}(l)(\beta)} : l \in n \rangle$ is not on the graph of the function f_{mk} as desired. Lemma 6.2.12 follows.

Towards the proof of Lemma 6.2.13, let α be an arbitrary countable ordinal. For definiteness I will deal with the transfinite Fubini iteration K of a sequence of σ-ideals of length $\omega \cdot \omega \cdot \alpha$ whose $(\omega \cdot \beta + n)$-th element is the ideal J_n. Fix several objects whose existence is provable in ZFC:

- An increasing sequence $\vec{\kappa} = \langle \kappa_\delta : \delta \in \omega_2 \rangle$ of regular cardinals below $\aleph_{\omega_2}$ such that the true cofinality of their product modulo the bounded ideal on ω_2 is ω_{ω_2+1}, from [63], Chapter II, Theorem 1.5. This means that there is a sequence $\vec{h} = \langle h_\gamma : \gamma \in \omega_{\omega_2+1} \rangle$ of functions in $\prod_\delta \kappa_\delta$ which is increasing and cofinal in the modulo bounded ordering. Fix such a sequence.
- A club guessing sequence $\vec{C} = \langle C_\gamma : \gamma \in \omega_{\omega_2+1} \rangle$ from Fact 6.2.5. The sequence will guess closed unbounded subsets of ω_{ω_2+1} by segments of length $\omega \cdot \omega \cdot \alpha$. For every ordinal $\gamma \in \omega_{\omega_2+1}$ let $C(\gamma, \beta)$ denote the β-th nonaccumulation point of the set C_γ.
- A club guessing sequence $\vec{D} = \langle D_\delta : \delta \in \omega_2 \rangle$ from Fact 6.2.5. The sequence will guess closed unbounded subsets of ω_2 by segments of length $\omega \cdot \omega \cdot \alpha$ again, with similar notational convention for as in the previous item, using the symbol $D(\delta, \beta)$ for the β-th nonaccumulation point of the set D_δ.
- Without harm we may assume that $\mathfrak{c} > \aleph_{\omega_2}$. So fix a sequence $\vec{s} = \langle s_\gamma : \gamma \in \omega_{\omega_2+1} \rangle$ of pairwise distinct elements of 2^ω.

Now suppose that $\gamma \in \omega_{\omega_2+1}$ and $\delta \in \omega_2$ are ordinals such that the ordertypes of the sets C_γ and D_δ are both $\omega \cdot \omega \cdot \alpha$. Define a $\omega \cdot \alpha$ sequence $\vec{r}_{\gamma\delta}$ by setting its $\omega \cdot \beta + n$-th element to be the point in the space $(2^\omega)^{n+1}$ whose k-th coordinate for every number $k \in n+1$ is the point on the s sequence indexed by the ordinal $h_{C(\gamma, \omega \cdot (\omega \cdot \beta+n)+k)}(D(\delta, \omega \cdot (\omega \cdot \beta+n)+n-k))$. I will show that the collection $\{\vec{r}_{\gamma\delta} : \gamma \in \omega_{\omega_2+1}, \delta \in \omega_2\}$ is K-positive, proving the lemma. This means that for every Player I's strategy σ in the iteration game I must find ordinals $\gamma \in \omega_{\omega_2+1}$ and $\delta \in \omega_2$ such that the sequence $\vec{r}_{\gamma\delta}$ is a legal counterplay against the strategy.

Fix a continuous increasing $\in$-tower $\langle M_\gamma : \gamma \in \omega_{\omega_2+1} \rangle$ of elementary submodels of large enough structure, each of them of size $\aleph_{\omega_2}$ and such that $M_\gamma \cap \omega_{\omega_2+1} \in \omega_{\omega_2+1}$. In particular, $\vec{\kappa}, \vec{h}, \vec{C}, \vec{D}, \sigma \in M_0$. Let $E = \{\gamma \in \omega_{\omega_2+1} : M_\gamma \cap \omega_{\omega_2+1} \in \omega_{\omega_2+1}\}$. Since this is a closed unbounded subset of ω_{ω_2+1}, there must be an ordinal γ such that the set $C_\gamma \subset \gamma$ is cofinal of ordertype $\omega \cdot \omega \cdot \alpha$ and $C_\gamma \subset E$.

Also, fix a continuous increasing $\in$-tower $\langle N_\delta : \delta \in \omega_2 \rangle$ of elementary submodels of large enough structure, each of them of size $\aleph_1$ and such that $N_\delta \cap \omega_2 \in \omega_2$. In particular, $\vec{\kappa}, \vec{h}, \vec{C}, \vec{D}, E, \sigma, \gamma \in N_0$. Let $F = \{\delta \in \omega_2 : N_\delta \cap \omega_2 = \delta\}$. Since this is a closed unbounded set, there must be an ordinal $\delta \in \omega_2$ such that the set $D_\delta \subset \delta$ is cofinal of ordertype $\omega \cdot \omega \cdot \alpha$ and $D_\delta \subset F$.

I claim that the sequence $\vec{r}_{\gamma\delta}$ is a legal counterplay against the strategy σ. To see this, consider the situation at round $\omega \cdot \beta + n$ for some ordinal $\beta \in \alpha$ and number $n \in \omega$. Suppose that up to this point, the sequence constituted a legal partial play; we want to see that it will provide a legal answer even in this round. The strategy σ commands Player I to play functions $\{f_{mk} : k \in n+1, m \in \omega\}$ such that for each $k \in n+1$ and each $m \in \omega$ the function maps $(2^\omega)^{n+1 \setminus \{k\}}$ to (2^ω). I must show that the $n+1$-tuple $\vec{\imath} = \vec{r}_{\gamma\delta}(\omega \cdot \beta + n) \in (2^\omega)^{n+1}$ is not on the graph of any of these functions, that is $\vec{\imath}(k) \neq f_{mk}(\vec{\imath} \upharpoonright (n+1 \setminus \{k\}))$.

To this end, fix integers $k \in n+1$ and $m \in \omega$. Define a function $g \in \prod_\eta \kappa_\eta$ by letting $g(\eta)$ to be the supremum of the set $\{\xi \in \kappa_\eta : s_\xi = f_{mk}(\vec{u}, \vec{v})$, where $\vec{u} \in (2^\omega)^k$ is a sequence all of whose entries are on the $\vec{s}$-sequence, and are indexed by ordinals $< \sup\{\kappa_{\eta'} : \eta' \in \eta\}$, and $\vec{v} \in (2^\omega)^{(n+1) \setminus (k+1)}$ is a sequence all of whose entries are on the $\vec{s}$-sequence and are indexed by the ordinals in the range of the functions $\{h_{\gamma'} : \gamma' \in C_\gamma \cap C(\gamma, \omega \cdot (\omega \cdot \beta + n) + k)\}\}$. It is immediate that this set has size $< \kappa_\eta$ and so the function g is well-defined. There are two important points.

- $g \in M_{C(\gamma, \omega \cdot (\omega \cdot \beta + n) + k)}$. This so happens because the function is defined from objects contained in the model, in particular from the set $C_\gamma \cap \gamma(\omega \cdot (\omega \cdot \beta + n) + k)$ which belongs to the model by the coherence of the C-sequence.
- $g \in N_{D(\delta, \omega \cdot (\omega \cdot \beta + n) + k)}$ by the same reason as in the previous item, this time using the coherence of the D-sequence.

By the first point, the function $g \in \prod_\delta \kappa_\delta$ is dominated by the function $h_{C(\gamma, \omega \cdot (\omega \cdot \beta + n) + k)}$ from some ordinal on. By the second point, this ordinal must be smaller than $D(\delta, \omega \cdot (\omega \cdot \beta + n) + k)$. By the definition of the function g and the sequence $\vec{\imath} = \vec{r}_{\gamma\delta}(\omega \cdot \beta + n)$ then, it must be the case that $\vec{\imath}(k) \neq f_{mk}(\vec{\imath} \upharpoonright (n+1 \setminus \{k\}))$ as desired. Lemma 6.2.13 follows. $\qquad \square$

6.3 Preservation theorems

The terminology used in the previous work is highly suitable for proving preservation theorems for the countable support iteration. The preservation theorems obtained are in spirit quite different from those obtained by Shelah [64], not only because they work only in a definable context and with large cardinal assumptions, but also because they cover a quite different scale of preservation properties.

Before I indulge in a list of theorems, I will recall some terminology and basic tricks used in their proofs. Suppose that I is an iterable σ-ideal on a Polish space X. The symbol I^ω stands for the ω-th iterated Fubini power of the ideal I. A nonempty tree $p \subset X^{<\omega}$ is an I, ω-tree if its levels are Borel and each of its nodes has an I-positive set of immediate successors. Recall that if the ideal I is iterable and satisfies the second universally Baire dichotomy then every universally Baire I^ω-positive set contains all branches of some I, ω-tree. I will need to relativize this to an arbitrary sequence $t \in X^{<\omega}$. The symbol $I^\omega(t)$ wil stand for the σ-ideal on the space of all sequences in X^ω which contain t as an initial segment defined by $A \in I^\omega \leftrightarrow \{t^\frown \vec{x} : \vec{x} \in A\} \in I^\omega(t)$. A tree $p \subset X^{<\omega}$ is an I, ω, t-tree if it contains t as its trunk, its levels are Borel, and every node of p past t splits into I-positively many immediate successors. Clearly the work of Section 5.1 shows that every universally Baire $I^\omega(t)$-positive set contains all branches of some I, ω, t-tree.

The following proposition explains the "bootstrapping" move, which makes it possible to prove the preservation theorem only for iterations of length ω and then argue abstractly for the general case.

Proposition 6.3.1. *(LC) Suppose that J is a σ-ideal on a Polish space Y and for every iterable ideal I, if $I \not\perp J$ then $I^\omega \not\perp J$. Then*

1. *for every iterable ideal I and every countable ordinal $\alpha \in \omega_1$, $I^\alpha \not\perp J$;*
2. *for every iterable ideal I and every ordinal γ, if for every Borel J-positive set $B \subset Y$ $P_I \Vdash \dot{B} \cap V \notin J$, then the countable support iteration of P_I of length γ forces the same;*
3. *if in addition the ideal J is homogeneous, for every iterable ideal I and every ordinal γ, if $P_I \Vdash \dot{Y} \cap V \notin J$ then the countable support iteration of length γ forces the same.*

Moreover, if the variable I ranges over $\mathbf{\Pi}^1_1$ on $\mathbf{\Sigma}^1_1$ ideals then the large cardinal assumption can be dropped.

And finally, the following proposition records the fusion argument for the countable support iteration.

Proposition 6.3.2. *Suppose that I is an iterable ideal, $B \notin I^\omega$ is a Borel set, and $f : B \to \omega^\omega$ is a Borel function. Then there is a I, ω-tree p such that $[p] \subset B$ and for $\vec{x} \in [p]$ the value $f(\vec{x})(n)$ depends only on $\vec{x} \upharpoonright n$.*

The proof is just a repetition of the standard proof of the preservation of properness in the countable support iteration and as such I omit it.

6.3.1 Uniformity of ergodic ideals

Theorem 6.3.3. *[83] (LC) Suppose that I is an iterable ideal on a Polish space X and J is an ergodic c.c.c. ideal on a Polish space Y. If suitable large cardinals exist and $I \not\perp J$ then $I^\omega \not\perp J$. If the ideal I is $\mathbf{\Pi}^1_1$ on $\mathbf{\Sigma}^1_1$ and the ideal J comes from a Suslin c.c.c. forcing then the large cardinal assumption can be dropped.*

Proof. I will treat the large cardinal version. Let E be a countable Borel equivalence relation on the space Y witnessing the assumption that the ideal J is ergodic. Suppose that $I \not\perp J$ holds. Towards the proof of $I^\omega \not\perp J$, choose Borel sets $B \subset X^\omega$, $C \subset Y$, and $D \subset B \times C$ such that the horizontal sections of the complement of D are in the ideal I; I must find a vertical J-positive section of the set D. There is a universally Baire collection of winning strategies $\tau_y : y \in C$ for Player I in the iteration games with payoff sets $(B \times C) \setminus D)^y) \subset B$. Since E is an equivalence relation with countable classes, without loss of generality I may suppose that the moves of the strategies are invariant under the equivalence relation E. I will find a sequence $\vec{x} \in B$ such that for a J-positive set $C' \subset C$, the sequence $\vec{x}$ is a legal sequence of answers of Player II against the strategy τ_y for all $y \in C'$. This will indicate that $C' \subset D_{\vec{x}}$ and $D_{\vec{x}}$ is the desired positive vertical section.

Find an I, ω-tree $p \subset X^\omega$ such that $[p] \subset B$. By induction on $n \in \omega$ build sequences $t_n \in X^{<\omega}$ such that

- $0 = t_0 \subset t_1 \subset \ldots, t_n \in p, |t_n| = n$;
- writing $C_n = \{y \in C : \text{ the sequence } t_n \text{ is a legal sequence of Player II's answers to the strategies } \tau_y\}$, the set C'_n is J-positive. Note that since the strategies are invariant under the equivalence relation E, so is the set C'_n, and by the ergodicity it must be the case that $C \setminus C_n \in J$.

The initial state $t_0 = 0$ of the induction satisfies the induction hypothesis. Suppose t_n has been obtained. Now $I \not\perp J$, the initial assumption, is equivalent to $I \not\perp_{uB} J$ by Proposition 3.2.4, since the ideal J is c.c.c. and therefore satisfies the first universally Baire dichotomy. Consider the Borel I-positive set $B_n = \{x \in X : t_n^\frown x \in p\}$, the universally Baire J-positive set C_n, and the universally Baire set $D_n \subset B_n \times C_n$ defined by $\langle x, y \rangle \in D_n$ if $x \notin \tau_y(t_n)$. The complement of this set has J-small horizontal sections by the definition of the strategies, and so $I \not\perp_{uB} J$ implies that there must be a point $x \in B_n$ such that $(D_n)_x \notin J$. The sequence $t_{n+1} = t_n^\frown x$ satisfies the induction hypotheses.

In the end, let $\vec{x} = \bigcup_n t_n \in [p] \subset B$. The main point of the argument now is that $C' = \bigcap_n C_n$ is a J-positive set, since for every number $n \in \omega$ it was the case that $C \setminus C_n \in J$. The theorem follows. $\qquad\square$

Corollary 6.3.4. *Under the above assumptions the statement $P_I \Vdash V \cap X \notin J$ is preserved under the countable support iteration of the suitably definable forcing. In particular, the following properties are preserved:*

1. *$V \cap \mathbb{R}$ is not meager;*
2. *$V \cap \mathbb{R}$ is not Lebesgue null;*
3. *$V \cap$meager is cofinal in* meager;
4. *$V \cap$null is cofinal in* null.

Proof. It is just necessary to observe that the properties enumerated above can be expressed as preservation of J-positivity for a suitable ergodic σ-ideal associated with a Suslin forcing. The first two properties are already in this form. The third property is equivalent to the preservation of positivity for the ideal associated with Hechler forcing and the usual amoeba forcing for measure will work for the last property. $\square$

The reader may want to compare these techniques and results to the parallel results of [41] and [42], which employ combinatorial approach to proper forcing.

The book [83] introduced a weakening of the ergodic property of ideals that is still sufficient for the above proof. The following definition restates Definition 5.4.4 in that book.

Definition 6.3.5. *Suppose that I is a σ-ideal on a Polish space X. I will call the ideal weakly ergodic if the partial order P_I^e is $\aleph_0$-distributive. Here, a condition $p \in P_I^e$ is a pair $\langle B_p, E_p \rangle$ where $B_p \subset X$ is a Borel I-positive set and E_p is a countable Borel equivalence relation on B_p such that E_p-saturations of I-small sets are I-small. P_I^e is ordered by $q \leq p$ if $B_q \subset B_p$, B_q is invariant under E_p, and $E_p \upharpoonright B_q^2 \subset E_q$.*

Note that if large cardinals exist and I is a universally Baire weakly ergodic ideal then the forcing P_I^e must in fact be strategically closed since the descending chain game is determined. Recall that in the *descending chain game* [30] Players I and II alternate, creating a descending chain $p_n : n \in \omega$ of conditions in the poset P_I^e, and Player II wins if the chain has a lower bound. In this particular case, this condition reduces to $\bigcap_n B_{p_n} \notin I$, since then the largest lower bound will be the pair $\langle B = \bigcap_n B_{p_n}, E = \bigcup_n E_{p_n} \upharpoonright B^2 \rangle$.

It is not difficult to see that P_I then embeds into $P_I^e * P_J$ where J is the σ-ideal generated by Borel sets disjoint from some set B_p where p comes from the P_I^e-generic filter. The forcing P_J is c.c.c. in all cases that I can see.

The extent of the family of weakly ergodic ideals is something of a mystery.

Proposition 6.3.6. *The following ideals are weakly ergodic.*

1. *Every c.c.c. ergodic ideal.*
2. *The Laver ideal.*

3. The c_{min} ideal.
4. The Mathias ideal.
5. The E_0-ideal.

Proof. In the case of a c.c.c. ergodic ideal I, Player II wins the descending chain game by simply finding a countable Borel equivalence relation E such that every E-invariant Borel set is either I-small or I-large and E-saturations of I-small sets are I-small, and then in his first move playing a condition $\langle B_1, E_1 \rangle$ such that $E \upharpoonright B_1 \subset E_1$. His remaining moves are completely irrelevant, since for every number $n \in \omega$ it must be the case that $B_1 \setminus B_n \in I$ by the ergodicity, and therefore $B \setminus \bigcap_n B_n \in I$ and $\bigcap_n B_n \notin I$ as required.

For the Laver ideal, as the descending chain game on P_I^e develops, Player II indicates conditions $\langle B_{2n+1}, E_{2n+1} \rangle : n \in \omega$ in such a way that

- $B_{2n+1} = [T_n]$ for some Laver tree T_n and the trees $T_n : n \in \omega$ form a fusion sequence;
- for every sequence $s \in \omega^{<\omega}$ and every branching node $t \in T_n$ let $f_{t,s,n} : O_s \to [T_n \upharpoonright t]$ be the natural homeomorphism; then E_{2n+1} is required to be a Borel countable equivalence relation containing all the pairs $(x, y) : \exists s_0, t_0, s_1, t_1 \ f_{t_0,s_0,n}^{-1}(x) = f_{t_1,s_1,n}^{-1}(y)$.

It is not difficult to see that this is possible. Suppose B_{2n+1}, E_{2n+1}, T_n have been obtained, and B_{2n+2}, E_{2n+2} is the answer of Player I. There is a node $s \in \omega^{<\omega}$ such that the set B_{2n+2} contains all branches of some Laver tree S with trunk s. Since B_{2n+2} is invariant under the equivalence relation E_{2n+1}, the second item above implies that in fact it contains all branches of some Laver trees with trunks equal to the nodes at the n-th splitting level of the tree T_n. Let T_{n+1} be the Laver tree combining all of them, find the homeomorphisms as in the second item above, and let B_{2n+3} be the E_{2n+2}-saturation of the set $[T_{n+1}]$ and E_{2n+3} be the equivalence relation generated by E_{2n+2} and the pairs in the second item above. This concludes the inductive step.

In the end, the condition $\langle B, E \rangle$ defined by $B = \bigcap_n B_n$ and $E = \bigcup_n E_n \upharpoonright B$ is a lower bound of the chain obtained. Note that the set B contains all branches of the Laver tree $T = \bigcap_n T_n$.

For the c_{min}, Mathias and E_0-ideal proceed in the same way as above, with a notion of fusion suitable for the associated forcings. $\qquad\square$

Theorem 6.3.7. *(LC) Suppose that J is a universally Baire weakly ergodic ideal satisfying the second dichotomy. Whenever I is an iterable ideal and $I \not\perp_{uB} J$ then $I^\omega \not\perp_{uB} J$.*

Proof. The argument is exactly the same as in Theorem 6.3.3 except that on the J side it is necessary to simulate a run of the descending chain game in the forcing P_J^e. □

There are several attractive corollaries. The reader should note that all cases below in which the ideal J satisfies the first dichotomy can be actually obtained from other preservation theorems in this section. The novel cases are those in which the ideal J does not satisfy the first dichotomy. Theorem 6.3.7 is the only theorem in this section that can handle preservation theorems of this kind. Note that in these cases it is not clear if $\perp_{uB}$ is the same as $\perp$.

Corollary 6.3.8. *(LC) The following statements are preserved under the countable support iteration of suitably definable proper forcings.*

1. *$2^\omega \cap V$ is c_{min}-positive.*
2. *No unbounded reals are added.*
3. *No splitting or unbounded reals are added.*
4. *$2^\omega \cap V$ cannot be covered by a universally Baire set without a Borel E_0-positive subset.*

If the iterated ideals are Π_1^1 on Σ_1^1 then the statement "$2^\omega \cap V \notin E_0$-ideal" is preserved.

6.3.2 Uniformity of porosity ideals

Theorem 6.3.9. *Suppose that I is an iterable ideal on a Polish space X and J is a σ-ideal generated by a universally Baire abstract porosity. If suitable large cardinals exist and $I \not\perp J$ then $I^\omega \not\perp J$. If the abstract porosity function is coanalytic then the large cardinal assumptions can be dropped.*

Proof. I will treat the large cardinal case. Suppose that the σ-ideal J is generated by some porosity function $\text{por} : \mathcal{P}(U) \to \mathcal{B}(Y)$ for some collection U of Borel subsets of the space Y. Suppose for contradiction that $I \not\perp J$ and $I^\omega \perp J$, and fix Borel sets $B \subset X^\omega$, $C \subset Y$ and $D \subset B \times C$ witnessing the latter statement. Thus the vertical sections of the set D are in the ideal J while the horizontal sections of its complement are in the ideal I^ω.

Using the standard P_I^ω-uniformization arguments find an I, ω-tree $p \in P_I^\omega$ and Borel functions $f_n : n \in \omega$, $f_n : [p] \to \mathcal{P}(U)$, such that for every sequence $\vec{x} \in [p]$, $D_{\vec{x}} \subset \bigcup_n (\text{por}(f_n(\vec{x})) \setminus \bigcup f_n(\vec{x}))$. Find a universally Baire collection $\tau_y : y \in C$ of strategies winning for Player I in the iteration games with payoff sets $(B \times C) \setminus D)^y$. I will find a point $\langle \vec{x}_{end}, y_{end} \rangle \in [p] \times C$ such that

- $\vec{x}_{end}$ is a legal sequence of Player II's answers to the strategy $\tau_{y_{end}}$;
- $y \notin \bigcup_n (\text{por}(f_n(\vec{x}_{end})) \setminus \bigcup f_n(\vec{x}_{end}))$.

This will be a contradiction as such a point $\langle \vec{x}, y \rangle$ can be neither in the set D nor in its complement.

Fix a winning strategy σ for the Nonempty player in the Borel precipitous game with the ideal J. By induction on n build sequences $t_n \in X^n$, conditions $C_n \in P_J$ and I, ω, t_n-trees $p_n \subset p$ such that

- $0 = t_0 \subset t_1 \subset t_2 \subset \ldots$ are sequences in the tree T; in the end $\vec{x}_{end} = \bigcup_n t_n$. Also $p = p_0 \supset p_1 \supset \ldots$
- $C = C_0 \supset C_1 \supset \ldots$ are conditions i P_J such that for some $C_0' \supset C_1' \supset \ldots$ the sequence $C_0, C_0', C_1, C_1' \ldots$ forms a legal play against the strategy σ in the Borel precipitous game; so in the end $\bigcap_n C_n$ must be nonempty and the point y_{end} will be any of its elements;
- for every point $y \in C_n$ the sequence t_n will be a legal sequence of Player II's answers against the strategy τ_y;
- the set C_{n+1} is disjoint from all sets $\text{por}(f_n(\vec{x})) \setminus \bigcup f_n(\vec{x})$ for $\vec{x} \in [p_{n+1}]$.

The first two items concern the actual construction of the point $\langle \vec{x}_{end}, y_{end} \rangle \in [p] \times C$, the last two items are present to guarantee the required properties of that point. The proof will be complete once I show how to perform the induction.

The inductive construction is straightforward. The initial setup $0 = t_0, p = p_0, C = C_0$ satisfies the induction hypothesis. Suppose that t_n, C_n, T_n have been constructed. Let $C_n' \subset C_n$ be the set dictated to player Nonempty by the strategy σ in the Borel precipitous game. Consider the set $B_n' = \{x \in X : t_n^\frown x \in p_n\}$ and the set $D_n' \subset B_n' \times C_n'$ given by $\langle x, y \rangle \in D_n'$ if $x \notin \tau_y(t_n)$. The complement of the universally Baire set D_n' has I-small horizontal sections. The porosity ideals satisfy the first universally Baire dichotomy by Theorem 4.2.3, thus $I \not\perp J \leftrightarrow I \not\perp_{uB} J$ by Proposition 3.2.4, and there must be a point $x \in B_n'$ such that the vertical section $(D_n')_x$ contains a J-positive Borel set $C_n'' \subset C_n'$. Let $t_{n+1} = t_n^\frown x$.

Let $a = \{u \in U : C_n'' \cap u \in J\}$. By the countable additivity of the ideal $I^\omega(t_{n+1})$ and the dichotomy 5.1.12, there is a I, ω, t_{n+1} tree $p_{n+1} \subset p_n$ such that either for all sequences $\vec{z} \in [p_{n+1}], f_n(\vec{z}) \subset a$, or there is a set $u \notin a$ such that for all sequences $\vec{z} \in [p_{n+1}], u \in f_n(\vec{z})$. In the former case, let $C_{n+1} = (C_n'' \setminus \bigcup a) \setminus \text{por}(a)$, in the latter case let $C_{n+1} = C_n'' \cap u$. The induction hypothesis continues to hold. $\square$

Corollary 6.3.10. *The following properties are preserved in the countable support iteration of suitably definable forcings:*

1. $V \cap \mathbb{R}$ *is not meager;*
2. $V \cap \omega^\omega$ *is unbounded;*
3. $V \cap 2^\omega$ *cannot be decomposed into countably many* c_{min}*-homogeneous sets;*
4. *any ground model Borel non-σ-continuous function remains non-σ-continuous on* $V \cap \mathbb{R}$*;*
5. $V \cap B$ *is not metrically σ-porous for every Borel set* $B \subset \mathbb{R}$*.*

Proof. The first three statements are associated with ideals generated by closed sets, treated in Section 4.1; every σ-ideal generated by closed sets is a porosity ideal. The last two ideals have been discussed in Section 4.2. Note that the first four ideals are homogeneous and therefore it is enough to speak about the preservation of positivity of the whole space. I do not know if the metric porosity ideal is homogeneous.   □

6.3.3 Outer regular capacities

The standard notion of a capacity and outer regular capacity was defined in Section 4.3.

Theorem 6.3.11. *[85] (LC) Suppose that I is an iterable σ-ideal and ϕ is an outer regular capacity. If suitable large cardinals exist and ZF+DC+AD+ proves that the capacity ϕ is continuous in increasing well-ordered unions and $I \not\perp \phi$ then $I^\omega \not\perp \phi$. If the ideal I is $\mathbf{\Pi}^1_1$ on $\mathbf{\Sigma}^1_1$, the capacity ϕ is ZFC-correct, and coanalytic sets are capacitable for it, then the large cardinal assumption can be dropped.*

It is remarkable that neither the LC nor ZFC arguments uses the dichotomies from Section 5.1. Instead they use a complicated manipulation of the capacities. In the next section, the reader can find a theorem with a nontrivial overlap with the present theorem, whose proof uses the dichotomies.

Proof. Suppose ϕ is an outer regular capacity on a Polish space Y. Fix a basis $\mathcal{O}$ for the space Y closed under finite unions. Suppose that I is an iterable σ-ideal on a Polish space X such that $I \not\perp \phi$. Suppose that $p \in P^\omega_I$ is a condition and $\dot{O}$ is a name for an open subset of the space Y of capacity $\leq \varepsilon$. By Proposition 6.3.2, there is a I, ω-tree $q \subset p$ and a Borel function $f : [q] \to \mathcal{O}^\omega$ such that $q \Vdash \dot{O} = \bigcup_n \dot{f}(\vec{x}_{gen})(n)$ and for a sequence $\vec{x} \in [q]$ the value $f(\vec{x})(n)$ depends only on $\vec{x} \restriction n$; for a sequence $t \in q$ of length $\geq n$ the symbol $f(t)(n)$ will stand for the unique possible value of $f(\vec{x})(n)$ where $t \subset \vec{x}$.

It will be enough to show that $\phi(\{y \in Y : q \Vdash \check{y} \in \dot{O}\}) \leq \varepsilon$. In order to do this, for a sequence $t \in q$ write $B_t = \{x \in X : t^\frown x \in q\}$ and by induction on ordinal α define sets $A(t, \alpha) \subset Y$ for all $t \in p$ simultaneously:

- $A(t, 0) = \bigcup_{n \in |t|} f(t)(n)$;
- $A(t, \alpha + 1) = \int_{B_t} D(t, \alpha) \, dI$ where $D(t, \alpha) \subset B_t \times Y$ is the set of all pairs $\langle x, y \rangle$ such that $y \in A(t^\frown x, \alpha)$;
- $A(t, \alpha) = \bigcup_{\beta \in \alpha} A(t, \beta)$ for limit ordinals α.

It is easy to simultaneously argue by induction on the ordinal α that $s \subseteq t$ and $\beta \leq \alpha$ imply $A(s, \beta) \subseteq A(t, \alpha)$. After that, it is again easy to argue by induction on the ordinal α for all sequences $t \in p$ simultaneously that $\phi(A(t, \alpha)) \leq \varepsilon$: this follows by the assumption $I \not\perp \phi$ on the successor step, and by the assumption that the capacity is continuous in wellordered increasing unions under AD+. Note that the whole inductive process takes place inside some inner model of AD+ all of whose sets are universally Baire – Fact 1.4.5.

By the Replacement Axiom, the whole inductive process must stabilize at some ordinal Ω. I claim that $A(0, \Omega) = \{y \in Y : q \Vdash \check{x} \in \dot{O}\}$. For the left-to-right inclusion argue by induction on the ordinal α for all sequences $t \in p$ simultaneously that $A(t, \alpha) \subset \{y \in Y : \{\check{x} \in [q] : y \notin \bigcup \mathrm{rng}(f)\} \in I^\omega(t)\}$. For the (more important) opposite inclusion note that if $y \notin A(0, \Omega)$ then the universally Baire tree $r \subset p, r = \{t \in p : y \notin A(t, \Omega)\}$ is I-positively branching and so by Corollary 5.1.15 contains a condition $s \in P_I^\omega$. The setup of the sets $A(t, 0)$ then implies that $y \notin \bigcup_{\check{x} \in [s]} \bigcup_n f(\check{x})(n)$ and therefore $s \Vdash \check{y} \notin \dot{O}$.

The previous two paragraphs in conjunction show that $\phi(A(0, \Omega)) = \phi(\{y \in Y : q \Vdash \check{y} \in \dot{O}\}) \leq \varepsilon$ as desired.

The ZFC case requires several nontrivial patches of the above argument. First, the Borel function $f : q \to \mathcal{O}$ is obtained through a reference to the ZFC-correctness of the capacity ϕ. After that, consider the operator Γ on the space $q \times Y$ defined by $(t, y) \in \Gamma(A) \leftrightarrow (t, y) \in A \vee \{x \in X : t^\frown x \in q \wedge (t^\frown x, y) \notin A\} \in I$, for every set $A \subset q \times Y$. An important point is that the ideal I is $\mathbf{\Pi}^1_1$ on $\mathbf{\Sigma}^1_1$ so that this formula defines a monotone coanalytic operator. By a theorem of Cenzer and Mauldin [5], 1.6, given a coanalytic set $A \subset q \times Y$, the transfinite sequence given by the description $A = A_0$, $A_{\alpha+1} = \Gamma(A_\alpha)$ and $A_\alpha = \bigcup_{\beta \in \alpha} A_\beta$ for limit ordinals α, stabilizes at ω_1 in a coanalytic set $A_{\omega_1} = \Gamma(A_{\omega_1})$ such that for every analytic set $C \subset A_{\omega_1}$ there is an ordinal $\alpha \in \omega_1$ such that $C \subset A_\alpha$. Now consider the set $A = \{(t, y) \in q \times Y : y \in f(t)\}$, its associated coanalytic fixed point A_{ω_1} and the coanalytic set $A' = \{y \in Y : (0, y) \in A_{\omega_1}\}$. If $y \in Y$ is a point such that $y \notin A'$ then the tree $q_y = \{t \in q : (t, y) \notin A_{\omega_1}\} \subset q$ is an analytic I-positively splitting tree and therefore it contains an I, ω-subtree $r \subset q_y$. Clearly, $r \Vdash \check{y} \notin \dot{O}$. Thus $q \Vdash \dot{O} \cap V \subset \dot{A}' \cap V$ and it will be enough to show that the set A' has ϕ-mass $\leq \varepsilon$.

This is still a nontrivial statement, and the coanalytic capacitability assumption is used here. First argue by induction on $\alpha \in \omega_1$ for all nodes $t \in q$ simultaneously that $\phi(\{y \in Y : (t, y) \in A_\alpha\}) \leq \varepsilon$. This is clear at $\alpha = 0$, and at limit stages of the induction this follows from the continuity of the capacity ϕ under countable increasing unions. The successor step $\alpha = \beta + 1$ follows from the assumption that $I \not\perp \phi$, but extra care is required. Let $t \in q$ be a node, write $B_t = \{x \in X : t^\frown x \in q\}$ and write $D \subset B \times Y$ for the set $\{(x, y) : x \in B, (t^\frown x, y) \in A_\beta\}$. I must show that $\phi(\int_B D \, dI) \leq \varepsilon$. This would directly follow from $I \not\perp \phi$ if the set D was analytic. However, D is coanalytic and another patch is called for. Suppose for contradiction that the integral has ϕ-mass $> \varepsilon$. Let $D = \bigcup_{\gamma \in \omega_1} D_\gamma$ be an increasing union of Borel sets such that every Borel subset of D is included in one of D_γ's as a subset. I will reach the contradiction by showing that the integral of one of the sets D_γ has ϕ-mass $> \varepsilon$. Since the ideal I is $\mathbf{\Pi}_1^1$ on $\mathbf{\Sigma}_1^1$, the set $\int_B D \, dI$ is coanalytic. By the capacitability of the coanalytic sets, there is a Borel set $C \subset \int_B D \, dI$ of the same ϕ-mass. Since the ideal I is $\mathbf{\Pi}_1^1$ on $\mathbf{\Sigma}_1^1$, the analytic set $(B \times C) \setminus D$ with I-small horizontal sections has a Borel superset with I-small horizontal sections; denote its Borel complement in $B \times C$ by the letter D'. Now $D' \subset D$ and so $D' \subset D_\gamma$ for some ordinal $\gamma \in \omega_1$. This brings about the desired contradiction and concludes the successor stage of the induction. Finally, in the end note that the set A' is coanalytic and by the capacitability of coanalytic sets it must have a Borel subset $A'' \subset A'$ of the same ϕ-mass, and this must be a subset of one of the sets $\{y \in Y : (0, y) \in A_\alpha\}$ by the aforementioned theorem of Cenzer and Mauldin. If $\phi(A') > \varepsilon$, then $\varepsilon > \phi(A'') \geq \phi(\{y \in Y : (0, y) \in A_\alpha\})$. However, I have just proved that the latter set has capacity $\leq \varepsilon$, contradiction. Thus $\phi(A') \leq \varepsilon$ as desired. $\square$

Corollary 6.3.12. *Under the above assumptions, the preservation of the following capacities is preserved under the countable support iteration of definable forcings:*

1. *outer Lebesgue measure;*
2. *Newtonian capacity and all capacities ocurring in potential theory ;*
3. *all strongly subadditive capacities ;*
4. *the $c_{f,h}$ capacity introduced in Example 4.3.46.*

The preservation theorem resulting from the last item above is connected to the f, h-bounding preservation theorems of Shelah [2], 7.2.18, 7.2.19.

6.3.4 Pavement submeasures

Theorem 6.3.13. *[85] Suppose that I is an iterable σ-ideal on a Polish space X and ϕ is a pavement submeasure on a Polish space Y derived from a countable collection of Borel pavers. If suitable large cardinals exist and $I \not\perp \phi$ then $I^\omega \not\perp \phi$. If the ideal I is $\mathbf{\Pi}^1_1$ on $\mathbf{\Sigma}^1_1$ and every coanalytic set has a Borel subset of the same ϕ-mass then the large cardinal assumptions can be dropped.*

The assumption on the coanalytic sets in the last sentence is parallel to the same requirement in Theorem 6.3.11. Consult Theorem 4.5.6 to see that the assumption is satisfied in many situations. If for some reason it is necessary to work in a situation where the assumption is not satisfied, the conclusion of the theorem will be weaker: the iteration will not preserve ϕ-masses of all sets but just the masses of sets of the form $B \cap V$ where B is a Borel set coded in the ground model V.

Proof. Fix a countable collection U of Borel sets and a weight function $w : U \to \mathbb{R}^+$ which give rise to the pavement measure ϕ.

Suppose for contradiction that $I \not\perp \phi$ and $I^\omega \perp \phi$. This means that there are a real number $\varepsilon \geq 0$, a Borel I^ω-positive set $B \subset X^\omega$ and a Borel set $D \subset B \times Y$ ϕ-masses of whose vertical sections are bounded below ε and the set $C = \int_B D \, dI^\omega \subset Y$ has ϕ-mass $> \varepsilon$. By the usual P_I^ω-uniformization argument and Proposition 6.3.2 this means that there are a condition $p \in P_I^\omega$ such that $[p] \subset B$ and a function $f : [p] \to U^\omega$ such that for every sequence $\vec{x} \in [p]$ the value $f(\vec{x})(n)$ depends only on $\vec{x} \restriction n$ and moreover $\Sigma_n \Sigma \{w(u) : u \in f(\vec{x})(n)\} \leq \varepsilon$, together with a universally Baire system $\tau_y : y \in C$ of winning strategies for Player I in the iteration games with payoff sets $\{\vec{x} \in [p] : y \notin \bigcup_n f(\vec{x})(n)\}$.

Fix a winning strategy σ for the Nonempty player in the Borel precipitous game with the ideal I_ϕ. In order to reach a contradiction, I will build by induction the following objects:

- Sequences $0 = t_0 \subset t_1 \subset \ldots$ in the tree p, with $|t_n| = n$. In the end I will put $\vec{x}_{end} = \bigcup_n t_n$.
- I, ω, t_n-trees p_n such that $p = p_0 \supset p_1 \supset \ldots$
- Borel sets $C \supset C_0 \supset C_0' \supset C_0'' \supset C_1 \supset C_1' \supset C_1'' \supset C_2 \supset \ldots$ of positive ϕ-mass such that the sets $C_0', C_0'', C_1', C_1'' \ldots$ form a play of the Borel precipitous game according to the strategy σ. The intersection $\bigcap_n C_n$ will contain a singleton $y_{end} \in Y$.
- For every point $y \in C_n$ the sequence t_n will be a correct sequence of Player II's responses to the strategy τ_y.

- Finite sets $Q_n \subset U$ such that $\{f(\vec{x})(m) : m \in n, \vec{x} \in [p_n]\} \subset Q_n$, $C_n \cap \bigcup Q_n = 0$ and and the numbers $\Sigma\{w(u) : u \in (\text{rng}(f(\vec{x})) \setminus Q_n)$ for $\vec{z} \in [p_n]$ are bounded below $\phi(C_n)$.

Once this construction is complete, consider the point $\langle \vec{x}_{end}, y_{end} \rangle \in [p] \times C$. By the fourth item above, the sequence $\vec{x}_{end}$ is a correct sequence of Player II's answers to the strategy $\sigma_{y_{end}}$ and therefore it should be the case that $\langle \vec{x}_{end}, y_{end} \rangle \in D$. On the other hand, the fifth item implies that $y_{end} \notin \bigcup \text{rng}(f(\vec{x}_{end}))$, and this is the desired contradiction.

The initial setup $p = p_0$, $0 = t_0$, $0 = Q_0$, $C = C_0$ satisfies the induction hypothesis. Now suppose the objects t_n, p_n, C_n, Q_n have been defined. By induction on $k \in \omega$ build a descending sequence of I, ω, t_n-trees q_k and finite sets $R_k \subset U \setminus Q_n$ such that for every number k and every sequence $\vec{z} \in [q_k]$ it is the case that $R_k \subset \text{rng}(f(\vec{z}))$ and $\Sigma\{w(u) : u \in (\text{rng}(f(\vec{z})) \setminus Q_n) \setminus R_k\} < 2^{-k}$. This is easily possible using the countable completeness of the ideal I^ω and the dichotomy 5.1.12. In the end, $\Sigma\{w(u) : u \in \bigcup_k R_k\} < \phi(C_n)$ by the last item of the induction hypothesis, so the set $C'_n = C_n \setminus \bigcup_k \bigcup R_k$ has positive ϕ-mass. Let C''_n be the strategy σ's answer to C'_n. Let $k \in \omega$ be a number so large that $2^{-k} < \phi(C''_n)$. Let $B_n = \{x \in X : t_n^\frown x \in q_k\}$ and consider the universally Baire set $D_n \subset B_n \times C''_n$ given by $\langle x, y \rangle \in D_n \leftrightarrow x \notin \tau_y(t_n)$. Note that the horizontal sections of the complement of D_n are I-small. By the assumptions, $I \not\perp \phi$, by the dichotomy in Theorem 4.5.6 and Proposition 3.2.9 this is equivalent to $I \not\perp_{uB} \phi$ and there must be a point $x \in B_t$ such that $\phi((D_n)_x) = \phi(C''_n)$. Use Theorem 4.5.6 to find a Borel set $C_{n+1} \subset (D_n)_x$ of the same ϕ-mass, let $t_{n+1} = t_n^\frown x$ and $p_{n+1} = q_k \restriction t_{n+1}$ and continue with the induction process.

The ZFC argument requires several patches again. First, use the ZFC-correctness of the submeasure ϕ to find the function $f : [p] \to (U)^\omega$. Note that since the ideal I^ω is $\mathbf{\Pi}^1_1$ on $\mathbf{\Sigma}^1_1$, the set $C \subset Y$ must be coanalytic, and use the assumptions to replace it with a Borel subset of the same ϕ-mass. Use the proof of Theorem 5.1.9 to show that the strategies $\tau_y : y \in C$ can be chosen in such a way that there is a real r such that for every $y \in C$ and every finite sequence $t \in p$, if the iteration game reaches the situation where Player II will have played the sequence t, the strategy τ_y responds with the union of all $\mathbf{\Sigma}^1_1(r, y)$ sets in the ideal I. The inductive process then proceeds as above, with Proposition 3.2.9 and $\perp_{uB}$ replaced with 3.2.10 and $\perp_a$. $\square$

Corollary 6.3.14. *The following statements are preserved under the countable support iteration of suitably definable forcings:*

1. *preservation of outer Lebesgue measure;*
2. $V \cap \omega^\omega$ *is dominating;*
3. *preservation the* $c_{f,h}$ *capacity introduced in Example 4.3.46.*

6.3.5 Other preservation theorems

There is a large class of preservation theorems for properties that cannot be in an obvious way stated as Fubini properties. Their proofs use the old-fashioned algebraic approach to the countable support iteration as in [2], plus the determinacy results of Section 3.10. It is always necessary to find the appropriate determined game, and after that, the arguments follow the same pattern. I do not see a uniform way for finding the determined games, and consequently, I hesitate to state a general theorem. Instead, I will list a number of preservation theorems, prove one of them, and let the others to the interested readers. The necessary determined games appear in Section 3.10.

Theorem 6.3.15. *(LC) Suppose that J is an analytic P-ideal, and I is an iterable σ-ideal on a Polish space. If P_I forces $\dot{J} \cap V$ to be cofinal in $\dot{J}$ under inclusion, then so do all of its countable support iterations. If in addition I is a $\mathbf{\Pi}_1^1$ on $\mathbf{\Sigma}_1^1$ ideal and the forcing P_I is bounding, the large cardinals are not necessary.*

Theorem 6.3.16. *(LC) Suppose that I is an iterable ideal on a Polish space X such that P_I has the continuous reading of names. Then all of its iterations have the continuous reading of names.*

Theorem 6.3.17. *(LC) Suppose that ϕ is a pavement submeasure on 2^ω defined from a countable set of clopen pavers. Suppose I is an iterable ideal on a Polish space X. If P_I strongly preserves ϕ then so do all of its countable support iterations. If, in addition, the ideal I is $\mathbf{\Pi}_1^1$ on $\mathbf{\Sigma}_1^1$ and P_I is bounding, then the large cardinal assumptions are not necessary.*

Theorem 6.3.18. *(LC) Suppose that ϕ is a strongly subadditive capacity on the Cantor space 2^ω. Suppose I is an iterable ideal on a Polish space X. If P_I strongly preserves ϕ then so do all of its countable support iterations. If, in addition, the ideal I is $\mathbf{\Pi}_1^1$ on $\mathbf{\Sigma}_1^1$ and P_I is bounding, then the large cardinal assumptions are not necessary.*

This list is in no way exhaustive. I will prove the last theorem. Recall that a forcing P is said to strongly preserve a universally Baire submeasure ϕ on a Polish space Y if for every real number $\varepsilon > 0$ and every set $A \subset Y$ in the extension with $\phi(A) < \varepsilon$ there is a ground model coded Borel set $B \subset Y$ such that $A \subset B$ and $\phi(B) < \varepsilon$. Strong preservation of submeasures is a common concern in forcing arguments; in this book it appears for example in Theorem 4.3.13.

To cut the notational clutter, I will consider only iterations of length ω. The general length case follows by a familiar bootstrapping argument. The iteration of a forcing P_I of length $n \leq \omega$ will be denoted by P_{I^n}. I will use the old fashioned algebraic approach to the iteration, so $P_{I\omega}$ consists of ω-sequences $\vec{p}$ such that $\forall n \in \omega \; \vec{p} \upharpoonright n \Vdash_{P_{In}} \vec{p}(n) \in \dot{P_I}$.

I will need extra notation to handle the strong preservation game of Section 3.10 efficiently. Suppose that P is a forcing and τ is a (finite or infinite) play of the game. The symbols $\tau(p_{ini})$, $\tau(\varepsilon_{ini})$, $\tau(\delta_{ini})$, and $\tau(U_{ini})$ denote Player I's initial choices, for every number $k \in \omega \; \tau(k, I)$ denotes k-th move of Player I, $\tau(k, II)$ denotes the k-th nontrivial move of Player II, $\tau(\omega, I) = \bigcup_k \tau(k, I)$ and $\tau(\omega, II) = \bigcup_k \tau(k, II)$. Thus, $\tau(\omega, I)$ is a P-name for an open set of ϕ-mass $\leq \tau(\varepsilon_{ini})$, $\tau(\omega, II)$ is an open set of mass $\leq \tau(\delta_{ini})$, and the result of the play τ, if nonzero, forces in P that $\tau(\omega, I) \subset \tau(\omega, II)$.

The following is a key concept, and it reveals the idea behind the definition of the game. Suppose $P * \dot{Q}$ is a two step forcing iteration. I will call a pair $\langle \tau_P, \tau_Q \rangle$ *coherent past* m if τ_P is a play of the strong preservation game for P and $\tau_P(p_{ini}) \Vdash \tau_Q$ is a play of the strong preservation game for $\dot{Q}$, and for every $k \in \omega$, $\tau_P(m + k, I) = \tau_Q(k, II)$; I will call the pair *coherent* if there is a number m such that it is coherent past m. Note that $\tau_Q(k, II)$ is a P-name for a clopen set and as such a suitable move for player I in the strong preservation game for P. Note that if the pair is coherent and Player II won both plays, with the resulting conditions $p, \dot{q}$, then $\langle p, \dot{q} \rangle \Vdash \tau_Q(\omega, I) \subset \tau_P(\omega, II)$. I will use the obvious variation of this definition for iterations of arbitrary finite length. More importantly, I will also use it in the case where $\tau_P, \tau_{\dot{Q}}$ are finite plays. In such a case, I require that $\tau_P(p_{ini}) \Vdash$ Player II makes exactly k many nontrivial moves in the play $\tau_{\dot{Q}}$, where $m + k$ is the length of the play τ_P

One way to obtain a coherent pair of plays proceeds like this: let $\sigma_P, \dot{\sigma}_Q$ be winning strategies in the game for $P, \dot{Q}$ respectively, let $0 < \varepsilon < \delta$ be real numbers, let $m \in \omega$ be a natural number, and let τ be a finite play against the strategy σ_P such that τ has length m, with $\dot{U}$ as the last name played by Player I. Let $\varepsilon' > 2^{-m}\varepsilon$, $\delta' = 2^{-m}\varepsilon$ let $\dot{q}$ be a name for a condition in $\dot{Q}$, and let $\dot{U}_n : n \in \omega$ be $P * \dot{Q}$-names for clopen sets such that $\langle p, \dot{q} \rangle \Vdash \dot{U} \subset \dot{U}_0 \subset \dot{U}_1 \subset \ldots, \phi(\dot{U}_n) - \phi(\dot{U}) < \gamma, \phi(\dot{U}_n) - \phi(\dot{U}_k) < 2^{-k}\varepsilon'$ for $k \in n$. Then let $\tau_{\dot{Q}}$ be the play resulting from the application of the strategy σ_Q to the initial choices $\dot{q}, \varepsilon', \delta', \dot{U}$, and the names $\dot{U}_n : n \in \omega$, and τ_P is the play that is obtained from τ by adding names for the nontrivial moves of Player II in the play τ_Q as moves of Player I, and letting the strategy σ_P act on them.

Towards the proof of the theorem, suppose that I is an iterable σ-ideal on a Polish space X, and suppose that the forcing P_I strongly preserves ϕ. I will prove that $P_{I\omega}$ strongly preserves ϕ. Let $\vec{p} \in P_{I\omega}$ be a condition, $0 < \varepsilon < \delta$ be real numbers, and $\dot{O} \subset 2^\omega$ be a name for an open set of ϕ-mass $< \varepsilon$. I will find a condition $\vec{q} \leq \vec{p}$

and an open set $R \subset 2^\omega$ of ϕ-mass $< \delta$ such that $\vec{q} \Vdash \dot{O} \subset \dot{R}$. This will prove the strong preservation property of the iteration.

By a universally Baire absoluteness argument, the forcing P_I strongly preserves ϕ in all forcing extensions, and by the results of Section 3.10.6, there are P_{I^n}-names $\dot{\sigma}_n$ for a winning strategy for Player II in the strong preservation game for P_I, this for all $n \in \omega$. Also, there are names $\dot{U}_n : n \in \omega$ for clopen sets forming an inclusion-increasing sequence such that $\vec{p} \Vdash \dot{O} = \bigcup_n U_n$, and $\phi(\dot{U}_m) - \phi(\dot{U}_n) < 2^{-n-1}\varepsilon$ whenever $n \in m \in \omega$.

I will construct a descending sequence $\vec{p}_n : n \in \omega$ of conditions in the iteration below $\vec{p}$ and, for every $n \in \omega$, P_{I^n}-names τ_n so that

- $\vec{p}_n \restriction n+1 = \vec{p}_{n+1} \restriction n+1$; thus the sequence of conditions has a lower bound $\vec{p}_\omega$ given by $\vec{p}_\omega(n) = \vec{p}_n(n)$;
- $\vec{p}_{n+1} \Vdash \dot{U}_n$ is in fact a $P_{I^{n+1}}$-name;
- $\vec{p}_n \restriction n \Vdash \langle \tau_n, \tau_{n+1} \rangle$ is a coherent pair of plays against the strategies $\dot{\sigma}_n, \dot{\sigma}_{n+1}$ respectively, $\tau_n(p_{ini}) = \vec{p}_n(n)$, and $\dot{U}_n \subset \tau_{n+1}(U_{ini})$. Moreover, $\tau_0(\varepsilon_{ini}) = \varepsilon$ and $\tau_0(\delta_{ini}) = \delta$.

After this is done, consider the condition $\vec{q} \in P_{I^\omega}$ defined by $\vec{q} \restriction n \Vdash \vec{q}(n) =$ the result of the play τ_n. Since the strategies $\dot{\sigma}_n : n \in \omega$ are forced to be winning, this is indeed a definition of a condition in the iteration. Let $R = \tau_0(\omega, II)$. Since $\tau_0(\delta_{ini}) = \delta$ and σ_0 is a winning strategy, $R \subset 2^\omega$ is an open set of ϕ-mass at most δ. By the coherence requirement, for every $n \in \omega$ it is the case that $\vec{q} \Vdash \dot{U}_n \subset \tau_n(\omega, I) \subset \tau_0(\omega, II) = R$ and therefore $\vec{q} \Vdash \dot{O} \subset \dot{R}$ as required.

To construct the plays and the conditions as above, by induction on $n \in \omega$ build conditions $\vec{p}_n$ and P_{I^n}-names $\dot{m}_n$ for nonzero natural numbers, and P_{I^k}-names $\tau_{k,n}$ for $k \leq n$ so that:

- $\vec{p}_n \restriction n = \vec{p}_{n+1} \restriction n$;
- $\tau_{k,n}$ is a P_{I^k}-name for a finite play of the strong preservation game against the strategy $\dot{\sigma}_k$;
- $\vec{p}_k \restriction k$ forces τ_{k,n_0} to be a proper initial segment of τ_{k,n_1} whenever $n_0 \in n_1$ are natural numbers, and $\tau_{k,n}$ and $\tau_{k+1,n}$ cohere above $\dot{m}_k$;
- writing $\dot{l}_n = \Sigma_{k \in n} \dot{m}_k$, it is the case that $\vec{p}_{n+1} \Vdash \dot{U}_l$ is in fact a P_{I^n}-name and it is the last move of Player I in the play $\tau_{n,n}$. Moreover, $\tau_{0,n}(\varepsilon_{ini}) = \varepsilon$, $\tau_{0,n}(\delta_{ini}) = \delta$, and for $k > 0$, $\tau_{k,n}(\varepsilon_{ini}) = 2^{-l_k-1}\varepsilon$ and $\tau_{k,n}(\delta_{ini}) = 2^{-l_k}\varepsilon$.

To initiate the construction, let $\vec{p} = \vec{p}_0$. Now suppose that $n \in \omega$ is a natural number and $p_n, \tau_{k,n-1} : k \in n$, as well as $\dot{m}_k : k \in n$ have been constructed. Let $\dot{l} = \Sigma_{k \in n} \dot{m}_k$. Find a decreasing sequence $\vec{p}_n \geq \vec{r}_0 \geq \vec{r}_1 \geq \dots$ so that for every $h \in \omega$,

$\vec{p}_n \upharpoonright n+1 = \vec{r}_h \upharpoonright n+1$ and $\vec{r}_h \Vdash \dot{U}_{l+h+1}$ is in fact a $P_{j^{n+1}}$-name. Consider the P_{I_n}-name $\vec{\tau}_n = \vec{\tau}$ for an infinite play of the strong preservation game against the strategy $\dot{\sigma}_n$ such that $\vec{\tau}(p_{ini}) = \vec{p}_n(n)$, $\vec{\tau}(\varepsilon_{ini}) = 2^{-l}\varepsilon$, $\vec{\tau}(\delta_{ini}) = 2^{-l-1}\varepsilon$ (except at $n = 0$ where these two values will be ε and δ), $\vec{\tau}(U_{ini}) = \dot{U}_l$, and $\vec{\tau}(h, II) = \dot{U}_{l+h+1}$. By downward induction on $k \in n$ find the P_{j^k}-names $\vec{\tau}_k$ for infinite plays against the strategies $\dot{\sigma}_n$ extending $\tau_{k,n-1}$ such that the sequence $\langle \vec{\tau}_k : k \leq n \rangle$ coheres. By upward induction on $k \leq n$ find P_{j^k}-names $\dot{g}_k$ for natural numbers so that $g_0 =$the length of $\tau_{0,n-1} + 1$, and $\dot{g}_k =$the index of the round of the play $\vec{\tau}_k$ in which Player II makes the $\dot{g}_{k-1}$-th nontrivial move. Let $\dot{m}_n = \dot{g}_n$, let $\tau_{k,n} = \vec{\tau}_k \upharpoonright \dot{g}_n$ for all $k \leq n$, and let $\vec{p}_{n+1} \in P_{j^\omega}$ be the condition given by $\vec{p}_{n+1} \upharpoonright n+1 = \vec{p}_n \upharpoonright n+1$ and the remainder of $\vec{p}_{n+1}$ is forced to be equal to $\vec{r}_{\dot{m}_n} \upharpoonright [n+1, \omega)$. The induction hypotheses continue to hold, and the theorem follows.

6.3.6 Cichoń's diagram

What pentagram is to heavy metal, Cichoń's diagram is to set theory. It depicts ZFC-provable inequalities between several cardinal invariants, with the assertion that no other inequalities between them can be proved.

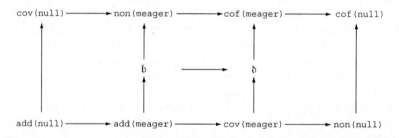

In this section I will illustrate how the techniques of this book can be used to show that none of the inequalities documented in the diagram can be reversed. This duplicates the work of [2], using forcings that naturally come up in abstract analysis.

Start with the upper right corner. Consider the packing measure ϕ on the metric space described in Example 4.1.27. The associated σ-finite forcing $P_{I_{a\phi}}$ is proper, bounding, and preserves outer Lebesgue measure and Baire category. All of these features proliferate through the countable support iteration. I do not know if it is possible to replace the packing measure with a packing measure on a Euclidean space.

Move down to non(null). Consider the metric porosity ideal defined in Example 4.2.10. The forcing is bounding and preserves category by the results of Section 4.2. Again, the resulting model has $\mathrm{cof(null)} = \mathrm{non(null)} = \aleph_2$

while the other invariants remain small. I do not know if it is possible to replace the metric space above with a Euclidean space.

In order to increase non(meager) while keeping the other invariants small, let ψ be the two-dimensional Hausdorff measure on the three-dimensional Euclidean space and let $I_{\sigma\psi}$ be the σ-ideal generated by sets of finite mass. By the results of Section 4.4 the resulting forcing is bounding and does not add a splitting real, in particular it does not add random reals. It also preserves outer Lebesgue measure since the ideal $I_{\sigma\psi}$ is polar as shown in Example 3.6.4. These properties are preserved under the countable support iteration. The forcing increases non(meager) by Corollary 3.5.8 or by a direct computation.

In order to increase $\mathfrak{b}$ while keeping the other invariants small, consider the Laver ideal I with the attendant Laver forcing equivalent to P_I. Since $\mathfrak{b} = \mathrm{cov}(I)$, this increases $\mathfrak{b}$. The forcing preserves outer Lebesgue measure as well as many other outer regular capacities by Theorem 3.6.11. An interesting argument offers itself to show that Laver forcing and its iteration do not add random reals. Consider the ideal J of non-σ-splitting sets from Section 4.1.7. The forcing P_J is bounding which is equivalent to $I \not\perp J$. Since the ideal J is σ-generated by closed sets this feature persists through the countable support iteration. On the other hand, Proposition 4.1.29 shows that $\mathrm{null} \perp J$ and so neither P_I nor its iterations can add random reals. Thus in the resulting model $\mathfrak{b} = \aleph_2$ while $\mathrm{non}(\mathrm{null}) = \mathrm{cov}(\mathrm{null}) = \aleph_1$.

In order to increase $\mathfrak{b}$ and non(null) keeping the other invariants small, it is common to use Mathias forcing and its countable support iteration. Recall that the Mathias forcing is associated with the ideal I of sets nowhere dense in the algebra $\mathcal{P}(\omega)$ mod fin. I must prove that in the resulting model $\mathrm{cov}(\mathrm{null}) = \mathrm{cov}(\mathrm{meager}) = \aleph_1$, in other words the iteration adds no Cohen or random reals. Consider any of the capacities ϕ from Section 4.3.5. Since ϕ is Ramsey, $I \not\perp \phi$ by Theorems 4.3.23 and 4.3.25, and this feature persists through the countable support iteration by Theorem 6.3.11 – in particular the set of ground model reals is of full capacity in the extension. On the other hand, ϕ is an outer regular capacity so meager $\perp \phi$ by Corollary 3.5.8 or a direct computation, and the construction of the capacity ϕ implies that $\mathrm{null} \perp \phi$ by Proposition 4.3.47. Thus no Cohen or random reals are added in the iteration.

In order to increase $\mathfrak{b}$, non(null) and cov(null) it is customary to iterate Mathias and Solovay forcings. I must show that in the resulting model cov(meager) is small, in other words the iteration adds no Cohen reals. Consider any of the Ramsey capacities ϕ from Section 4.3.6. Since the capacity ϕ is Ramsey, $I \not\perp \phi$, and since it is strongly subadditive, $\mathrm{null} \not\perp \phi$, and this feature persists through the iteration by Theorem 6.3.11. On the other hand, ϕ is an outer regular capacity, therefore meager $\perp$ ϕ by Corollary 3.5.8 or a direct computation, and so the iteration can add no Cohen real.

Increasing $\mathfrak{b}$ and cov(null) is commonly achieved by iterating Miller forcing and Solovay forcing. I must show that in the resulting model $\mathfrak{b} = \mathrm{non(null)} = \aleph_1$. Neither Miller forcing nor Solovay forcing adds a dominating real, and this feature persists through the countable support iterations of definable forcings by Theorem 6.3.9. Both Miller and Solovay forcing preserve the outer Lebesgue measure, which again persists through the iteration by Theorem 6.3.11.

The only problematic possibility left out in the above discussion is whether it is possible to increase $\mathfrak{b}$, cov(null), and non(null) while keeping cov(meager) small. I will leave this part to the interested readers.

7

Questions

7.1 Basics

Question 7.1.1. Let $A \subset 2^\omega \times 2^\omega \times 2^\omega$ be an analytic (coanalytic, etc.) set. For every point $x \in \mathbb{R}$ let I_x be the σ-ideal generated by the sets $A_{x,y} : y \in 2^\omega$. What is the complexity of the set $\{x \in 2^\omega : P_{I_x}$ is proper$\}$?

Question 7.1.2. Suppose that $I = \bigcap_n I_n$ is an intersection of a decreasing nonstabilizing sequence of universally Baire σ-ideals. Does the forcing P_I collapse $\mathfrak{c}$ to $\aleph_0$?

Note that Proposition 2.2.6 shows that the forcing P_I adds a countable set of ordinals which cannot be covered by a ground model countable set. With some definability this should imply that $\mathfrak{c}$ is collapsed.

Question 7.1.3. Prove that some of the forcings presented in this book are not homogeneous.

A typical case is that of I generated by sets of finite two-dimensional Hausdorff measure in $\mathbb{R}^3$. There appears to be no reason why the forcing P_I should be homogeneous, but at the same time, I have no idea how to argue for inhomogeneity. In fact, for every forcing mentioned in this book, either it is obviously homogeneous or else I think it cannot be homogeneous and at the same time I have no proof of the inhomogeneity.

7.2 Properties

Question 7.2.1. Is there a pair of universally Baire σ-ideals I, J such that P_I, P_J are proper and $I \perp J$ is not equivalent to $I \perp_{uB} J$? How about the special case of $J = E_0$-ideal?

Question 7.2.2. (CH) Is there a definable σ-ideal I such that the forcing P_I is proper and preserves Ramsey ultrafilters and does not preserve all P-points?

Question 7.2.3. Suppose that I is a universally Baire σ-ideal on a Polish space X such that the forcing P_I is proper. Suppose that P_I is bounding and preserves outer Lebesgue measure. Must there be a polar ideal J on X such that I and J contain the same Borel sets?

Section 3.5 proves a variation of this with the Lebesgue measure replaced by Baire category. Section 3.6 contains the proof of this in the special case that $I = I_\phi$ where ϕ is some submeasure which is outer regular on compact sets.

Question 7.2.4. Let I be the σ-ideal on $[0, 1] \times [0, 1]$ associated with the side-by-side product of Solovay forcing with itself. Is I $\mathbf{\Pi}_1^1$ on $\mathbf{\Sigma}_1^1$?

Example 3.11.9 shows that writing J for the Lebesgue null ideal, MRR(J, J) fails badly. Since $P_J \times P_J$ is a definable proper forcing, the question arises, what is its associated ideal? At this point I do not have a definition of this ideal which does not involve the forcing relation. An answer to the question should help resolve this problem.

Question 7.2.5. Is $\mathfrak{sn} \leq \mathfrak{d}$ provable in ZFC?

Simple complexity calculations show that the σ-ideal connected to any proper bounding forcing consisting of an analytic collection of finitely branching trees must be $\mathbf{\Pi}_1^1$ on $\mathbf{\Sigma}_1^1$ and therefore the forcing and its iterations preserve $\mathfrak{sn}$. Thus a bounding poset increasing $\mathfrak{sn}$ must have a somewhat unusual form.

Question 7.2.6. Suppose that I is a σ-ideal such that P_I is c.c.c. Is it true that every positive Borel set contains a positive closed set modulo the ideal I?

Question 7.2.7. (AD) Suppose that I is a σ-ideal on a Polish space such that there is a countable ordinal α such that every Borel set either has a $\mathbf{\Sigma}_\alpha^0$ I-positive subset or a $\mathbf{\Sigma}_\alpha^0$ I-small subset. Does then every (not necessarily Borel) set have this property?

This question is driven by the simple fact that all ideals satisfying the first dichotomy that I can produce have a simply describable basis consisting of Borel sets of limited complexity; in fact, such ideals in this book are always generated by $G_{\delta\sigma}$ sets.

7.3 Examples

Question 7.3.1. Suppose that I is a universally Baire porosity ideal. Is there a universally Baire porosity generating I?

The membership of a σ-ideal I in all other classes of ideals considered in this book are recognizable in the model $L(\mathbb{R})[I]$. The general question is whether the existence of generating porosity is recognizable in $L(\mathbb{R})[I]$.

Question 7.3.2. Is there a countable ordinal α such that Σ^0_α sets are dense in the forcing P_I for every porosity σ-ideal I?

Question 7.3.3. Let I be the σ-ideal generated by the metrically porous subsets of the unit interval. Does the forcing P_I preserve outer Lebesgue measure?

[27] proved that the ideal of metrically porous sets is not polar. The associated forcing is bounding by [80]. The question relates to Question 7.2.3.

Question 7.3.4. (ZF+DC+AD+) Is there a subadditive outer regular capacity c which is not continuous under increasing wellordered unions?

Note that Theorems 4.3.21 and 4.5.6 show that most capacities in fact are so continuous.

Question 7.3.5. Is the Newtonian capacity restricted to the closed unit ball Ramsey?

This is motivated by the effort to separate the various capacity forcings from each other, in particular from the Solovay forcing. If the Newtonian capacity is Ramsey then the associated forcing does not add splitting reals and thus is quite different from the Solovay forcing. The larger program is to classify capacities by the forcing properties of their associated forcings.

Question 7.3.6. Is the theory ZF+"every subset of $\mathbb{R}^3$ has a Borel subset of the same Newtonian capacity" equiconsistent with ZFC?

Question 7.3.7. Can a pavement submeasure forcing add a Cohen real?

Note that Theorem 4.5.13 says that the only possible intermediate extensions are given by c.c.c. reals. The bounding c.c.c. reals cannot be excluded because they are all generated by a pavement submeasure forcing. The question seeks to eliminate any other possibility; recall that every definable c.c.c. forcing which is not bounding adds a Cohen real. The known pavement submeasure forcings fail to add a Cohen real, each for a very different reason.

Question 7.3.8. Let X be a separable Banach space and I be the σ-ideal of Gauss null sets on X [10]. Is the forcing P_I proper?

7.4 Operations

Question 7.4.1. Characterize those suitably definable σ-ideals I on ω^ω such that it is provable in ZFC that if $0^\#$ does not exist then there is a forcing extension $V[G]$ and in it a point $x \in \omega^\omega$ such that $V[G] \models L[x] \cap \omega^\omega \notin I$.

This question is motivated by the proof of Theorem 5.1.16.

Question 7.4.2. Let I be the ideal associated with the eventually different real forcing. Is I^{ω^*} a nontrivial ideal?

Question 7.4.3. Is there a σ-ideal I such that the σ-ideal I^{ω^*} is nontrivial while the σ-ideal $I^\mathbb{Q}$ is trivial?

Question 7.4.4. Let I be the ideal of Lebesgue null sets. Is the forcing $P_{I^{\omega^*}}$ proper?

The forcing would be a sort of a countable support illfounded iteration of the random forcing, and the proof should generalize to the capacity forcings.

Question 7.4.5. Suppose that I, J are σ-ideals on the same Polish space X such that P_I, P_J are both proper. Let K be the σ-ideal generated by $I \cup J$. Is it true that P_K is proper? If P_I, P_J are bounding, is P_K bounding? If P_I, P_K are bounding and do not add splitting reals, is P_K such?

7.5 Applications

Question 7.5.1. Consider the CPA game G for the ideal I of countable sets. Is the existence of a winning strategy for Player I in the game G equivalent to CH?

Question 7.5.2. Which variations of the Ciesielski–Pawlikowski axioms prove $\mathfrak{c} = \aleph_2$?

Question 7.5.3. Is there a useful axiomatization of the side-by-side product models parallel to the CPA axioms?

Let I denote the countable side-by-side product of the ideal of the countable sets. It is not difficult to employ Σ_1^2 absoluteness to show that under suitable large cardinal assumptions, if $\mathfrak{x}$ is a tame invariant and $\mathfrak{x} < \mathrm{cov}(I)$ can be forced, then it holds in the long side-by-side product of the Sacks forcing. In the product model, $\mathfrak{x} = \aleph_1$ and $\mathrm{cov}(I)$ can be pushed arbitrarily high. However, it is not clear whether any reasonable sentence in the product model is responsible for this effect.

Bibliography

[1] David R. Adams and Lars Inge Hedberg. *Function Spaces and Potential Theory*. New York, Springer-Verlag, 1996.

[2] Tomek Bartoszynski and Haim Judah. *Set Theory. On the Structure of the Real Line*. Wellesley, MA, A. K. Peters, 1995.

[3] Andreas Blass and Saharon Shelah. Near coherence of filters. III. A simplified consistency proof. *Notre Dame Journal of Formal Logic*, 30:530–538, 1989.

[4] Joerg Brendle, Greg Hjorth, and Otmar Spinas. Regularity properties for dominating projective sets. *Annals of Pure and Applied Logic*, 72:291–307, 1995.

[5] Douglas Cenzer and R. Daniel Mauldin. Inductive definability: measure and category. *Advances in Mathematics*, 38:55–90, 1980.

[6] G. Choquet. Theory of capacities. *Ann. Inst. Fourier*, 5:131–295, 1953.

[7] Jens Peter Reus Christensen. Some results with relation to the control measure problem. In *Vector Space Measures and Applications*, Lecture Notes in Mathematics, **645**, pages 27–34. New York, Springer-Verlag, 1978.

[8] Jacek Cichon, Michal Morayne, Janusz Pawlikowski, and Slawomir Solecki. Decomposing Baire functions. *Journal of Symbolic Logic*, 56:1273–1283, 1991.

[9] Krzystof Ciesielski and Janusz Pawlikowski. *The Covering Property Axiom CPA. A Combinatorial Core of the Iterated Perfect Set Model*. Cambridge, Cambridge University Press, 2004.

[10] Marianna Csörnyei. Aronszajn null and gaussian null sets coincide. *Israel Journal of Mathematics*, 111:191–201, 1999.

[11] Roy O. Davies and C. Ambrose Rogers. The problem of subsets of finite positive measure. *Bulletin of London Mathematical Society*, 1:47–54, 1969.

[12] Gabriel Debs. Polar σ-ideals of compact sets. *Transactions of American Mathematical Society*, 347:317–338, 1995.

[13] Carlos DiPrisco and Stevo Todorcevic. Perfect set properties in $L(R)[U]$. *Advances in Mathematics*, 139:240–259, 1998.

[14] Natasha Dobrinen. Games and general distributive laws in Boolean algebras. *Proceedings of American Mathematical Society*, **131**:309–318, 2003.

[15] Ilijas Farah. *Analytic Quotients*. Memoirs of AMS 702. Providence, RI American Mathematical Society, 2000.

[16] Ilijas Farah and Boban Velickovic. Maharam algebras and Cohen reals. Unpublished, 2006.

[17] Ilijas Farah and Jindřich Zapletal. Between Maharam's and von Neumann's problems. *Mathematical Research Letters*, **11**:673–684, 2004.

[18] Ilijas Farah and Jindřich Zapletal. Four and more. *Annals of Pure and Applied Logic*, **140**:3–39, 2006.

[19] J. Feldman and C. C. Moore. Ergodic equivalence relations, cohomology and von Neumann algebras. *Transactions of American Mathematical Society*, **234**:289–324, 1977.

[20] Qi Feng, Menachem Magidor, and Hugh Woodin. Universally Baire sets of reals. In Haim Judah, W. Just, and Hugh Woodin, editors, *Set Theory of the Continuum*, number 26 in MSRI publications, pages 203–242. New York, Springer-Verlag, 1992.

[21] David Fremlin. On a construction of Steprāns. 2002. Preprint.

[22] Stefan Geschke, Martin Goldstern, and Menachem Kojman. Continuous Ramsey theory and covering the plane by functions. *Journal of Mathematical Logic*, **4**:109–145, 2004.

[23] Stefan Geschke, Menachem Kojman, Wieslaw Kubiś, and Rene Schipperus. Convex decompositions in the plane and continuous pair colorings of the irrationals. Preprint, 2001.

[24] Moti Gitik and Saharon Shelah. Forcings with ideals and simple forcing notions. *Israel Journal of Mathematics*, **68**:129–160, 1989.

[25] Moti Gitik and Saharon Shelah. More on simple forcing notions and forcings with ideals. *Annals of Pure and Applied Logic*, **59**:219–238, 1993.

[26] J. D. Howroyd. On dimension and on the existence of sets of finite positive Hausdorff measure. *Proc. London Math. Soc.*, **70**:581–604, 1995.

[27] P. D. Humke and D. Preiss. Measures for which sigma-porous sets are null. *Journal of London Mathematical Society*, **32**:236–244, 1985.

[28] Stephen Jackson. The weak square property. *J. Symbolic Logic*, **66**:640–657, 2001.

[29] Thomas Jech. *Set Theory*. Academic Press, San Diego, 1978.

[30] Thomas Jech. More game-theoretic properties of Boolean algebras. *Annals of Pure and Applied Logic*, **26**:11–29, 1984.

[31] Sven Jossen and Otmar Spinas. A two-dimensional tree ideal. In *Logic Colloquium 2000*, Lecture Notes in Logic 19, pages 294–322. Urbana, Association for Symbolic Logic, 2005.

[32] Helen Joyce and David Preiss. On the existence of subsets of finite positive packing measure. *Mathematika*, **42**:15–24, 1995.

[33] Haim Judah, Andrzej Roslanowski, and Saharon Shelah. Examples for Souslin forcing. *Fundamenta Mathematicae*, **144**:23–42, 1994. math.LO/9310224.

[34] Vladimir G. Kanovei. *Varia*. ArXiv preprint.

[35] Alexander Kechris and Alain Louveau. The structure of hypersmooth Borel equivalence relations. *Journal of American Mathematical Society*, **10**:215–242, 1997.

[36] Alexander Kechris, Alain Louveau, and Hugh Woodin. The structure of σ-ideals of compact sets. *Transactions of American Mathematical Society*, **301**:263–288, 1987.

[37] Alexander Kechris and Alain Louveau. *Descriptive Set Theory and the Structure of Sets of Uniqueness*. Cambridge, Cambridge University Press, 1989.

[38] Alexander Kechris, Slawomir Solecki, and Stevo Todorcevic. Borel chromatic numbers. *Advances in Mathematics*, **141**:1–44, 1999.

[39] Alexander S. Kechris. On a notion of smallness for subsets of the Baire space. *Transactions of American Mathematical Society*, **229**:191–207, 1977.

[40] Alexander S. Kechris. *Classical Descriptive Set Theory*. New York, Springer-Verlag, 1994.

[41] Jakob Kellner. Preserving non-null with Suslin+ forcing. math. LO/0211385, 2002.

[42] Jakob Kellner and Saharon Shelah. Preserving preservation. *Journal of Symbolic Logic*, **70**:914–945, 2005.

[43] Jakob Kellner and Saharon Shelah. Saccharinity. math. LO/0511330, 2005.

[44] N. S. Landkof. *Foundations of Modern Potential Theory*. New York, Springer-Verlag, 1972.

[45] Paul B. Larson. *The Stationary Tower Forcing*. University Lecture Series 32. Providence, RI, American Mathematical Society, 2004. Notes from Woodin's lectures.

[46] Richard Laver. On the consistency of Borel's conjecture. *Acta Mathematica*, **137**:151–169, 1987.

[47] Alexander S. Kechris Leo Harrington and Alain Louveau. A Glimm–Effros dichotomy for Borel equivalence relations. *Journal of the American Mathematical Society*, **3**:903–928, 1990.

[48] Anthony Martin. An extension of Borel determinacy. *Annals of pure and applied logic*, **49**:279–293, 1990.

[49] D. Anthony Martin. A purely inductive proof of Borel determinacy. In A. Nerode and R. A. Shore, editors, *Recursion theory*, number 42 in Proceedings of Symposia in Pure Mathematics, pages 303–308. Providence, RI American Mathematical Society, 1985.

[50] K. Mazur. F_σ-ideals and $\omega_1 \omega_1*$ gaps in the Boolean algebra $P(\omega)/I$. *Fund. Math*, **138**:103–111, 1991.

[51] Arnold W. Miller. Covering 2^ω with ω_1 many disjoint closed sets. In *The Kleene Symposium*, pages 415–421. New York, North Holland, 1980.

[52] Arnold W. Miller. On relatively analytic and Borel subsets. *Journal of Symbolic Logic*, **70**:346–352, 2005.

[53] Mahendra G. Nadkarni. On the existence of a finite invariant measure. *Proceedings of Indian Academy of Sciences, Math. Sci.*, **100**:203–220, 1990.

[54] Kanji Namba. Independence proof of a distributive law in complete Boolean algebras. *Commentarii Mathematici Universitatis Sancti Pauli*, **19**:1–12, 1970.

[55] Itay Neeman. *The Determinacy of Long Games*. de Gruyter Series in Logic and its Applications 7. Berlin, Walter de Gruyter, 2004.

[56] C. A. Rogers. *Hausdorff Measures*. Cambridge, Cambridge University Press, 1970.

[57] Andrzej Roslanowski and Saharon Shelah. Sweet & sour and other flavours. *Acta Mathematica*. Submitted.

[58] Andrzej Roslanowski and Saharon Shelah. Norms on possibilities I: Forcing with trees and creatures. *Memoirs of the American Mathematical Society*, **141**, 1999.

[59] Saharon Shelah. Analytical Guide and updates for "Cardinal Arithmetic." Shelah [Sh:E12].

[60] Saharon Shelah. *Non-structure Theory*, Oxford: Oxford University Press, Accepted.

[61] Saharon Shelah. Properness without elementaricity. *Journal of Applied Analysis*. Accepted.

[62] Saharon Shelah. Vive la différence I: Nonisomorphism of ultrapowers of countable models. In *Set Theory of the Continuum*, volume 26 of *Mathematical Sciences Research Institute Publications*, pages 357–405. Berlin, Springer Verlag, 1992. math.LO/9201245.

[63] Saharon Shelah. *Cardinal Arithmetic*. Oxford, Clarendon Press, 1994.

[64] Saharon Shelah. *Proper and Improper Forcing*. New York, Springer-Verlag, second edition, 1998.

[65] Saharon Shelah. Borel sets with large squares. *Fundamenta Mathematicae*, **159**:1–50, 1999.

[66] Saharon Shelah and Otmar Spinas. The distributivity numbers of $P(\omega)$/fin and its square. *Transactions of the American Mathematical Society*, accepted.

[67] Saharon Shelah and Jindřich Zapletal. Duality and the pcf theory. *Mathematical Research Letters*, **9**:585–595, 2002.

[68] Roman Sikorski. On the representation of Boolean algebras as fields of sets. *Fundamenta Mathematicae*, **35**:247–258, 1948.

[69] Slawomir Solecki. Covering analytic sets by families of closed sets. *Journal of Symbolic Logic*, **59**:1022–1031, 1994.

[70] Slawomir Solecki. Decomposing Borel sets and functions and the structure of Baire class 1 functions. *Journal of American Mathematical Society*, **11**:521–550, 1998.

[71] Slawomir Solecki. Analytic ideals and their applications. *Annals of Pure and Applied Logic*, **99**:51–72, 1999.

[72] Otmar Spinas. Analytic countably splitting families. *J. Symbolic Logic*, **69**:101–117, 2004.

[73] Juris Steprāns. *Cardinal invariants associated with Hausdorff capacities,*. pages 174–184.

[74] Juris Steprāns. Many quotient algebras of the integers modulo co-analytic ideals. Preprint.

[75] Michel Talagrand. Maharam's problem. math. FA/0601689, 2006.

[76] Jan Van Mill. *Infinite-Dimensional Topology, Prerequisites and Introduction.* New York, North Holland, 1989.

[77] Boban Velickovic and W. Hugh Woodin. Complexity of the reals in inner models of set theory. *Annals of Pure and Applied Logic,* **92**:283–295, 1998.

[78] Peter Walters. *An Introduction to Ergodic Theory.* New York, Springer-Verlag, 1975.

[79] Hugh Woodin. Supercompact cardinals, sets of reals, and weakly homogeneous trees. *Proceedings of National Academy of Sciences,* **85**:6587–6591, 1988.

[80] Luděk Zajíček and Miroslav Zelený. Inscribing compact non-sigma-porous sets into analytic non-sigma-porous sets. 2003. Unpublished.

[81] Luděk Zajíček. Porosity and sigma-porosity. *Real Analysis Exchange,* **13**:314–350, 1987/88.

[82] Jindřich Zapletal. Forcing with ideals of closed sets. *Commentationes Mathematicae Universitatis Carolinae,* **43**:181–188, 2002.

[83] Jindřich Zapletal. *Descriptive Set Theory and Definable Forcing.* Memoirs of American Mathematical Society. Providence, RI American Mathematical Society, 2004.

[84] Jindřich Zapletal. Potential theory and forcing. 2005. math. LO/0502394.

[85] Jindřich Zapletal. Two preservation theorems. 2005. math. LO/0502047.

[86] Jindřich Zapletal. Proper forcing and rectangular Ramsey theorems. *Israel Journal of Mathematics,* **152**:29–47, 2006.

[87] Jindřich Zapletal and Michael Hrušák. Forcing with quotients. 2005. Preprint.

Index